DR. C. G. JUSTUS, PROFESSOR
SCHOOL OF GEOPHYSICAL SCIENCES
GEORGIA INSTITUTE OF TECHNOLOGY
ATLANTA, GA.   30332

RADIATION IN A CLOUDY ATMOSPHERE

Radiation in a Cloudy Atmosphere

Edited by

E. M. FEIGELSON
Institute of Atmospheric Physics, Academy of Sciences, U.S.S.R.

D. Reidel Publishing Company

A MEMBER OF THE KLUWER ACADEMIC PUBLISHERS GROUP 

Dordrecht / Boston / Lancaster

Library of Congress Cataloging in Publication Data

Radiatsiia v oblachnoi atmosfere. English.
      Radiation in a cloudy atmosphere.

      (Atmospheric sciences library)
      Translation of: Radiatsiia v oblachnoi atmosfere.
      Bibliography: p.
      Includes index.
      1. Solar radiation. 2. Cloudiness. I. Feigelson, E. M. (Eva Mikhailovna) II.
   Title. III. Series.
QC911.R2913   1984   551.5'7   84-9761
ISBN 90-277-1803-2

Published by D. Reidel Publishing Company
P.O. Box 17, 3300 AA Dordrecht, Holland

Sold and distributed in the U.S.A. and Canada
by Kluwer Academic Publishers,
190 Old Derby Street, Hingham, MA 02043, U.S.A.

In all other countries, sold and distributed
by Kluwer Academic Publishers Group,
P.O. Box 322, 3300 AH Dordrecht, Holland

Translated from the Russian by U.A.T. Ltd.

TABLE OF CONTENTS

Radiative heat transfer is a fundamental factor in the
energetics of the terrestrial atmosphere: the system consisting
of the atmosphere and the underlying layer is heated by the Sun,
and this heating is compensated, on the average, by thermal radiation. Only over a period of 1-3 days from some specified initial
moment can the dynamic processes in the atmosphere be considered
to be adiabatic. Global dynamic processes of long duration are
regulated by the actual influxes of heat, one of the main ones
being the radiative influx.

Radiation must be taken into account in long-term weather
forecasting and when considering the global circulation of the
atmosphere, the theory of climate, etc. Thus it is necessary to
know the albedo of the system, the amount of solar radiation
transmitted by the atmosphere, the absorptivity of the atmosphere
vis-à-vis solar radiation, and also the effective radiation flux,
the divergence of which represents the radiative cooling or heating.
All these quantities have to be integrated over the wavelength
spectrum of the solar or thermal radiation, and they must be
ascertained as functions of the determining factors. The relationships between the indicated radiation characteristics, the optical
quantities directly determining them, the optically active components of the atmosphere, and the meteorological fields will be
discussed in this book.

Since cloudiness is clearly the main regulator of the radiation processes in the atmosphere, we will focus most of our attention on cloud-related effects in the radiative heat transfer. A
cloudless atmosphere is taken to be only a particular case or
else is considered to be the medium beyond the clouds.

In order to ensure an expert description of the cloud properties used to calculate the radiation parameters, specialists in
cloud physics (the authors of monographs [1, 4, 7]) were called
in to participate in writing this book. In this respect, the
present work differs from the previous two books on radiation in
a cloudy atmosphere [5, 6], in which the parts concerning the
properties of clouds were not written on a professional (from the
point of view of cloud physics) level.

The present publication also differs from the works cited
above in some other important ways, even though their ideological
similarity cannot be denied. First of all, here the theory is
kept to a minimum, while the factual data are accorded as great
a volume as possible. In particular, the radiation regimes of the
polar and tropical regions are described quite fully. Secondly,
since this monograph represents a collective effort by many
Soviet experts on the subject of radiation in a cloudy atmosphere,

its content is quite comprehensive. At the same time, the editors were faced with enormous difficulties in making the results, methods, and styles of the various authors consistent.

The idea of preparing a collective monograph originated from a suggestion by the International Commission on Radiation, under the auspices of the International Association of Meteorology and Atmospheric Physics (IAMAP), to devise a radiation model of a cloudy atmosphere. The first fruits of the work on this model were two papers entitled "A tentative radiation model of a cloudy atmosphere" [2, 3]. Subsequently, however, the authors stopped trying to create such a model, since no single model can encompass the various problems involved.

The material presented in this book gives the information needed to construct various models and to achieve various parametrizations of the radiative fluxes.

The authors are: I. P. Mazin (Central Aerological Observatory), Chaps. 1, 2; Yu.-A. R. Mullamaa (Institute of Astrophysics and Atmospheric Physics, Estonian Academy), Chaps. 3, 11; L. D. Krasnokutskaya (Institute of Atmospheric Physics, USSR Academy), Chaps. 4, 5; L. M. Romanova (Institute of Atmospheric Physics, USSR Academy), Chaps. 6, 7; E. M. Feigel'son (Institute of Atmospheric Physics, USSR Academy), Chaps. 8, 9; N. I. Goisa (Ukrainian Scientific Research Institute, Chap. 10; A. S. Ginzburg (Institute of Atmospheric Physics, USSR Academy), Chaps. 12, 14; G. N. Kostyanoi (Central Aerological Observatory), Chap. 13; N. A. Zaitseva (Central Aerological Observatory), Chaps. 15, 16.

PART I. CLOUD STRUCTURES AND OPTICAL CHARACTERISTICS OF CLOUDS

INTRODUCTION

In existing books and surveys dealing with cloud physics
attention is usually focused on the physics of the origin and
development of clouds and the physics of the processes determining
their microstructure. However, the data of the numerous experimental observations are still fragmentary and are scattered over
individual papers and reports, being cited in books only as exam-
ples and illustrations. During the 1970's, it is true, some works
did appear in which their authors tried to generalize the accumu-
lated empirical material and to arrive at definite statistical
and climatological results on the basis of it. The most important
publication of this sort was the *Aeroclimatological atlas-handbook
of the USSR* [1]. We have made considerable use of this handbook,
together with publications generalizing the experimental studies
(for instance, Refs. [4, 21] to Chap. 1 and Ref. [17] to Chap. 4),
during the writing of this part of the book.

In Part I an attempt is made to generalize experimental data
on the physical structure of clouds and their optical characteristics. Due to the limited size of the book, however, it is not
possible to describe in detail the collected material. Therefore,
we have just attempted to present to the reader, in a convenient
form, data on the main microphysical and macrophysical characteristics of clouds (mean values, variability, etc.). As far as possible, use has been made of empirical results based on reliable
measurements. In some instances, because of imperfect measurement
techniques, variability of the characteristics in question, or
the small number of measurements carried out, the results obtained
have had to be regarded as merely estimates.

In other cases they constitute reliable, statistically valid
data. For further details on the actual measurements, the reader
is referred to the original works, cited in the reference lists
for each chapter.

In Part I our main interest will be those cloud parameters
and characteristics which may be needed to estimate and calculate
the optical and radiation properties of clouds, to devise diverse
numerical descriptions of various atmospheric processes, taking
into account the optical and radiation properties of clouds, and
so forth. Consequently, in addition to direct empirical data on
cloud structure (Chaps. 1-3), we will present along with the
measured optical characteristics of clouds calculations of the
optical parameters of clouds (Chap. 4) and of the atmosphere as
a whole (Chap. 5).

CHAPTER 1

SPATIAL STRUCTURE OF CLOUDS

1.1. Cloud frequencies

When studying any mean characteristics of clouds over a time t, one should bear in mind that these characteristics relate only to the part of the period during which cloudiness was observed.

Frequencies of total cloud cover, lower-level clouds, and clouds of vertical extent over continents and oceans can be found in climate handbooks [17] and atlases [16, 29]. Since these data were obtained from observations on the ground, information about the frequencies of clouds in the upper and middle levels are practically nonexistent. Only by incorporating the results of aircraft soundings is it possible to estimate the frequencies of clouds in the upper and middle levels [1, 12]. For large regions the total cloud cover and the amount of any cloud type present can be estimated with the aid of satellite measurements [1]. The cloud frequency is closely related to the general circulation of the atmosphere, the nature of the underlying surface, and the insolation. In addition, the cloud frequency is affected by site elevation, orography, and other factors.

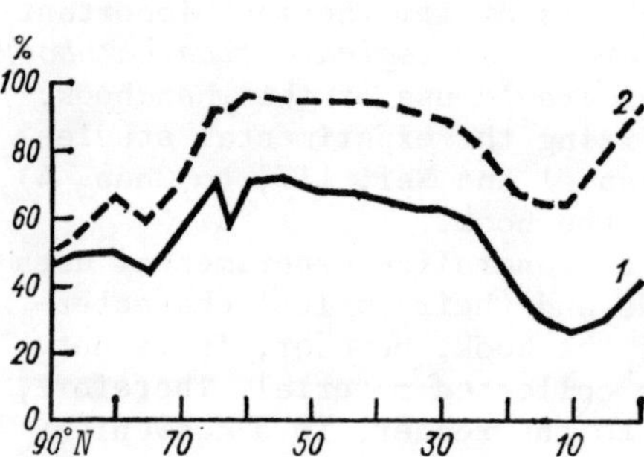

Fig. 1.1. Latitude and dependence of frequencies of various kinds of weather along 20°W meridian. January. 1) frequency of overcast skies (8-10), 2) total frequency of overcast and partly cloudy (3-7) skies. Region above curve 2 characterizes frequency of clear skies (0-2).

The global distribution of the cloud frequency can be divided into four climatic zones: a polar zone from 70 to 90° latitude; a temperature zone from 30 to 70°; a tropical (subtropical) zone from 10 to 30°; an equatorial zone from 0 to 10°. These zones for January in the Northern Hemisphere are evident from an inspection of Fig. 1.1 [14].

In summer there is a northward shift in the zone boundaries, because the circulation conditions vary and the underlying surface becomes moistened. In the polar zone there is a sharp increase in both the total cloudiness and the low-lying cloudiness. The

4

southern part of the temperate zone moves downward, under the influence of subtropical anticyclones, and accordingly the amount of cloud cover there is sharply reduced. The intertropical convergence zone shifts northward, increasing the area of the equatorial zone (there is a simultaneous decrease in the area of this zone in the Southern Hemisphere).

Consequently, in summer in the Northern Hemisphere three large zones are discernible in the distribution of the cloudiness frequency: a polar zone (combined with the temperate zone), a tropical zone, and an equatorial zone. A definite tendency can be detected in the relative variation of the frequencies of stratiform and cumulus cloudiness: with the approach to the equator, the relative frequency of cumuli increases (Fig. 1.2).

In the Southern Hemisphere as a whole, there is also a zonal distribution of the cloud frequency, which is less affected by the presence of continents than in the Northern Hemisphere [15, 19].

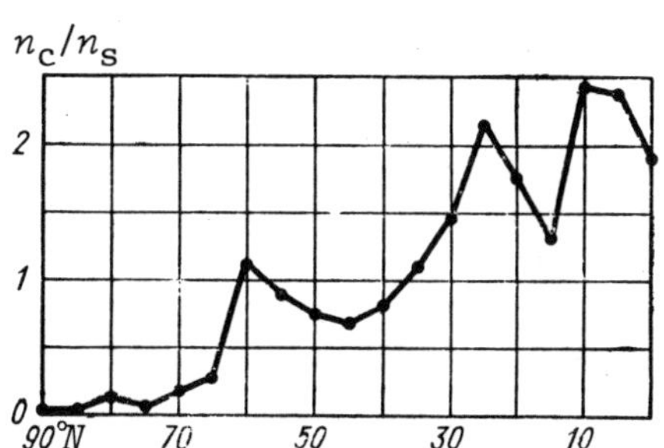

Fig. 1.2. Ratio of frequency of cumulus (n_c) to frequency of stratiform cloudiness (n_s) along 20°W meridian. January (according to [14]).

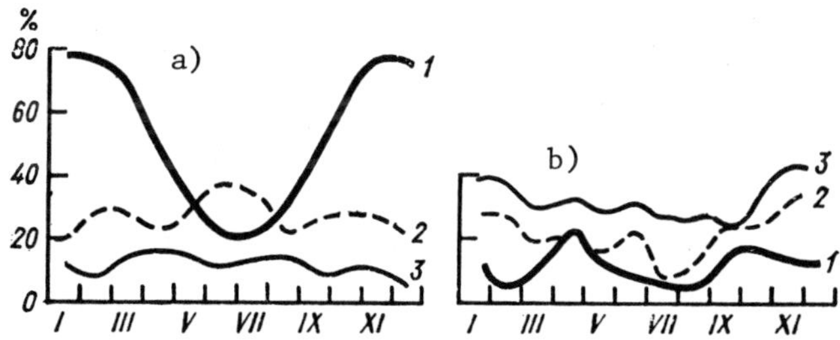

Fig. 1.3. Annual variation of frequency of overcast weather (cloud cover of 8–10), according to the data of aircraft sounding over Minsk (a) and Krasnoyarsk (b).
1) lower-level clouds; 2) middle-level clouds; 3) upper-level clouds.

Figure 1.3 shows the annual variation of the frequency of clouds in the upper, middle, and lower levels over continents, according to data of aircraft soundings at two places in the USSR [9]. During the cold season the frequency of a solid cloud cover in the lower level, in the middle and high latitudes of the European USSR, is 2 or 3 times the frequency of middle-level cloudiness and 4 to 6 times the frequency of upper-level cloudiness.

Over the Asian part of the USSR, east of the 90° meridian, the opposite pattern is observed: cloudiness in the upper and middle levels is predominant.

1.2. Structure of stratiform clouds

1.2.1. Definitions

Here stratiform clouds are defined as clouds whose horizontal dimensions are orders of magnitude greater than their vertical dimensions. These include: stratus (St), stratocumulus (Sc), nimbostratus (Ns), altocumulus (Ac), altostratus (As), and cirrostratus (Cs), as well as Ns-As, As-Cs, and Ns-As-Cs frontal cloud systems.

1.2.2. Height distribution of clouds

Table 1.1 shows the predominant heights of the lower, middle, and upper cloud levels in the various geographical zones.*)

Table 1.1. Heights (km) of cloud levels

Level	Zone		
	polar	temperate	tropical
Lower	<2	<2	<3
Middle	2... 4	2... 7	3... 8
Upper	3... 8	5... 13	7... 18

The height of the lower boundary z_{1b} of stratiform clouds increases, on the average, from north to south. For instance, at various places the annual mean heights of St clouds are: Amderma (Arctic) 0.23 km [11], Moscow 0.47 km, and Simla (India) 1.4 km [23]. The mean heights of As clouds are: at Amderma 2.2 km [11],

---

*) These height values differ somewhat from those given in the International Cloud Atlas [26], since results of recent observations were taken into account [7]. Some intervals overlap (for instance, 2-7 km and 5-13 km) because the Instructions specify that clouds are to be placed in a given level on the basis of the height of their lower boundary. Thus clouds with their base in a lower-lying level may also extend up into a higher stratum.

at Moscow 2.9 km, and at Batavia (Java) 5.4 km [23]. In summer z_{1b} is generally greater than in winter. For instance, in summer in Moscow for Sc clouds z_{1b} = 1.2 km, while in winter it is 0.7 km [25]. However, in some regions, such as the Maritime Territory, z_{1b} is lower in summer than in winter, because of the arrival of moist air from the sea. There are also meridional variations in height, related to the conditions of the atmospheric circulation. Moreover, a diurnal variation in the frequency of the lower-boundary height also exists.

The distribution of the cloud depth (thickness) H is described well by the exponential function [6]

$$f(H) = \frac{1}{H^*} e^{-H/H^*}, \qquad (1.1)$$

where H^* is the mean cloud depth. The mean depths $\overline{H}$ given in Tables 1.2 and 1.3 can be used for the quantity H^*.

Table 1.2. Mean values (km) of heights of $\overline{z}_{1b}$ and $\overline{z}_{ub}$, vertical depth $\overline{H} = z_{ub} - z_{1b}$ (numerator), and corresponding standard deviations (denominator), according to data of aircraft soundings carried out from 1957 to 1963 over Moscow (Vnukovo) in the daytime.

Cloud type	$\overline{z}_{1b}/\sigma_{1b}$		$\overline{z}_{ub}/\sigma_{ub}$		$\overline{H}/\sigma_{\Delta H}$	
	Winter	Summer	Winter	Summer	Winter	Summer
St	0.25 / 0.16	0.29 / 0.19	0.55 / 0.32	0.58 / 0.39	0.30 / 0.32	0.29 / 0.36
Sc	0.85 / 0.51	1.26 / 0.57	1.14 / 0.54	1.59 / 0.60	0.29 / 0.25	0.33 / 0.28
As	3.80 / 1.14	3.93 / 1.17	4.73 / 1.16	4.83 / 1.06	0.93 / 0.90	0.90 / 0.77
Ac	3.58 / 1.11	3.56 / 1.01	3.80 / 1.04	3.78 / 1.02	0.22 / 0.23	0.22 / 0.27
Ns — As	0.59 / 0.52	0.76 / 0.56	2.34 / 1.53	2.11 / 1.01	1.75 / 1.46	1.35 / 0.72

The data in Table 1.2 [10] give us an idea of how the heights of the upper and lower cloud boundaries and the cloud depth fluctuate. The relative variability of the cloud-boundary heights in the lower level $\sigma/\overline{z}$ is seen to be 1.5 to 2 times the variability of the cloud-boundary heights in the middle level. On the whole, the standard deviations σ_{lb} and σ_{ub} for each type of cloud are comparatively small and they do not vary much from season to season.

Table 1.3. Depth characteristics of stratiform clouds (km) at various places in the USSR. Autumn.

Station or region	Cloud type									
	St		Sc		As		Ac		Ns (Ns-As-Cs)	
	$\bar{H}$	H_{90}	$\bar{H}$	H_{90}	$\bar{H}$	H_{90}	$\bar{H}$	H_{90}	$\bar{H}$	H_{90}
Riga	0.44	0.7	0.26	0.6	1.42	2.4	0.20	0.5	1.68	3.4
Moscow	0.28	0.6	0.32	0.7	1.06	2.8	0.20	0.5	1.71	4.1
Omsk	0.32	0.6	0.31	0.6	1.34	2.0	0.26	0.5	2.35	3.8
Yakutsk	—	—	0.30	0.6	1.18	1.9	0.46	0.7	—	—
Khabarovsk	—	—	0.36	0.6	1.18	1.8	0.42	0.9	2.26	2.8
Northern Kazakhstan	0.47	—	0.50	—	0.91	—	0.36	—	1.76	—
Tashkent	—	—	—	—	1.03	2.0	0.37	1.0	2.55	4.1
Arctic (Laptev Sea)	0.40	0.6	0.33	0.6	0.28	0.8	0.29	0.5	—	—

In addition to the mean values, Table 1.3 gives the 90th percentile values (H_{90}) of the cloud depths (that is, the probability that in 90% of the cases the cloud depth will be less than the given value) for various places in the USSR, according to aircraft sounding data. These figures indicate that even in different climatic zones of the USSR the mean cloud-depth characteristics for clouds of a given type do not vary greatly.

Table 1.4. Frequency of boundary heights of Cs clouds over European USSR (2138 ascents for z_{lb} and 2386 ascents for z_{ub}).

Height, km	Frequency, %		Height, km	Frequency, %	
	z_{lb}	z_{ub}		z_{lb}	z_{ub}
<5	5		8.0 . . . 8.9	16	19
5.0 . . . 5.9	14		9.0 . . . 9.9	8	25
6.0 . . . 6.9	30		10.0 . . . 10.9	3	21
<7		4	11.0 . . . 11.9	1	14
7.0 . . . 7.9	23	12	>12.0	0.4	5

Tables 1.4 and 1.5 give, respectively, boundary-height frequencies and depth frequencies for cirrostratus clouds. All the data were obtained during flights of weather-reconnaissance aircraft [2] or special flights of high-flying sounding aircraft [12] from 1953 to 1959. Only those data were processed for which Cs clouds were observed as individual strata lying mainly above 6 km, that is, an As-Cs system would not appear in these statistics.

Table 1.5. Frequency of depths of Cs clouds over European USSR (510 ascents).

Depth, km	Frequency %	Depth, km	Frequency %
< 0.5	14.7	3.1. . . 4.0	14.9
0.5. . . 1.0	11.6	4.1. . . 5.0	6.7
1.1, . . 2.0	25.9	5.1. . . 6.0	4.3
2.1. . . 3.0	21.3	> 6.0	0.6

According to the data of [2], z_{ub} and z_{lb} for upper-level clouds depend on the season and the latitude and are determined by the regime of the tropopause. Boundary z_{ub} is generally 0.5-2 km below the level of the tropopause. The highest upper cloud boundary is observed in the summer at lower latitudes. For instance, over Southeast Asia for Ci and Cs this boundary lies above 15 km in more than 50% of the cases, whereas in the Arctic $z_{ub} < 10$ km 98% of the time [4].

1.2.3. Horizontal extent

Stratiform clouds can stretch over huge areas. Satellite photographs of clouds have revealed that fields of solid cloud cover can in the Northern Hemisphere (including the USSR) encompass areas as great as about 50 million km^2 [1]. The mean area of solid cloud cover fluctuates from 5 to 15 million km^2. The mean, and also the maximum, areas of the cloudless space in these same regions are about twice as large.

In extratropical latitudes the cyclonic activity and the frontal cloud systems associated with it are one of the main factors influencing the frequency of stratiform clouds and their horizontal extent. Satellite photographs [27] show that the width of a frontal zone in Central Europe can be as great as 1000 km, the length of the zone being 7000 km.

Table 1.6 [5] gives the frequency of the width of the upper-level cloud zone for warm and cold fronts over the European USSR, according to data obtained during flights of Tu-104 and Il-18 aircraft from 1959 to 1961. A study [22] based mainly on ground-level observations indicated that the areas of frontal zones of cirrus range from 1 to 2.8 million km^2 in 70% of the cases, while areas greater than 4 million km^2 are encountered only 6% of the time.

Table 1.6. Frequency (%) of zone width for high-level frontal clouds over European USSR

Zone width, km	Front type		Zone width, km	Front type	
	cold	warm		cold	warm
$<$ 50	2.4	—	501. . . 600	—	25.0
51. . . 100	12.2	—	601. . . 700	—	20.8
101. . . 200	26.8	—	701. . . 800	—	11.7
$<$ 200	—	1.3	801. . . 900	—	6.2
201. . . 300	29.3	6.5	901. . . 1000	—	2.6
301. . . 400	22.0	10.4	1001. . . 1500	—	2.6
401. . . 500	7.3	12.9	No. of fronts	41	77

Table 1.7. Frequencies (%) of cloudy regions and clear spaces of various extents in low-level frontal clouds [13].

Extent, km	Front type			
	cold	warm	cold	warm
	Cloudy regions		Clear spaces	
$<$ 10	20	14	30	10
10. . . 20	28	19	32	20
20. . . 30	19	19	14	17
30. . . 50	18	21	14	20
50. . . 75	9	11	6	12
75. . . 100	4	8	3	6
100. . . 150	0	3	1	6
150. . . 200	1	4	—	7
200. . . 300	1	1	—	2

A frontal zone does not form a single band over its entire extent. The cloud zones are divided into macrocells a few hundred kilometers in size, and these in turn consist of cloud bands or a continuous field having cells of cloudiness of nonuniform density tens of kilometers in size [5]. Table 1.7 gives frequencies of cloud regions of different sizes, and also of intervening clear spaces, for low-level frontal clouds.

Frontal clouds include all clouds situated not more than

200 km from the front line near the ground. Inspection of Table 1.7
shows that in regions of low-level frontal cloudiness the frequencies of clear and cloudy spaces are practically the same, and
also that the cloudy parts are generally less than 50 km in extent
(85% for cold fronts and 73% for warm fronts). However, it should
be kept in mind that high-level and middle-level clouds may well
be present at altitudes above the flight level, over both the
cloudy and clear parts. Similar data on the frequencies of cloud
sizes are obtained from nephanalysis maps for all clouds, not
just for frontal clouds [1].

The mean lifetime of a solid cloud cover over the European
USSR is 13-15 h in winter and about 5 h in summer [1]. The maximum continuous duration of solid cloudiness in winter is 200 h,
and in summer 50-60 h. Table 1.8 [1] gives the probability that
a cloud cover of 10 will last for various times over the European
USSR (on the basis of data for 10 sites).

Table 1.8. Probabilities (%) of various durations of cloud cover
of 10 over European USSR.

Season	Hours				
	1	3	6	12	24
Winter	93	87	83	78	74
Summer	80	64	52	41	35

In winter the probability that a solid cloud cover will persist is very high, and even for a duration of a day and a night it
is 74%; in summer this probability drops sharply. The probability
that clear skies will continue for the first hour in summer is
about 80%, but for 24 h the probability is only 35%.

1.3. Clouds of vertical extent

1.3.1. Definitions

Clouds with considerable vertical extent include convective
clouds of various depths: Cu hum., Cu med., Cu cong., and cumulonimbus (Cb). The depths of these clouds decreases in winter and
increase in summer, and they become 1-2 km deeper upon transition
from temperate to tropical latitudes. In summer in the middle
latitudes these convective clouds have the following typical
depths: Cu hum., up to 1.0 km; Cu med., 1-2 km; Cu cong., up to
2-3.5 km; Cb, more than 3-4 km. Finally, we should also mention
fractocumulus (Cu fr.) and trade cumulus (Cu pass.), which in [7]

were classified as low-level clouds; these have depths of hundreds of meters.

1.3.2. *Shapes and typical sizes*

The shapes of cumulus clouds are extremely capricious and diverse. However, if we consider a field made up of a large number of clouds, we can refer to the mean shape, the mean size ratio, and so forth. Then it is convenient to speak of the axial symmetry of cumuli and of the ratio between the diameter L and the depth H of the clouds. The horizontal extent of clouds which are not deep ($H < 1.5$ km) is usually greater than the vertical extent. Deeper clouds, on the other hand, have a tendency to tower upward.

On the average, within the main bulk of the cloud the diameter varies negligibly; it decreases considerably only in the summit part of the cloud, which accounts for about a quarter of the total depth. Consequently, in the following we will find it convenient to use the mean diameter L over the entire cloud depth and the mean relative diameter $\delta = L/H$. The median value of the relative diameter is related as follows to the cloud depth:*)

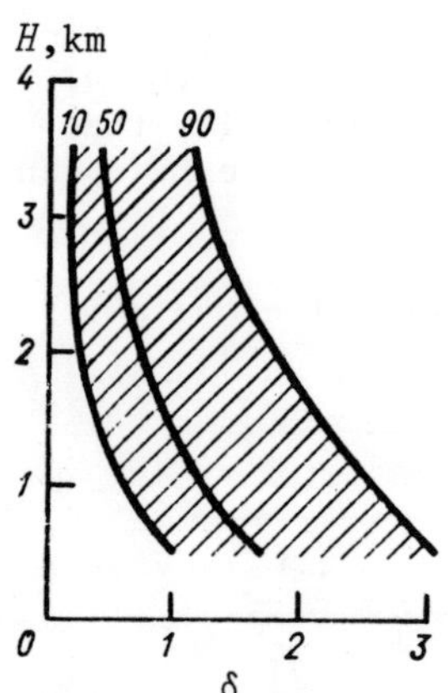

Fig. 1.4. Relative cloud diameter δ = = L/H plotted against cloud depth H.
Numbers above curves denote corresponding portions of frequency. Shaded region includes 80% of all observations.

$$\delta_{\mathrm{med}} = 1.5 H^{-1.1}. \qquad (1.2)$$

In 80 to 90% of the cases the observed values of δ do not differ from δ_{med} by more than a factor of 2–2.5 (Fig. 1.4).

Aircraft sounding data, including the results of measuring the sizes of clouds more than 50 m in diameter, give a mean diameter of 1.5 km for the clouds in a field of internal Cu, and they indicate that their distribution can be approximated by the following relation for $n(L)$, where $n(L)$ is the fraction of the total cloud cover having horizontal extent between L and dL:

---

*) In this subsection we are making use of empirical material obtained mainly on the experimental meteorological test area (at Dnepropetrovsk) by F. Ya. Voit and A. I. Furman [21].

$$n(L) = \begin{cases} 3.05 \cdot 10^{-3} L \left(1 - \dfrac{L}{3.75}\right)^{4.35} & \text{for } L < 3.75 \text{ km} \\[2mm] 0 & \text{for } L > 3.75 \text{ km} \end{cases} \qquad (1.3)$$

Here $n(L)$ is normalized to unity, that is, $\int\limits_0^\infty n(L)\,dL = 1$. More than 50% of the clouds have diameters from 0.3 to 1.2 km, and only in 5% of the cases is $L > 3.5$ km. The rare instances of clouds with $L > 5$ km were not processed; it was assumed by the authors of [21] that these data pertain to conglomerates rather than to individual clouds.

Frequently clusters (groups) of Cu hum. and Cu med. about 3 km in diameter are observed, the diameter of the main cloud being about 1.5 km and those of the surrounding small clouds in the cluster ranging from about 0.2 to 0.3 km. The distances between clouds in a group are 0.2–0.3 km, and groups are 3 to 10 km apart, on the average, this separation becoming greater as evening approaches. Fields of cumuli have areas ranging from 10^4 to 10^5 km^2. Often anvils of dense cumulus (Cb inc.) are seen to spread out over tens of kilometers in the direction of the wind gradient, and then, expanding even more, to form cirrostratus clouds covering vast areas.

1.3.3. *Degree of sky cover and cloud duration*

Because of the relative proximity of clouds to the Earth, standard ground-based meteorological observations of the degree of sky cover give distorted results. According to the data of [24, 28], observations made on the ground give sky covers p for convective cloudiness which are too high, on the average, by 0.2–0.3 (or by 2–3 on the 10-point scale). The distribution curve for the degree of sky cover can be approximated by a beta distribution:

$$n(p) = \frac{\Gamma(m+n)}{\Gamma(m)\,\Gamma(n)}\, p^{m-1}(1-p)^{n-1}, \qquad (1.4)$$

where $\Gamma(x)$ is the gamma function. For instance, for the Ukraine, $m = 4.59$ and $n = 10.23$. The mode of the distribution, that is, the most probable degree of sky cover, is in this case $p_0 = 0.28$, and the mean value is

$$\bar{p} = \int\limits_0^1 p\,n(p)\,dp = 0.31. \qquad (1.5)$$

Convective clouds persist for periods from tens of minutes (Cu hum., Cu med.) to several hours (Cb). The deeper (denser) the clouds, the longer their lifetimes. The dissipation of one cloud is often accompanied by the creation of a new one, situated 10--20 km away from the dissipating cloud.

1.4. Regional peculiarities of clouds in the Arctic, Antarctic, and tropics

Most of the information about clouds presented in the previous sections pertains to well-documented clouds of the middle

Table 1.9. Mean heights of lower boundaries z_{1b} and depths H of stratiform clouds in Arctic

Cloud type	z_{1b}, m		H, m	
	Winter	Summer	Winter	Summer
St	350	170	150	400
Sc	650	450	400	600
Ns	500		1500	
As, Ac	1600	2500	500	

latitudes. Polar and tropical clouds possess some peculiarities of morphological structure, and we will consider these now.

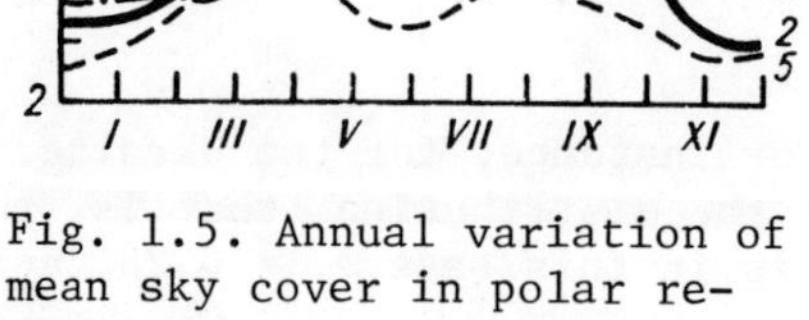

Table 1.9 gives the heights of clouds of various types in the polar latitudes. In the Arctic stratiform cloudiness predominates, due to the high stability of the air masses there. The salient features of arctic clouds are their shallow depth and low height. Especially thin clouds are observed during the cold season (100–150 m), and the warm-season clouds are the deepest (up to 1000 m). In contrast to the middle latitudes, clouds in the Arctic are the most low-lying in summer. For instance, the lower boundaries of summer stratus clouds descend to 180--200 m over the arctic seas and to 150–170 m in the Central Arctic.

Fig. 1.5. Annual variation of mean sky cover in polar regions.
1) Atlantic, 2) arctic coast, 3) Central Arctic, 4) Antarctic coast, 5) Central Antarctica.

In summer the cloudiness is extremely uniform and stretches over
enormous areas. The mean horizontal extent of stratus is 460 km
[8], and in individual instances can be 2000 km. The annual variation of the sky cover in the Central Arctic is 4.5-5. In winter
there is only half as much cloudiness there as in summer (Fig. 1.5).
     The cloud-formation conditions in Antarctica differ from those
in the Arctic. The least amount of cloud cover (1-3) is observed
over the central part of the antarctic plateau (elevation more
than 3000 m above sea level). These antarctic regions rise above
low-level clouds, and sometimes even above clouds of the middle
level [15]. On the antarctic coast the amount of sky cover increases to 6-7. For Antarctica a rather poorly defined annual
variation of the cloud cover is typical: the amount of cloudiness
in summer is only 1 or 2 points more than in winter. The frequency
of overcast skies along the coast in summer is 60-70%, and in
winter 50-60%. The lower cloud boundary is higher than in the
Arctic, the most probable height of the clouds above the coast
being from 600 to 1500 m. The clouds are below 300 m in only 1-3%
of all cases.
     In the tropics convective clouds and typical trade cloudiness
are predominant. The lower boundary of the main layer of cumuli
lies at a height of 1.5-3 km, that is, somewhat higher than for
clouds in the middle latitudes. Trade cumuli are situated below
1 km. Tropical cumuli are 1-2 km deeper than similar clouds in the
middle latitudes, and sometimes even more than this [7]. Clouds of
the upper and middle levels in the tropics are in many cases also
genetically related to cumuliform clouds, being formed by the
spreading out of deep cumulus and cumulonimbus clouds at certain
levels.
     Trade cumuli, like thin fractocumuli (up to 0.5 km in depth),
often take the form of regular lines and cover 0.2-0.4 of the
sky (cloud cover of 2-4).

CHAPTER 2

PHYSICAL CHARACTERISTICS OF CLOUDS

2.1. Temperature and phase

2.1.1. *Temperature characteristics of clouds*

Cloud temperatures usually differ from the temperature of
the surrounding air by some tenths of a degree. Only in the case
of dense cumuli can this difference be more than a degree. On
the whole, convective clouds tend to heat up the surrounding atmosphere somewhat. Within the cloud mass itself, however, the temperature fluctuations may be considerable, reaching 1 to 3°C. On the
average, these clouds stay a little (0.2-0.5°C) cooler than the
ambient air [9]. Inside the clouds the temperature gradients are,
as a rule, close to moist-adiabatic.

Many physical characteristics of clouds, in particular the
cloud phase makeup and water content, are temperature dependent,
and it is precisely this dependence which frequently determines
the seasonal and latitude variations of the physical parameters.
Consequently, it is important to have an idea of the mean atmospheric temperatures. Mean air temperatures for the hemispheres and
mean global temperatures are given in [6], and mean temperatures
at various latitudes in the Northern Hemisphere are given in [2].

2.1.2. *Cloud phase*

At positive (above-freezing) temperatures clouds can naturally be assumed to be droplet clouds. During the fall of snowflakes
and crystalline particles, the melting layer usually does not lie
more than several hundred meters below the zero isotherm. However,
in such cases as the fall of hail, ice particles are also present
in the atmosphere at high positive temperatures.

At negative (subfreezing) temperatures one can have droplet,
mixed, or crystal clouds. The proportion of drops and crystals
(according to mass or according to number of particles) in a mixed
cloud generally varies in time and from place to place. Individual
parts of a cloud may consist wholly of droplets or only of crystals,
while droplets and crystals coexist in other regions. However,
experimental data on the proportions of droplets and crystals in
clouds of various forms are practically nonexistent. The only
definite thing is that the relative proportion of crystals increases as the temperature drops. At the same time, the frequencies
of droplet (that is, consisting just of liquid droplets), crystal

(consisting just of crystals), and mixed (containing both droplets
and crystals) clouds have been studied in detail, on the basis of
data from regular aircraft soundings carried out over several years.
In all, observations of more than 41,500 clouds were analyzed [3].

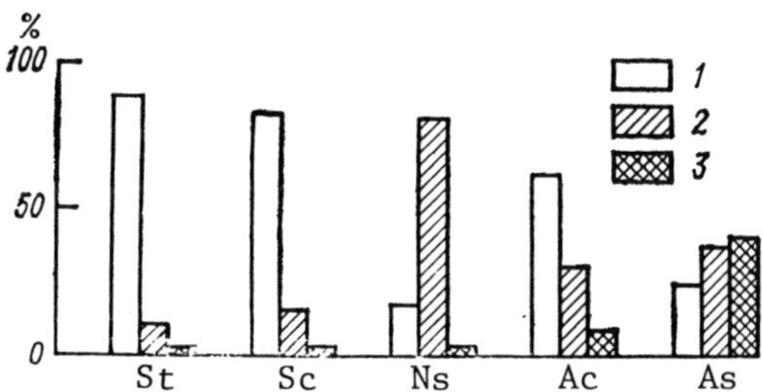

Fig. 2.1. Frequencies of phases in clouds of various types at a
negative temperature [3].
1) purely droplet supercooled clouds, 2) mixed clouds, 3) crystal
clouds.

It was established that the temperature is the main factor
determining the probability that supercooled droplets will be
present in a cloud. The concentration of ice nuclei, which determines the probability of the appearance of crystals in the cloud,
increases sharply with a drop in temperature. The droplet size is
the next most important factor: the smaller the droplets, the more
likely it is that unfrozen droplets will be present in the cloud
at the given temperature. Finally, important roles are also played
by the seeding of lower-lying layers with crystals falling from
higher layers at a lower temperature, as well as by the secondary
multiplication of crystals.

Table 2.1. Frequencies (%) of different phases of clouds as function of temperature (for USSR).

Phase of cloud	Temperature (from-to), °C											
	0 −2	−4 −6	−8 −10	−12 −14	−16 −18	−20 −22	−24 −26	−28 −30	−32 −34	−36 −38	−40 −42	−44 −46
Droplet	84	69	54	37	23	17	10	6	3	2	1	0
Mixed	14	26	35	42	40	35	31	25	17	11	7	6
Crystal	2	5	11	21	37	48	59	69	80	87	92	94

Table 2.1 gives the mean frequencies of the various phases
at various temperatures for clouds of all forms. These data are
practically independent of the season of the year.
The dependence of the phase on the cloud form is actually

determined by the temperature regime of the clouds. Figure 2.1 clearly illustrates the frequency of the phases in clouds of different forms in the middle latitudes. In these latitudes high-level clouds are practically always crystal clouds. The phase makeup of clouds in polar and tropical regions differs somewhat from the one shown (see Section 2.5).

2.2. Cloud water content

2.2.1. Definitions

The water content of a cloud, in the wide sense of the word, is defined as the mass of water in the condensed state (in the form of droplets and crystals) per unit mass of cloud air (specific water content) or per unit cloud volume (absolute water content). In the narrow sense, the water content is defined as just the liquid-droplet part, the part of the water present in the form of crystals being called the ice content.

The specific water content is measured in g/kg or g/g, that is, it is dimensionless; the absolute water content (often called simply the "water content") is measured in g/m^3 or g/cm^3.

In clouds, as a rule, the moisture content is close to saturation. Consequently, the amount of water in vapor form is determined solely by the temperature. The absolute saturation moisture content a decreases sharply with a drop in temperature, as is evident from Table 2.2.

Table 2.2. Temperature dependence of absolute saturation moisture content a (a_w = saturation with respect to water; a_i = saturation with respect to ice).

$t,°C$......	20	10	0	−10	−20	−30
$a_w,g/m^3$...	17.6	9.7	4.8	2.3	1.0	0.45
$a_i,g/m^3$...	−	−	4.8	2.15	0.88	0.34

The amount of water in a cloud is usually considerably less than the amount of vapor, that is, the water content $w < a$. The characteristic values of the water content decrease with temperature like the saturation moisture content, only considerably less sharply. The water content of a cloud, however, depends on the height of the point above the lower cloud boundary, as well as on the temperature. This is demonstrated especially clearly by the behavior of the so-called adiabatic water content (see Fig. 2.4).

2.2.2. *Relation of water content to temperature*

On the average, for all forms of clouds, disregarding the dependence on the location in the cloud, in the middle latitudes the frequency of the water-content values is described well for different temperatures by the empirical relation [3]

$$N(w) = 100\left[1 - \exp\left(-\frac{w - \varepsilon}{w_0}\right)\right]. \qquad (2.1)$$

Here $N(w)$ is the frequency (in %) of water-content values lower than w; parameters w_0 and ε are temperature dependent, as is evi-

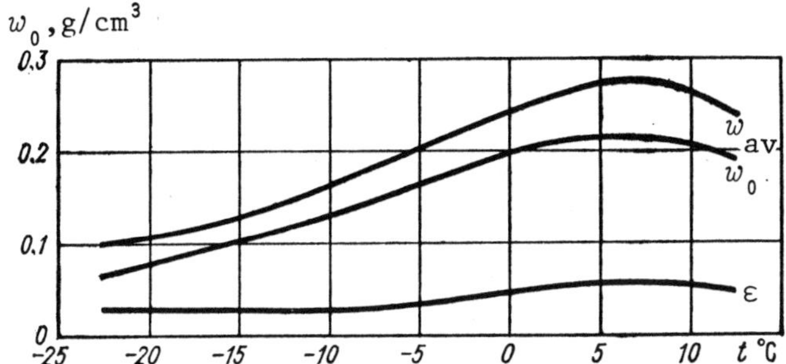

Fig. 2.2. Temperature dependence of characteristic values of water content.

dent from Fig. 2.2, which also shows the mean water content. It follows from formula (2.1) that

$$w_{\text{av}} = w_0 e^{\varepsilon/w_0} \approx w_0 + \varepsilon; \qquad (2.2)$$

the nth percentile w_n is*)

$$w_n = \frac{w_0}{\lg e}\left[2 - \lg(100 - n)\right] + \varepsilon. \qquad (2.3)$$

Thus, for instance, $w_{90} \approx 2.3\, w_0 + \varepsilon$ and $w_{99} \approx 4.6\, w_0$.

The $w_0(t_0)$ relation differs somewhat for different forms of clouds. As inspection of Fig. 2.3 shows, the lowest values of w_0 correspond to Ac and As clouds, and the highest to Ns clouds.

---

*) By definition, $w \leq w_n$ in n% of the cases.

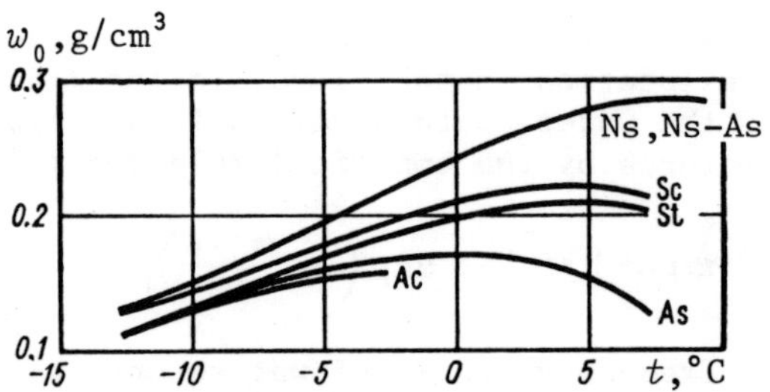

Fig. 2.3. Temperature dependence of w_0 for clouds of various types.

2.2.3. *Water content of convective clouds*

In vertically extensive clouds the mean water content $\overline{w}$
over the cloud section increases with the distance from the cloud
base, reaching the highest value in the upper part of the cloud
(at a height of about $0.8H$, where H is the depth of the cloud).

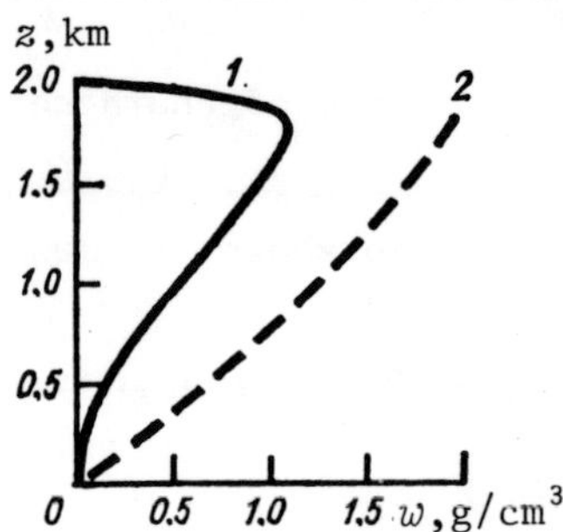

Figure 2.4 shows a typical $w(z)$ distribution for a cloud of depth
$H = 2$ km, the variation of the adiabatic water content being
plotted as well.*) The vertical variation of the water content
with increasing height is described well by the expression**)

$$\eta\,(\xi) = \frac{\overline{w}\,(\xi)}{\overline{w}_{\,\text{max}}} = \left(\frac{\xi}{\xi_0}\right)^m \left(\frac{1-\xi}{1-\xi_0}\right)^n, \qquad (2.4)$$

where $\xi = z/H$, and ξ_0 is the relative level at which the maximum
value of the water content is reached, that is, $\eta(\xi_0) = 1$.
For the steppe regions of the Ukraine [3], $m = 2.8$ and $n =
0.57$; the scatter of the values is about 30% for n ($\sigma_n = 0.8$)

---

*) The adiabatic water content is defined in [7].
**) In this notation, $\eta(\xi)$ is obviously not normalized to unity.

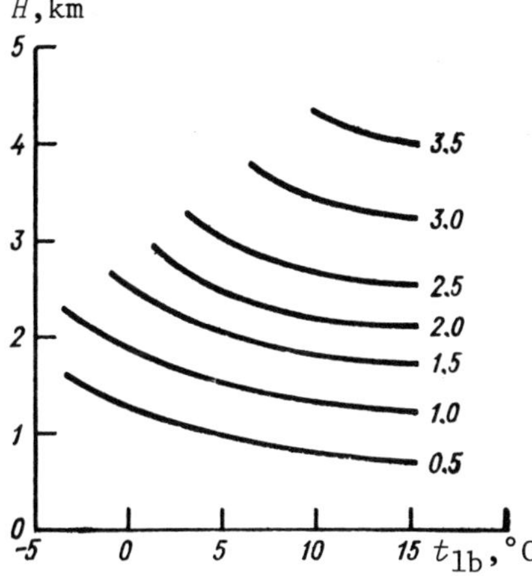

Fig. 2.5. Dependence of w_{max} on cloud depth and temperature t_{1b} on cloud base.
Numbers by curves indicate values of $\overline{w}_{max}$ in g/m^3.

and 50% for m ($\sigma_m = 0.25$); $\xi = 0.83 \pm 0.1$. The values $\overline{w}(\overline{\xi}_0) = w_{max}$ depend on the cloud depth and on the temperature. For the Ukraine [9] a good approximation of w_{max} (error generally not exceeding 30%) can be obtained with the aid of the diagram in Fig. 2.5. The following relation [9] can be used to evaluate the mean water content over the entire cloud:

$$w_{.av} = 0.34H. \tag{2.5}$$

Here H is in kilometers and w is in g/m^3.

2.2.4. Spatial variability

The depths of stratiform clouds of types St and Sc do not exceed 500–600 m, and the mean water content increases with height right up to the cloud summit. For deeper clouds, beginning at a height of 600–700 m, the mean water content gradually decreases (Fig. 2.6). In the case of Ns and Ns-As, the mean water content of the droplet fraction of the clouds*) at first increases with height, but then, beginning at 200–300 m, remains practically unchanged. In Ac clouds the vertical variation of the water content is similar to that in Fig. 2.6, except that the gradients in this case are lower by a factor of 1.5 to 2. In As clouds the height variation of the droplet part of the water content is the same as in Ns, that is, the mean water content is practically independent of the height.

The water content is a quite variable quantity, especially in clouds of convective origin. Possible values of the water content can be arrived at on the basis of relation (2.1), which suggests that 10% of the time the water content may be 1/4 of the mean value or twice as great as the latter. The spatial variability can

---

*) Data on the crystal fraction are as yet too scanty for statistical generalizations.

be typified by the maximum length of the interval L_w over which
the water content varies by no more than
a factor of two. It is also possible to
use the correlation radius L_{0w} as such
a nonuniformity scale (for more details
about the spatial nonuniformity, see
Section 4.9). The range of fluctua-
tion of scale L_w can be found from
Table 2.3, which gives the mean values
of L_w, L_{0w}, and also the region $L_1 - L_2$
including 80% of the measurements of L_w.
The values in Table 2.3 are actually only
tentative, however, and in a specific

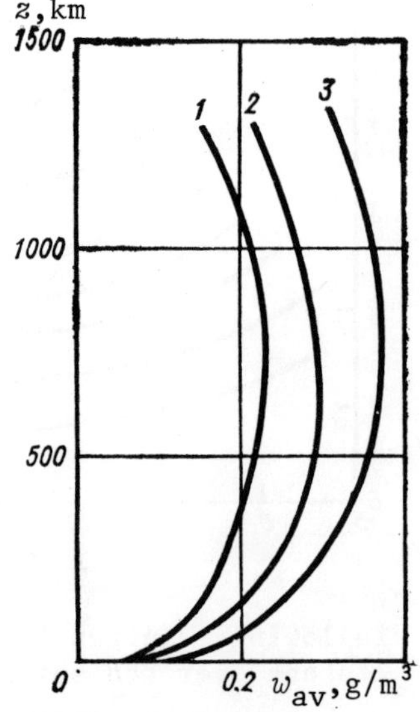

Fig. 2.6. Average variation of water
content with height in St and Sc
clouds for different temperatures.
1) –10 to –5°C; 2) –5 to +5°C;
3) +5 to +10°C.

cloud characteristic values $\overline{L}_w$ and $\overline{L}_{0w}$ may differ from those
tabulated by as much as 100%.

Table 2.3. Mean values of characteristic nonuniformity scales
for distribution of water content in various types of clouds.

Charac- teristic	Type of cloud					
	St	Sc	Ns	Ac	As	Cs
$\overline{L}_w$, km	0.6	0.6	1.1	0.7	0.9	0.6
$\overline{L}_{0w}$, km	1.0	1.5	1.5	1.4	1.4	1.4
$L_{w_1} - L_{w_2}$	0.05...4.5	0.03...3.0	0.05...4.0	0.1...1.6	0.1...1.6	

2.3. Droplet-size distribution

The concentration and size of cloud droplets can vary
greatly, but if we consider averages over large volumes of cloud
and over many cases then definite regularities show up in the
droplet-size distribution.

The radii of most cloud droplets range from a few microns
to some tens of microns. The droplet concentration varies from
tens per cm^3 to thousands per cm^3. The concentration of large
droplets (radii greater than 100 μm) is usually 1/1000 to
1/10,000 as great, being 0.1–10 per liter. In the presence of

precipitation, the rain drop concentration (radii greater than 100-200 µm) may be 10-100 per liter.

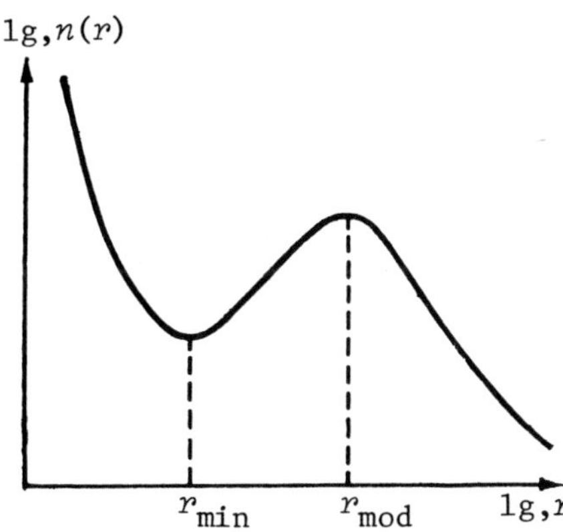

Fig. 2.7. Density of droplet-size distribution.

Along with cloud droplets, clouds also contain water-coated condensation nuclei; these take the form of tiny droplets about 1 µm or less in radius which are in equilibrium with the cloud material. Since in a cloud the moisture content is close to 100%, the radii of these droplets are 5 to 10 times the radii of the nuclei dissolved in them.

On the whole, the averaged distributions (size spectra) of the droplets in a cloud have the form shown in Fig. 2.7. In clouds of the middle latitudes r_{min} usually ranges from 0.5 to 2.0 µm, and r_{mod} from 2.5 to 7 µm. The left-hand part of the curve in Fig. 2.7, from $r = 0.1$ µm to r_{min}, characterizing the water-coated condensation nuclei, can be approximated by an exponential relation [1], where $r_0 = 1$ µm:

$$f(r) = A (r_0/r)^\nu. \qquad (2.6)$$

Table 2.4 gives values of A and ν, and also r_{min}, obtained from measurements with aircraft, carried out by Yudin in stratus clouds over the central part of the European USSR and the Baltic region, with the aid of a photoelectric counter [11].

Table 2.4. Generalization of eight series of observations in stratus clouds over central part of European USSR and over Baltic region (autumn).

Part of cloud	$A\mu m^{-1} \cdot cm^{-3}$	ν	r_{min} µm
Upper	45	3.2	2.0
Middle	50	3.0	1.7
Lower	75	3.7	1.2

Exponent ν varied slightly inside a cloud and from cloud to cloud (standard deviation $\sigma_\nu = 20$-30%). Parameter A fluctuated somewhat more ($\sigma_A = 30$-50%).

The mean droplet radius $\bar{r}$ observed during these flights ranged from 3.5 to 5.0 μm,*) the relative standard deviation $\sigma_r/\bar{r}$ being from 0.3 to 0.4.

In four series of measurements in stratus clouds over Arctic seas, parameter A was found to vary, from the upper part of the cloud to the lower, from 40 to 5 $\mu m^{-1} \cdot cm^{-3}$, while exponent ν varied from 3.7 to 5.5 and r_{min} from 2.5 to 1.1 μm. The standard deviations of the parameters increased accordingly, from 30% in the upper part of the cloud to 80 or 100% in the lower part. In stratocumulus clouds, 30% of the time r_{min} lay beyond the range of the instrument, that is, it was less than 1 μm. In growing cumuli $r_{min} < 1$ μm in 95% of the cases, while in dissipating cumuli outside of upcurrents $r_{min} > 1$ μm in 90% of the cases.

The region to the right of r_{min} is generally a unimodal curve, which can be approximated by a lognormal or gamma distribution. For simplicity of analysis and convenience of processing, a Γ distribution is more often used. Normalized to unity and defined in the region $(0, \infty)$, the Γ distribution has the form

$$ n(r) = \frac{1}{\Gamma(\alpha+1) r_0^{\alpha+1}} r^{\alpha} e^{-r/r_0}. \qquad (2.7) $$

The concentration $N(r)$ of droplets greater than r in radius is

$$ N(r) = N_0 \int_r^{\infty} n(r)\, dr = N_0 [1 - \gamma_{\alpha}(r/r_0)], \qquad (2.8) $$

where

$$ \gamma_{\alpha}(x) = \frac{1}{\Gamma(\alpha+1)} \int_0^{x} t^{\alpha} e^{-t}\, dt \qquad (2.9) $$

is the incomplete gamma function of index α.

The distribution parameters N_0, α, and r_0 depend on the cloud type, the averaging scale, and other factors. Let us recall that the mode radius r_{mod} is related as follows to the distribution parameters:

$$ r_{mod} = \alpha r_0, \qquad (2.10) $$

and the mean radius r_1 is related as follows:

____

*) The mean radius is defined over an interval $r_{min} - r_2$ where r_2 is so chosen that the concentration of larger droplets does not exceed 1-2 cm^{-3}.

$$r_1 = (\alpha + 1)\, r_0, \qquad\qquad (2.11)$$

while the root-mean-square radius is

$$r_2 = r_0 \sqrt{(\alpha + 1)(\alpha + 2)}, \qquad\qquad (2.12)$$

and the root-mean-cube radius is

$$r_3 = \sqrt[3]{(\alpha + 1)(\alpha + 2)(\alpha + 3)}. \qquad\qquad (2.13)$$

Accordingly, the standard deviation is

$$\sigma = r_0 \sqrt{\alpha + 1}, \qquad\qquad (2.14)$$

and the relative standard deviation is

$$\frac{\sigma}{r_1} = \frac{1}{\sqrt{\alpha + 1}}. \qquad\qquad (2.15)$$

Before considering some specific values of the distribution parameters, let us cite a few properties of a more general nature. Firstly, the typical sizes of droplets in stratiform clouds (St, Sc) increase with increasing height z above the lower boundary in proportion to $z^{1/3}$, on the average. The variation with height of the typical droplet sizes in convective clouds is not so definite, however. Although on the whole the sizes of these droplets also increase with height, mixing with dry ambient air, especially in the lower half of the cloud, may cause a great number of tiny droplets to appear, leading in a number of cases to a bimodal distribution [12]. The onset of precipitation may also appreciably change the size distribution of the cloud droplets and the height variation of the distribution parameters.

Finally, for small averaging scales, of the order of tens of meters, the droplet-size spectrum is quite narrow.*) Distribution parameter α, which according to formula (2.14) is a measure of the width of the spectrum in these cases, may be 6 or even more. The droplet-size spectrum is broadened considerably, while parameter α is reduced sizably, if the averaging scales are increased and if samples pertaining to different levels, different clouds, different times, etc., are averaged.

As we pointed out in [10], and subsequently verified by numerous studies carried out in various countries, when many cases are

---

*) According to the data of Aleksandrov and Yudin [1].

averaged, the droplet-size spectra for clouds of various types
in the radius range from 2-3 to 20-30 µm are quite broad, being
described by a gamma distribution with $\alpha = 2$. In the range of
droplet sizes from $r = 50\text{-}100$ µm to some maximum value $r_{\max}$, which
in nonprecipitating clouds may range from 100 to 1000 µm, the
droplet distribution is described by an exponential relation:

$$N(r) = \begin{cases} N_{100}\left(\dfrac{100}{r}\right)^{\beta-1} & \text{for } r < r_{\max}, \\ \approx 0\,(<1\text{ м}^{-3}) & \text{for } r > r_{\max}. \end{cases} \qquad (2.16)$$

Here r is in microns, and N_{100} is the concentration of droplets
(drops) larger than 100 µm in
radius. Usually N_{100} varies
from a few per m^3 to thousands
per m^3, and β from 3 to 15.

   The authors of [3]
generalized collected data on
the microstructure of droplet
clouds to obtain a diagram
characterizing the possible
droplet concentration in a
cloud. With some minor modifications, the diagram is presented in Fig. 2.8. In an
overwhelming majority of cases
real cloud-droplet concentrations do not lie outside of
the shaded region and are
practically never beyond the
upper curve. If precipitation
falls from the clouds, then
the concentration of large
droplets usually lies in the
unshaded region between the
curves.

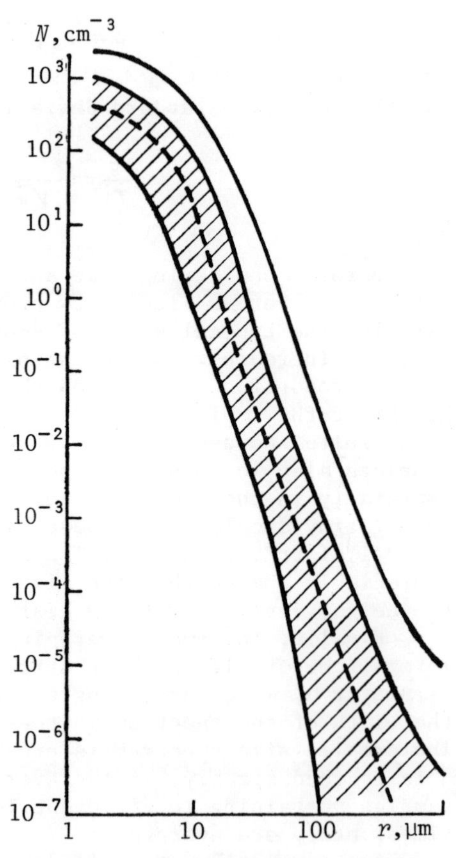

Fig. 2.8. Generalized diagram
of possible distribution of
cloud droplets.
Dashed curve gives droplet
distribution for $\alpha = 2$, $w =$
$= 0.4$ g/m^3, $r_1 = 4.5$ µm, $\beta =$
$= 6$, $N_{100} = 10^2$ m^{-3}, and
$r_{\max} = 400$ µm.

   The following can be recommended as model spectra: (2.6) in
the region of r from 0.1 to 1 µm; (2.7) in the region from $r_{\min} \approx$
≈ 1 µm to 20 µm. In the region from 20 µm to $r_{\max}$ the sum of

distributions (2.8) and (2.16) is recommended. For many optical and radiation problems the second term can be neglected.

Formula (2.17) represents the proposed model distribution:

$$
N(r) = \begin{cases}
\dfrac{A}{\nu - 1}\,\dfrac{1}{r^{\nu-1}} & \text{for } 0.1 \ \mu m < r < r_{min}, \\[2ex]
N_0\,[1 - \gamma_\alpha(r/r_0)] & \text{for } r_{min} < r < 20 \ \mu m, \\[2ex]
N_0\,[1 - \gamma_\alpha(r/r_0)] + N_{100}\left(\dfrac{100}{r}\right)^{\beta-1} & \\[1ex]
& \text{for } 20 \ \mu m \leqslant r < r_{max}, \\[2ex]
0 & \text{for } r > r_{max}
\end{cases}
\qquad (2.17)
$$

Correspondingly, the model size spectrum can be represented as

$$
n(r) = \begin{cases}
\dfrac{Ac}{r^{\nu}} & \text{for } 0.1 \ \mu m < r < r_{min}, \\[2ex]
\dfrac{c}{\Gamma(\alpha+1)\,r_0^{\alpha+1}}\,r^\alpha e^{-r/r_0} & \text{for } r_{min} < r < 20 \ \mu m. \\[2ex]
\dfrac{c}{\Gamma(\alpha+1)\,r_0^{\alpha+1}}\,r^\alpha e^{-r/r_0} + \dfrac{(1-\beta)\,N_{100}c}{100}\left(\dfrac{100}{r}\right)^\beta & \\[1ex]
& \text{for } 20 \ \mu m \leqslant r < r_{max}, \\[2ex]
0 & \text{for } r > r_{max}.
\end{cases}
\qquad (2.18)
$$

Here c is a normalization constant, found from the condition

$$
\int\limits_{0.1}^{\infty} n(r)\,dr = 1.
$$

When averaging over a great number of cases for clouds of the middle latitudes, the following parameter values are recommended: $\alpha = 2$ and $N_0 = 10^7 w/r_1^3$, where the water content w is in g/m^3 and the mean radius r_1 is in microns. According to (2.11), parameter $r_0 = r_1/3$. Table 2.5 gives some r_1 values.

Table 2.5. Mean radii r_1 for various cloud types.

Cloud type	St	Sc	Ns	Ac	Cu hum.	Cu med.	Cu cong.
r_1, μm	4.5 3.9; 4.5; 5.3	5.0 4.0; 4.8; 5.5	6.5	4.7	3.0	4.0	6.0

For St and Sc the values of r_1 in the lower, middle, and upper parts of the clouds, respectively, are given, as well as the mean value.

The dashed curve in Fig. 2.8, corresponding to parameters $\alpha = 2$, $w = 0.4$ g/m^3, $r_1 = 4.5$ μm, $N_{100} = 10^2$ m^{-3}, $\beta = 6$, and $r_{\max} = 400$ μm, can be taken to be typical for clouds.

The distribution parameters of extra-large droplets (N_{100}, β, $r_{\max}$) are quite variable and nonuniform in space. For instance, in St and Sc N_{100} varies from 0 in very thin clouds to 10^3 m^{-3}, while β ranges from 3 to 12. On the average, $\beta = 6$ for St and 5 for Sc, while $r_{\max}$ does not exceed 500 μm for St and 400 μm for Sc. In Cu hum. $N_{100} \to 0$ and in Cu med. $N_{100} \approx 10$ m^{-3}. In Cu cong. and Ns $N_{100} = 10^2$–10^3 m^{-3}, β is about 4 or 5, and $r_{\max}$ is of the order of 10^3 μm. Parameter A ranges from 40 to 80, while ν is generally 3 to 4.

2.4. Crystals in clouds

In contrast to the case with droplets, when considering crystals in clouds it is important to know their shape and orientation as well as their size and number. Ice particles in clouds can be divided into three groups, according to shape: rounded, laminar, and acicular. The first group includes particles whose dimensions in any direction are about the same; the second and third groups include particles which are appreciably smaller in one direction than in a perpendicular direction. If the crystals are small, so that the corresponding Reynolds number $Re = lu/\nu$ is less than one (here u is the fall speed of the crystal, l is a characteristic dimension, and ν is the kinematic viscosity of air), then as a first approximation the particles can be assumed to be oriented randomly in space. If, one the other hand, $Re > 1$, then it can be assumed that crystals of the first group are oriented randomly in clouds, crystals of the second group have their symmetry axis oscillating about the vertical, and crystals of the third group have their longitudinal axis close to the horizontal.

2.4.1. Shapes of cloud crystals

Ice crystals in the atmosphere have extremely diverse shapes, but as a rule they all possess a definite hexagonal symmetry, which

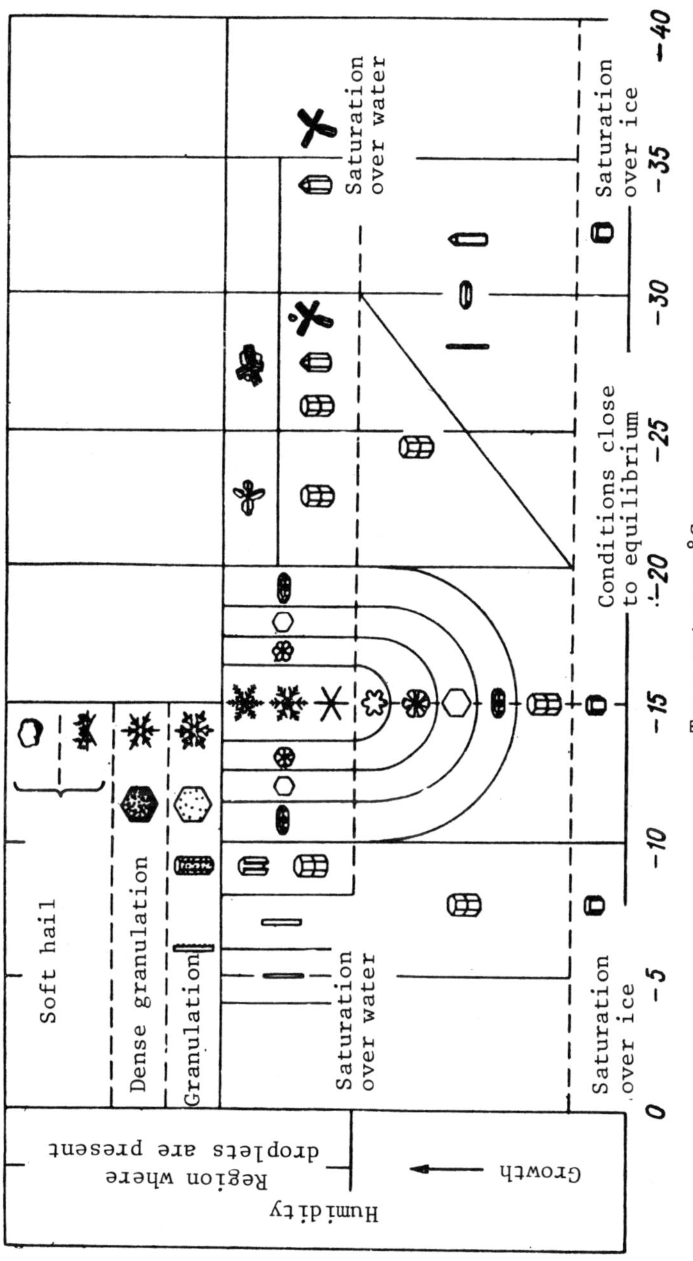

Fig. 2.9. Generalized diagram (after Magono and Lee [12]) showing relation of shape of ice crystals to conditions of crystal growth (temperature and humidity).

manifests itself as a variety of hexagonal prisms, hexagonal
plates, and six-ended dendrites. Rounded crystals may contain
various air inclusions (voids). The shape of the cloud crystals
depends on the temperature and humidity. Since, when crystals
originate in clouds, the humidity is close to the saturation value
over water, the shape of a crystal being formed may, as a first
approximation, be related only to the temperature.

Figure 2.9 juxtaposes the shapes of ice crystals with the
conditions of their formation. Soviet [3] and non-Soviet experimental data show a good fit with this diagram. The correlation
of the crystal shape with the temperature and humidity is sub-
stantiated more fully in [4].

As we see from Fig. 2.9, in high-level clouds, where the
temperature is below $-18°C$, the crystals can be expected to take
the form of columns, prisms, thick plates, and bundles (often
hollow) of crystals. In the $(-15 \pm 3)°C$ region crystals usually
take the form of dendritic starlets or thin plates. At $(-10 \pm 2)°C$
columns (but no longer hollow) and thick plates reappear, while
around $-5°C$ the crystals have the form of needles.

2.4.2. *Crystal size and concentration*

The sizes of crystals in clouds are usually hundreds of
microns along the major axis, with plate or column (needle) thicknesses of tens of microns. The "ice content", that is, the mass
of ice in the crystals, ranges from 10^{-3} g/m^3 in high-level clouds
(except Ci unc., where it may be as high as 10^{-1} g/m^3) to 10^{-1} g/m^3
in St and Ns. The concentration ranges from some units or tens per
liter in high-level clouds to tens or hundreds per liter in Ns, As
systems. The density of the ice in the crystals can vary considerably depending on their size and shape. With an increase in crystal size from tens of microns to millimeters, the relative proportion of the air inclusions increases and the density decreases from 0.9 to 0.5 g/m^3.

2.5. Microstructural features of polar and tropical clouds

There is as yet comparatively little information about the
microstructure of clouds in the Arctic and the tropics, while
data on antarctic clouds are still practically nonexistent. Indi-
cations are that, on the whole, in marine cumuli of the tropics
the total droplet concentration is an order of magnitude lower
than the droplet concentration in clouds of the middle latitudes,
being some tens per cm^3 [8]. At the same time, the concentration
of droplets with radii larger than 100 μm in cumuli of the low
latitudes is an order of magnitude higher than in the middle
latitudes, being from 1 to 10 per liter. Typical sizes of droplets in marine tropical clouds are 2-4 times the analogous droplet

sizes in clouds of the middle latitudes.

The water contents of arctic clouds are on the average several tens of percents less than the water contents of middle- -latitude clouds at the same temperature. This is due mainly to the fact that clouds of the Arctic are not as deep.

Although the temperature dependence of the cloud phase in various latitudes has not yet been studied much, it can be assumed that the increase in droplet size upon transition from polar clouds to tropical clouds increases the probability that the droplets will freeze at a given temperature [8]. In other words, it is more likely that the supercooled phase will be encountered in polar clouds than in clouds of the middle latitudes at the same .temperature, while this phase is less likely in tropical clouds. Differences in the aerosol structure of the atmosphere may also be of definite importance.

With regard to cloud types, in the polar regions St, Sc, and Ac are predominantly mixed clouds, although frequently (especially in summer) pure droplet clouds are encountered; polar Ns clouds are mixed, while polar As are mostly crystal clouds. In the tropics clouds which are quite low-lying, such as Cu pass., Sc, and even Ac, are at altitudes where the air temperature is above 0°C, so that these are, naturally, droplet clouds. Finally, in these regions droplets (usually with radii no larger than a few microns) may also often be found in high-level clouds as well. Consequently, in tropical latitudes the mixed phase is possible in high-level clouds.

CHAPTER 3

SPACE-TIME STATISTICAL STRUCTURE OF A CUMULUS FIELD

When applying any of the methods for calculating the radiation regime and its variability for cumulus clouds, in addition to the optical characteristics of an individual cloud it is important to know something about the size distribution of the clouds and the structure of the cloud field.

Studies of the structure of cumulus fields using optical methods have indicated a suitable form for modeling cloud fields and also have revealed the relationships between the statistical characteristics of various cloud parameters [12]. The parameters on which the theory is based were determined experimentally, as were a number of important empirical relationships.

The theoretical model is based on experimental data obtained at the Institute of Astrophysics and Atmospheric Physics of the Estonian Academy [5, 8, 9, 12]. The experimental data presented in Section 3.2 also verify the applicability of the theoretical model.

3.1. A theoretical-experimental model of the statistical structure

In their analysis of the structure of cumulus fields, the authors used an indicator function for the presence of clouds $n(\theta, \varphi, x, y, t)$, which is equal to 1 in the presence of clouds and is equal to 0 when there are no clouds present in the direction θ, φ over a point x, y at a time t. By applying the mathematical apparatus of the theory of random processes to the indicator function, we arrive at various statistical characteristics of cumulus fields.

3.1.1. Mean characteristics of cumulus cloudiness

After statistically averaging the cloud presence $n(\theta, \varphi, x, y, t)$, we obtain the mean cloud cover:

1) of direction θ, φ over point x, y (averaged over time t);

2) of a section of cloud field along the y axis in direction θ, φ at time t (averaged over x);

3) of the territory being studied in direction θ, φ at time t (averaged over x, y);

4) of the almucantar of θ over x, y at time t (averaged over φ);

5) of the sky over point x, y at time t (averaged over θ, φ for the whole sky).

In addition to the foregoing, the mean amounts of cloud cover averaged over other combinations of variables can be used. In the subsequent analysis we will pay special attention to the mean cloud cover at the zenith ($\theta = 0$), denoted by $n(0)$, and the mean sky cover n.

Assuming statistical isotropy of the cloud field, we can equate the averaged cloud covers of the almucantars to the probability of the cumulus cover in the viewing direction $n(\theta)$. A comparatively large number of calculations of $n(\theta)$ have been carried out.*) Considering the variability of the cloud field as a normal random process, we get the following formula for $n(\theta)$ [12]:

$$n(\theta) = 1 - [1 - n(0)]\, e^{-n(0)\,[N(\theta)-1]}, \qquad (3.1)$$

where $N(\theta) \geq 1$ is the mean multiplicity of intersection of a random surface in the viewing direction θ:

$$N(\theta) = \frac{1}{2} + \frac{1}{2}\,\mathrm{erf}\,\frac{\mathrm{ctg}\,\theta}{\sqrt{2}\cdot\sigma_{z'}} + \frac{\sigma_{z'}}{\sqrt{2\pi}\,\mathrm{ctg}\,\theta}\, e^{-\mathrm{ctg}^2\theta/2\sigma_{z'}^2}. \qquad (3.2)$$

Here erf is the probability integral, and $\sigma_{z'}^2$, is the variance of the derivative of the normal random surface describing the configuration of the upper cloud boundary. For the Baltic region $\sigma_{z'} = 1.2$, while in the tradewind zone over the oceans $\sigma_{z'} = 0.6$ if $n(0) < 0.6$ and $\sigma_{z'} = 1.5$ if $n(0) > 0.6$.

The presence of a special function in formula (3.1) complicates the calculations somewhat. As shown in [9], instead of (3.1) we can use the comparatively simple empirical relation

$$n(\theta) = 1 - [1 - n(0)]\, e^{-n(0)\,(\sec\theta - 1)\,b}, \qquad (3.3)$$

where coefficient b depends on the mean vertical depth of the clouds.

Figure 3.1. shows the mean cloud covers of the almucantars $n(\theta)$, obtained on the basis of sky photos taken on land (Estonian SSR) and in the Atlantic west of the Canary Islands (27°N, 25°W). The pictures were divided into three groups, according to the amount of relative cloudiness. Table 3.1 gives the numbers of photos in the groups and the values of coefficient b.

A comparison of Fig. 3.1a with Fig. 3.1b shows that the rise of the curves depends only comparatively slightly on parameter b and that no qualitative differences in the increase in cloudiness toward the horizon under land and sea conditions are observed. Consequently, in the calculations we can set $b = 1$.

---

*) One of the first was that of Väisälä (1929).

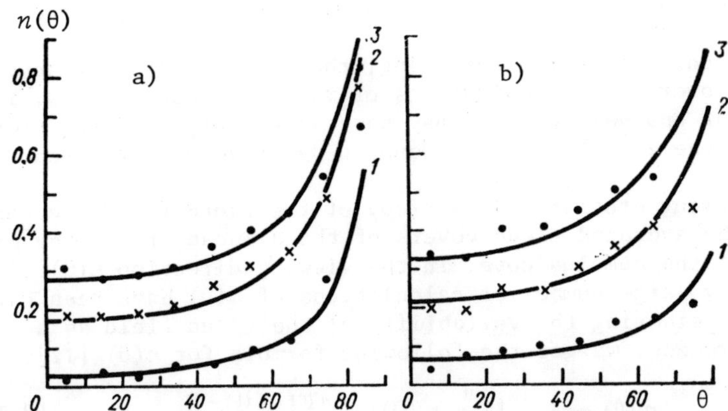

Fig. 3.1. Mean cloud covers of almucantars $n(\theta)$ as function of zenith angle θ, according to photos taken from land (a) and sea (b), grouped on the basis of amount of relative cloudiness. 1) n = 0-0.3; 2) n = 0.2-0.5; 3) n = 0.3-0.7. Approximation curves calculated using formula (3.3) and data of Table 3.1.

Table 3.1.

n	Land		Sea	
	No. of photos	b	No. of photos	b
0.0. . . 0.3	19	2.5	46	1.0
0.2. . . 0.5	76	1.0	35	1.0
0.3. . . 0.7	104	0.7	40	1.0

Now, by integrating $n(\theta)$ over the hemisphere, we obtain the mean cloud cover of the entire sky. Using formula (3.3) for $n(\theta)$, we get

$$n = \int_0^{\pi/2} n(\theta) \sin\theta\, d\theta = n(0) - bn(0)[1 - n(0)] e^{bn(0)} \mathrm{Ei}[-bn(0)],$$

$$(3.4)$$

where Ei is an integral exponential function.

The relation between the mean amount of cloudiness in the sky n and the mean amount at the zenith $n(0)$ can also be described by empirical formulas from [12]:

$$n = n(0) + 0.5[1 - n(0)]n(0) \qquad (3.5)$$

or

$$n = n\,(0) + 0.8\,[1 - n\,(0)]\,n^{0.8}\,(0). \qquad (3.6)$$

Experimental data [3] indicate that formula (3.5) is suitable for cloud depths $H \leq 0.5$ km, and formula (3.6) for $H > 1$ km. Because of the increase in cloudiness toward the horizon, the mean amount of cloudiness at the zenith is always less than the mean amount over the sky.

3.1.2. Statistics of cumulus parameters

The frequency of occurrence of clouds $\varkappa$ in a section of a cloud field, that is, the amount of cloud cover per unit time or per unit length in the normal random model, is given by the formula

$$\varkappa\,[n\,(0)] = \frac{1}{2\pi}\,\frac{\sigma_{z'}}{\sigma_z}\,\exp\{-[\arg \operatorname{erf}|\,1 - 2n\,(0)\,||^2\}, \qquad (3.7)$$

where σ_z and $\sigma_{z'}$ are the standard deviations of the normal random surface and its derivative from the mean [12]. For the middle

latitudes on land $\dfrac{1}{2\pi}\,\dfrac{\sigma_{z'}}{\sigma_z} = 0.45$ km^{-1}, while in the tradewind zone

the value is 0.96 km^{-1} if $n(0) < 0.6$ and 0.24 km^{-1} if $n(0) > 0.6$.

The mean lengths of the cloud sections s and the intervening spaces between clouds s_0 in a section of a cloud field depend on the frequency $\varkappa$ and the mean cloud cover $n(0)$:

$$\bar{s} = \frac{n\,(0)}{\varkappa}, \qquad \bar{s}_0 = \frac{1 - n\,(0)}{\varkappa}. \qquad (3.8)$$

The mean sizes of sections of both clouds and intervening spaces over land are somewhat larger than the corresponding mean sizes over the sea (Table 3.2).

The probability densities of sections of clouds and intervening spaces are approximated well by a lognormal law [5]

$$p\,(s) = \frac{1}{M\,\sqrt{2\pi}\cdot\sigma_w}\,\frac{1}{s}\,e^{-u^2/2}, \qquad (3.9)$$

where $u = (w - \bar{w})/\sigma_w$, $w = \lg s$, $\bar{w}$ is the mean value of the random quantity w, σ_w is the standard deviation of the random quantity w, and $M = 1/\lg e = 2.303$. The lognormal distribution is a two-parameter distribution. Quantities $\bar{w}$ and σ_w were used as the parameters in formula (3.9), but the mean value $\bar{s} = \bar{w}e^{M^2\sigma_w^2}$ and the

Table 3.2. Group averages of cloud frequency $\varkappa$ (km^{-1}), mean sizes of clouds $\bar{s}$ and intervening spaces $\bar{s}_0$ (km), and mean amounts of cloudiness $\bar{n}(0)$.

Measurements	$n(0) = 0-0.3$				$n(0) = 0.3-0.6$			
	$\bar{n}(0)$	$\bar{s}$	$\bar{s}_0$	$\varkappa$	$\bar{n}(0)$	$\bar{s}$	$\bar{s}_0$	$\varkappa$
Land	0.25	0.39	1.16	0.65	0.43	0.55	0.74	0.78
Sea	0.20	0.30	1.18	0.61	0.39	0.43	0.62	0.90

distribution mode $s_M = \bar{w}e^{-M^2\sigma_w^2}$ could just as well have been used instead.

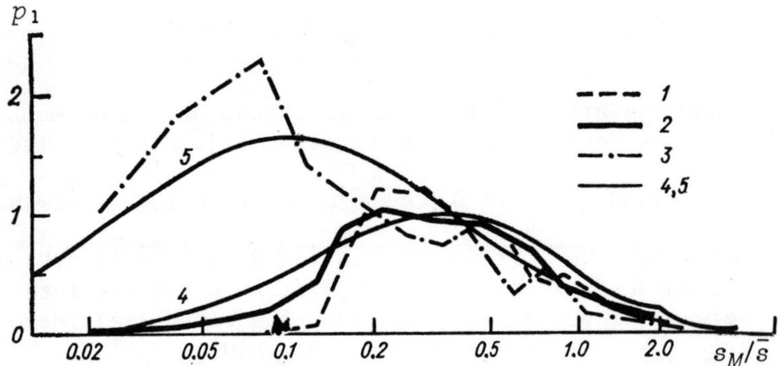

Fig. 3.2. Probability density of lengths of cumulus sections. 1) $n(0) = 0-0.3$; 2) $n(0) = 0.3-0.5$; 3) $n(0) = 0.6-1.0$;

4) lognormal distribution for $\lg \dfrac{s_M}{\bar{s}} = -0.6$; 5) for $\lg \dfrac{s_M}{\bar{s}} = -1.1$.

Figure 3.2 shows the lognormal and experimental distributions of the lengths of cumulus sections in the tradewind region [5].

Assuming that the cloud bases can be approximated by circles, let us now go from the distribution of sections to the distribution of diameters. The data of [1, 2, 5, 18] indicate that the distribution of cloud diameters on an area also conforms well to a lognormal law. The normalized mode of the diameter distribution $D_M/\bar{D}$ is a little larger than the mode of chords $s_M/\bar{s}$, that is, the distribution of sections is a little more asymmetric than the distribution of diameters. This explains the experimentally observed fact that for $s/\bar{s} > 0.2$ or $D/\bar{D} > 0.2$ as a first approximation the probability densities of both diameters and sections can be approximated satisfactorily by exponential or near-exponential functions

[7, 10, 12, 17, 19]. It should be noted that the energetics of clouds (including radiation processes) is proportional not to $p(D)$, but rather to $D^2 p(D)$, that is, to the relative area of the cloud cover, which has a diameter D. This means that the effects of clouds having relative diameters less than 0.2 can be ignored. Consequently, for energy calculations the maximum probability density (which occurs for $D/\bar{D} < 0.2$) does not have to be taken into account. Thus, as a first approximation the probability density of diameters or section lengths can also be approximated by an exponential function or by the following formula from [12], instead of by formula (3.9):

$$p\,[s,\,n\,(0)] = \frac{a\,[n\,(0)]}{\pi}\;\frac{e^{-a\,[n\,(0)]\,s}}{\{1 - e^{-2a\,[n\,(0)\,s]}\}^{3/2}}, \qquad (3.10)$$

where the dependence of parameter α on the mean cloud cover at the zenith $n(0)$ is described by the empirical expression

$$\alpha\,[n\,(0)] = 0.68 \pm 2.0\,[n\,(0) - 0.5]^2. \qquad (3.11)$$

In formula (3.11) a plus sign corresponds to $n(0) < 0.5$ and a minus sign to $n(0) > 0.5$. When formula (3.11) is applied to the tradewind region, for $n(0) < 0.6$ parameter $\alpha[n(0)]$ has to be multiplied by 1.8, while for $n(0) > 0.6$ it must be multiplied by 0.8.

The mean values $\bar{s}$ and $\bar{D}$ are related by the formula

$$\frac{\bar{s}}{\bar{D}} = \frac{\pi}{4}\left(\frac{D_M}{\bar{D}}\right)^{-2/3}, \qquad (3.12)$$

from which it follows that, if $D_M/\bar{D} < (4/\pi)^{-3/2}$, as is the case for the distribution of diameters of cumuli, then

$$\bar{s} > \bar{D}. \qquad (3.13)$$

This at first glance paradoxical result can be explained by the fact that the intersection of clouds with small diameters is less likely than the intersection of clouds with large diameters [5].

The frequency of clouds in an area $\varkappa_s$ and the frequency in a section $\varkappa$ are related by the formula [1]:

$$\varkappa_s = \frac{\varkappa}{\bar{D}}. \qquad (3.14)$$

3.1.3. Correlation and spectral characteristics of cumulus fields

The variance of the presence of clouds in direction θ is given by the simple expression

$$\sigma^2_{n\,(\theta)} = n\,(\theta)\,[1 - n\,(\theta)].\qquad(3.15)$$

The experimentally determined mean autocorrelation functions of the presence of clouds in a section of a cloud field for $n(0)$ from 0.1 to 0.9 can be approximated by the formula

$$r_{n\,(0)}(x) = \frac{2}{\pi}\,\arcsin\,e^{-\alpha\,[n\,(0)]\,|x|} =$$

$$= \frac{2}{\pi}\sum_{k=0}^{\infty}\frac{(2k-1)!}{(2k)!\,(2k+1)!}\,e^{-\alpha\,[n\,(0)]\,(2k+1)\,|x|},\qquad(3.16)$$

where the dependence of parameter α on $n(0)$ is given by formula (3.11) with a plus sign [9, 12]. The time function of correlation is obtained by multiplying $\alpha[n(0)]$ by 0.4 and replacing the distance x in kilometers by the time coordinate t in minutes (velocity of cloud motion 24 km/h). Since function (3.16) differs only slightly from the exponential, therefore as a first approximation the correlation function of the presence of clouds can also be approximated by an exponential function.

Using correlation function (3.16) in the form of a series, we obtain, respectively, the one-dimensional and two-dimensional spectral densities of the presence of clouds at the zenith:

$$S_{n\,(0)}(\omega) = \frac{1}{\pi}\int_0^{\infty} r_{n\,(0)}(x)\,e^{-j\omega x}\,dx =$$

$$= \frac{2\alpha}{\pi^2}\sum_{k=0}^{\infty}\frac{[(2k-1)!]^2}{(2k)!}\,\frac{1}{\alpha^2\,(2k+1)^2 + \omega^2},\qquad(3.17)$$

$$S_{2,\,n\,(0)}(\omega) = \frac{1}{2\pi}\int_0^{\infty} r_{n\,(0)}(x)\,J_0(\omega x)\,x\,dx =$$

$$= \frac{\alpha}{\pi^2}\sum_{k=0}^{\infty}\frac{(2k-1)!}{(2k)!}\,\frac{1}{[(2k+1)^2\,\alpha^2 + \omega^2]^{3/2}},\qquad(3.18)$$

where J_0 is a zero-order Bessel function.

A transition from the statistics of the cloud cover at the zenith to the correlation-spectral characteristics of the cover of the entire sky (sky cover) is possible provided the increase in cloudiness toward the horizon is neglected. Then the values obtained for the variance σ^2_n and the normalized correlation function r_n of the relative cloudiness will be somewhat high.

The frequency characteristic of the filter transforming the spectral densities of the zenith cloud cover $S_{2,n(0)}(\omega)$ into the spectral densities of the relative cloudiness $H^2_2(\omega)$ is a function only of the distance of the observer from the lower cloud boundary z:

$$H_2^2(\omega) = e^{-2z\omega}. \tag{3.19}$$

The variance of the relative cloudiness is defined by the integral

$$\sigma_n^2(z) = \sigma_{n\,(0)}^2 \, 2\pi \int\limits_0^\infty S_{2,\,n\,(0)}(\omega)\, e^{-2z\omega}\, \omega\, d\omega, \tag{3.20}$$

from which for $z = 0$ we have

$$\sigma_n^2(0) = \sigma_{n\,(0)}^2. \tag{3.21}$$

For $z = 1$ km, we get $\sigma_n^2(z) \approx 0.16^2 n(0)$, which fits well with the measured values.

The normalized correlation function of the relative cloudiness is calculated as the inverse Hankel transform of the spectral density (3.18), multiplied by (3.19):

$$r_n(x) = \frac{\sigma_{n\,(0)}^2}{\sigma_n^2(z)} \, 2\pi \int\limits_0^\infty e^{-2z\omega} S_{2,\,n\,(0)}(\omega)\, J_0(\omega x)\, \omega\, d\omega, \tag{3.22}$$

and it also depends on height z. For instance, if the cloud cover moves at a speed of 24 km/h, then the radius of correlation of the relative cloudiness increases from 5.5 to 8.5 min as distance z increases from 0.6 to 1.4 km.

3.1.4. The use of eigenvectors to describe the cloud cover of almucantars

The variability of the cloudiness from the zenith to the horizon is described by a system of eigenvectors of the cover of the almucantars. The latter were first calculated from sky photos [12].

In [9] the elements of the correlation matrix of the cloud cover of the almucantars were calculated theoretically, on the basis of the correlation function of the zenith cover (3.16)

$$K(\theta_i,\, \theta_j) = \frac{2\pi\sigma_{n\,(0)}^2}{\pi} \sum_{k=0}^\infty \frac{(2k-1)!}{(2k)!} \int\limits_0^\infty \frac{J_0(\omega R_i)\, J_0(\omega R_j)}{[(2k+1)^2\, a^2 + \omega^2]^{3/2}}\, \omega\, d\omega. \tag{3.23}$$

Despite the simplifications that were made (for instance, the increase in cloudiness from the zenith to the horizon was neglected), a good fit is observed between the experimental eigenvectors and the theoretical eigenvectors calculated from the correlation matrix (3.23). The first eigenvectors account for about 80% of the variance, the second about 10%, and the third about 5%. Therefore, the variability of the cover of the almucantars from the zenith to the horizon can be described reliably by three eigenvectors.

The eigenvectors derived above are applicable, for example, when reconstructing the zenith variation of the cloud cover of the almucantars or the viewing directions at a given time, according to a specified $n(0)$, and also when calculating the eigenvectors, averaged over the azimuth, of the long-wave radiation brightnesses [8, 9, 12].

3.2. An empirical model

In the previous section the statistical structure of a cumulus field was considered on the basis of a theoretical model of the spatial distribution of the clouds. Now we will present the results of a purely empirical study of this structure. Data of continuous recordings of the direct and total solar radiation were used, as well as data on the self-radiation of the sky in the region around the zenith, in the spectral interval from 8 to 13 μm.

The probability density $p(s)$ of cloud sections s on the line of sight of the instrument was evaluated, that is, the number of clouds per unit length of a section of the cloud field. The method used is described in [6, 15], and the main results are presented in [7, 11, 14, 16].

The total volume of data used comprises recordings of 195 hours duration or 7200 km in linear extent, collected at two points in the European USSR: Koltushi in the Leningrad Region and Tsimlyansk in the Rostov Region. All the observations were made on the ground. The time sections were converted to space sections with the aid of the wind-speed measurements carried out at the level of the lower cloud boundary. The cloud velocity is known to be 80-90% of the wind velocity v, if the latter exceeds 4 m/s. Since in 70% of the measurements v ranged from 8 to 13 m/s, the estimates of the linear sizes are 10-20% too high. All the measurement data were grouped into three gradations of the amount of cloudiness, for which the probability density of the sections $p(s)$ is approximated well by the formula

$$p(s) = \frac{0.43}{\sqrt{2\pi s \sigma_z}} e^{-u^2/2}, \qquad (3.24)$$

where

$$u = \frac{\lg s/\bar{s}}{\sigma_z}, \quad z = \lg s. \qquad (3.25)$$

The approximation parameters are given in Table 3.3.

Table 3.3. Approximation parameters.

Parameter	n					
	0-0.29		0.3-0.59		0.6-0.9	
	land	ocean	land	ocean	land	ocean
Mean cloudiness $\bar{n}$	0.20	0.22	0.44	0.43	0.80	0.75
Mean $\bar{s}$	0.50	0.48	0.80	0.53	0.90	0.65
σ_z	0.35	0.36	0.47	0.46	0.62	0.52

As n increases, the frequency of clouds with linear dimensions
from 0 to 1 km decreases (from 80% for $n = 0.2$ to 50% for $n = 0.8$-
-0.9), while the frequency of clouds with linear dimensions
greater than 3 km increases (from 3% for $n = 0.2$ to 20% for
$n = 0.8$-0.9).

The maximum frequency of clouds 1 to 2 km in size is observed for amounts of cumuli $n = 0.4$-0.6, being 25%, on the
average. Figure 3.3 shows the frequency $\varkappa$ and the mean size $\bar{s}$
as functions of the amount of cloudiness n (in Fig. 3.3b the scale
of s is 2 km).

These parameters have high variances. The mean curves of $s(n)$
and $\varkappa(n)$ practically do not differ at all from those obtained in
[12] (see preceding section).

Cloud regions can be divided into dense (d) or semitransparent
(t), depending on the amount of direct solar radiation transmitted
by the clouds [13]. Table 3.4 shows the corresponding amounts of
cloudiness n_d and n_t, as functions of $n = n_d + n_t$, according to
measurements carried out around Moscow.

Table 3.4. Relation between n_d and n_t for given n.

n	0.2	0.3	0.4	0.5	0.6	0.7	0.8	0.9
n_d . . .	0	0.05	0.16	0.22	0.32	0.40	0.58	0.84
n_t . . .	0.2	0.25	0.24	0.28	0.28	0.30	0.22	0.06

Cumulus fields in the northeastern and southeastern tradewind
regions of the tropical Atlantic were studied from research vessels, using the above-described methods. A generalization of the

volume of data obtained during a time of 300 hours (about 7600 km)
indicated the following. Over the tropical and northern parts of
the Atlantic, Cu hum. prevails, the amount being $n \le 0.6$. In the
equatorial Atlantic, the clouds are smaller than in the tropics,
and thus there are more of them. Regardless of the observation
region and the amount of cumuli, cloud nonuniformities up to

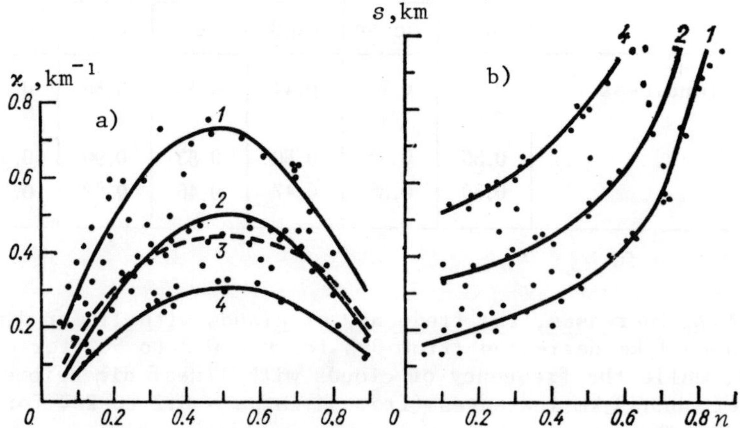

Fig. 3.3. Dependences of $\varkappa$ (curves a) and s (curves b) on n.
Dots refer to individual cases; 1 and 4 are envelopes; 2 are mean
values; 3 gives mean values according to data of [12].

800-900 m in size predominate. Individual clouds with a horizontal
extent of as much as 3 or 4 km are encountered. The probability
densities of the size distribution of cloud sections $p(s)$ can also
be approximated by formulas like (3.24) and (3.25), the approximation parameters of which are given in Table 3.3. The $s(n)$ and
$\varkappa(n)$ relations are the same as in Fig. 3.3, but the scatter of data
is greater.

Semitransparent regions in the tropical Atlantic occupy approximately the same area as in the middle latitude on land.

CHAPTER 4

OPTICAL PROPERTIES OF CLOUDS

INTRODUCTION

In this chapter the optical characteristics of clouds will be considered. These will be useful in subsequent chapters when calculating the albedo, transmission, and absorption. The characteristics of interest to us here are the coefficients of scattering σ, absorption α, and attenuation*) $\varepsilon = \sigma + \alpha$, and also the scattering (phase) function $\gamma(\varphi)$, where φ is the scattering angle. With respect to an individual spherical particle, all these quantities are calculated by solving Maxwell's equations using the method and formulas of Mie [2, 9].

When calculating for a single particle, we determine the effective cross sections or effectiveness factors

$$K_{\sigma,\alpha,\varepsilon}(\lambda, r) = \pi r^2 \widetilde{K}_{\sigma,\alpha,\varepsilon}(\lambda, r),$$

where r is the particle radius, and $\widetilde{K}$ is a dimensionless effective cross section of scattering, absorption or extinction.

If a unit volume of the medium contains N particles of radius r, then the "volume" (calculated per unit volume) coefficients $\sigma_\lambda(r) = K_\sigma N$, $\alpha_\lambda(r) = K_\alpha N$, $\varepsilon_\lambda(r) = K_\varepsilon N$, will have units of reciprocal length (L^{-1}). Finally, using the formula $\widetilde{\sigma} = \sigma/\rho$, we can find the "mass" (calculated per unit mass) coefficient of scattering (absorption, attenuation), which has units of $L^2 \cdot M^{-1}$. Here ρ is the amount of optically active substance per unit volume of air.

For a given size distribution of cloud particles $n(r)$ (see Chap. 2), the optical parameters of a polydisperse system are calculated:

$$\sigma_\lambda = \int_0^\infty \sigma_\lambda(r)\, n(r)\, dr \tag{1}$$

together with the analogous expressions for α_λ and ε_λ.

It is convenient to normalize to unity the function $\gamma(\varphi)$ describing the angular distribution of the scattered light:

---

*) For visible radiation the attenuation coefficient is known as the extinction coefficient (Translator).

$$\frac{1}{4\pi} \int_0^{2\pi} \int_0^{\pi} \gamma(\chi)\, d\psi \sin\theta\, d\theta = 1 \qquad (2)$$

and for spherical particles we can limit ourselves to the simple case of $\gamma(\mu)$, where $\mu = \cos\varphi$. Then normalization (2) becomes

$$\frac{1}{2} \int_{-1}^{1} \gamma(\mu)\, d\mu = 1. \qquad (3)$$

Knowing parameters σ, α, and ε, we can now determine the transmittance τ_ε entering directly into the equation of radiation transfer:

$$\tau_\varepsilon(z) = \int_{z_{lb}}^{z} \varepsilon(z)\, dz; \quad z \leqslant z_{ub}; \quad \tau_{0\varepsilon} = \tau_\varepsilon(z_{ub}), \qquad (4)$$

where z_{lb} and z_{ub} are the heights of the cloud boundaries; functions $\tau_{0\sigma}(z)$ and $\tau_{0\alpha}(z)$ are determined similarly.

Since, at any rate for $\lambda \leq 2$ μm, we know that $\lambda \ll \sigma$ in clouds, the effect of absorption can be neglected when evaluating $\tau_{0\varepsilon}$ and the total optical thickness of the cloud layer can be taken to be

$$\tau_0 = \tau_{0\,\varepsilon} = \tau_{0\,\sigma} = \int_{z_{ub}}^{z_{lb}} \sigma(z)\, dz. \qquad (5)$$

A second important parameter of the theory of radiation transfer is the probability of quantum survival or albedo of a unit volume:

$$\omega = \frac{\sigma}{\varepsilon} = \frac{\sigma}{\alpha + \sigma}. \qquad (6)$$

It would be unjustified, however, to set $\omega = 1$ on the basis of the foregoing inequality, since the radiation parameters are very sensitive to slight deviations of ω from unity (see Chap. 8).

Extensive tables have been compiled of all the above-mentioned quantities, covering wide ranges of wavelength and particle size. The main tables of this type are given in [9], and in the preceding three volumes of tables by the same authors. However, for our purposes the spectral resolution of the tables in [9] proved to be too low. As will be shown in Chap. 5, the spectrum of solar radiation is divided into 15 intervals, so as to provide an accurate identification of visible light, absorption bands of water vapor and liquid water, and the "transmission windows" between bands. At the same time, it was not in this case advisable to consider only two droplet-size distributions: "narrow" and "wide" (see Chap. 2). Consequently, the present chapter will begin with data of "precise" (that is, according to Mie's formulas) calculations of K_ε and K_σ for the indicated 15 wavelength intervals and two particle-size distributions.

Because of the complexity of these calculations, it is justified to use approximate methods, some of which are described in [2]. In Section 4.2 we will present some simple approximate formulas for calculating $\tilde{\varepsilon}$, $\tilde{\sigma}$, and $\tilde{\alpha}$; these are convenient, for instance, for applications in cloud-formation theory, where variations in droplet size and in the optical properties of droplets during cloud development have to be taken into account [1]. Finally, in Section 4.3 we will present a model of the optical parameters of a cloud layer, with a particle spectrum that varies with height. Particles grow in size with height, in accordance with the measurement data, and, in addition to the cloud particles proper, in these "precise" calculations the presence of a large number of water-coated condensation nuclei is taken into account (see Chap. 2).

This chapter includes diverse data of direct measurements of optical parameters of clouds. Section 4.4 surveys the main results of a comprehensive program of aircraft measurements of the horizontal transparency of clouds. These data are used to find ε and τ_0. The vertical transparency or τ_0 of thin clouds, which are transparent to direct solar light, is also discussed in Section 4.7.

Ice clouds, and especially their scattering functions, are considered in Section 4.5; some information about these clouds is also included in Sections 4.4 and 4.7.

The experimental findings led to an interesting evaluation in Section 4.6 of aerosol absorption in clouds, as indicated by aircraft measurements of reflected light.

Finally, the last two sections of this quite varied chapter, 4.8 and 4.9, present an assessment of the actual inhomogeneity of cloud layers. Within the framework of this monograph, stratiform clouds are assumed to be either homogeneous or horizontally stratified. The data of Sections 4.8 and 4.9 provide a basis for the construction of stochastic radiation-cloud models.

4.1. Spectrum of optical parameters

For 15 intervals in the short-wave part of the spectrum, the formulas of Mie's theory [2, 9] were used to calculate the optical parameters of an elementary volume of a cloudy medium: the attenuation cross section $K_{\varepsilon\lambda}$, the scattering cross section $K_{\sigma\lambda}$, the absorption cross section $K_{\alpha\lambda} = K_{\varepsilon\lambda} - K_{\sigma\lambda}$, and the scattering function $\gamma_\lambda(\varphi)$. The probability of quantum survival, or albedo, of a unit volume is $\omega_\lambda = K_{\sigma\lambda}/K_{\varepsilon\lambda}$.

All the above parameters are integrated over r with a

weighting factor $\bar{n}(r)$ (see formula (1) of the introduction to this chapter), where $\bar{n}(r) = n(r)/N_0$, $n(r)$ being the particle-size distribution and N_0 being the total number of particles. Therefore, the cross sections of attenuation, scattering, and absorption are normalized to a single particle and are, respectively, $\varepsilon_\lambda = N_0 K_{\varepsilon\lambda}$, $\sigma_\lambda = N_0 K_{\sigma\lambda}$, and $\alpha_\lambda = N_0 K_{\alpha\lambda}$. The scattering function is normalized with the aid of formula (3) of the introduction.

The mean cosines of the scattering function were also calculated, using the formula

$$\bar{\mu}_\lambda = \frac{1}{2} \int\limits_{-1}^{1} \gamma_\lambda(\mu)\, \mu\, d\mu \qquad (4.1)$$

as well as the weighted-mean (over the spectrum) function

$$\bar{\gamma}(\varphi) = \frac{\int\limits_0^\infty I_{0\lambda}\gamma_\lambda(\varphi)\, d\lambda}{\int\limits_0^\infty I_{0\lambda}\, d\lambda} \approx \frac{\sum\limits_{i=0}^{15} I_0^{(i)}\gamma_i(\varphi)}{I_0}, \qquad (4.2)$$

where $I_{0\lambda}$ or $I_0^{(i)}$ is the spectral component of the solar constant. The optical parameters were calculated for a "wide" ($\alpha = 2$, $1/r_0 = 0.4\ \mu m^{-1}$) and a "narrow" ($\alpha = 6$, $1/r_0 = 1.5\ \mu m^{-1}$) particle-size distribution (see (2.7)).

Table 4.1. Attenuation cross section K_ε and scattering cross sections K_σ in μm^2.

No., p/p	Δ_λ, μm	m	$\varkappa$	"Narrow" distribution K_ε	K_σ	"Wide" distribution K_ε	K_σ
1	0.40. . . 0.71	1.331	$0.017 \cdot 10^{-7}$	491	491	166	166
2	0.71. . . 0.76	1.333	$0.19 \cdot 10^{-6}$	494	494	169	169
3	0.76. . . 0.81	1.333	$0.17 \cdot 10^{-6}$	496	496	170	170
4	0.81. . . 0.86	1.331	$0.31 \cdot 10^{-6}$	497	497	170	170
5	0.86. . . 0.89	1.329	$0.47 \cdot 10^{-6}$	498	498	170	170
6	0.89. . . 1.00	1.325	$1.65 \cdot 10^{-6}$	499	499	171	171
7	1.00. . . 1.08	1.319	$1.56 \cdot 10^{-6}$	501	501	172	172
8	1.08. . . 1.21	1.317	$10.2 \cdot 10^{-6}$	503	502	173	173
9	1.21. . . 1.28	1.316	$13.3 \cdot 10^{-6}$	506	505	175	174
10	1.28. . . 1.54	1.314	$269 \cdot 10^{-6}$	508	495	176	173
11	1.54. . . 1.66	1.311	$125 \cdot 10^{-6}$	512	506	178	177
12	1.66. . . 2.08	1.303	$371 \cdot 10^{-6}$	517	502	180	177
13	2.08. . . 2.25	1.294	$471 \cdot 10^{-6}$	522	506	183	180
14	2.25. . . 3.00	1.245	$304 \cdot 10^{-5}$	533	464	197	184
15	3.00. . . 3.58	1.468	$111 \cdot 10^{-4}$	537	382	190	153

Table 4.2. Scattering functions $\gamma_i(\varphi)$ $i = 1, 2,\ldots,15$, in accordance with numbers of intervals $\Delta\lambda_i$ in Table 4.1 ("wide" distribution*).

φ_0	γ_1	γ_2	γ_3	γ_4	γ_5	γ_6	γ_7	γ_8
0	2381.81	1875.86	1750.02	1629.06	1535.56	1406.17	1196.14	1051.04
2	29.201	58.527	67.32	76.077	83.166	93.342	110.906	123.381
5	3.096	5.084	5.618	6.174	6.603	7.228	8.349	9.207
10	1.460	2.120	2.258	2.411	2.534	2.700	2.965	3.122
20	0.810	1.141	1.213	1.280	1.325	1.389	1.483	1.545
30	0.439	0.619	0.657	0.686	0.711	0.737	0.779	0.797
40	0.226	0.317	0.336	0.352	0.362	0.377	0.389	0.407
50	0.108	0.156	0.166	0.173	0.178	0.187	0.193	0.499
60	0.048 6	0.072 6	0.077 9	0.081	0.084	0.087	0.091 9	0.095
70	0.020 3	0.032 3	0.034 7	0.037 6	0.038 2	0.039 5	0.042 7	0.044 2
80	0.008 98	0.014 8	0.016 3	0.017 4	0.018 3	0.019	0.020 6	0.021 5
90	0.004 81	0.007 78	0.087 7	0.009 56	0.009 75	0.010 4	0.011 5	0.012 2
100	0.003 69	0.005 93	0.006 72	0.006 97	0.007 48	0.007 92	0.008 43	0.008 72
110	0.004 1	0.006 32	0.006 97	0.007 43	0.007 6	0.007 64	0.008 67	0.008 64
120	0.006 93	0.012 2	0.012 2	0.012 3	0.012 9	0.012 5	0.012 9	0.013 0
130	0.008 98	0.012 6	0.014 1	0.015 8	0.017 4	0.019 1	0.025 7	0.028 7
140	0.064 5	0.079	0.081	0.085 2	0.088 4	0.092 1	0.095 6	0.095 7
150	0.029 3	0.042 6	0.046 7	0.046 4	0.047 4	0.048 7	0.048 8	0.050 3
160	0.022 9	0.035 3	0.036 6	0.040 1	0.040 1	0.041 5	0.042 1	0.043 8
170	0.021 4	0.035 8	0.041 8	0.043 3	0.044 1	0.045 1	0.050 1	0.051 7
180	0.135	0.181	0.222	0.206	0.210	0.231	0.242	0.241
μ	0.942	0.918	0.912	0.908	0.905	0.901	0.895	0.891

*Scattering functions of "narrow" distribution given in [25].

φ	Y_9	Y_{10}	Y_{11}	Y_{12}	Y_{13}	Y_{14}	Y_{15}	$\overline{Y}$
0	889.23	747.38	583.27	455.58	351.21	271.52	208.1	1746.88
2	136.826	151.875	158.726	162.457	156.844	152.731	142.677	71.862
5	10.271	11.726	13.590	16.264	19.394	25.917	34.750	6.760
10	3.299	3.438	3.715	4.009	4.401	5.102	4.770	2.294
20	1.592	1.577	1.668	1.713	1.795	1.740	1.107	1.141
30	0.823	0.830	0.857	0.861	0.867	0.785	0.603	0.603
40	0.418	0.419	0.432	0.431	0.432	0.369	0.372	0.307
50	0.205	0.208	0.216	0.216	0.218	0.181	0.235	0.151
60	0.099 6	0.102	0.107	0.110	0.112	0.0915	0.15	0.071 1
70	0.047	0.050 4	0.0537	0.056	0.0583	0.0486	0.0968	0.032 6
80	0.024 1	0.025 5	0.0281	0.0298	0.0316	0.0277	0.0637	0.015 8
90	0.014	0.014 3	0.0163	0.018	0.0192	0.0174	0.0433	0.008 83
100	0.010	0.010 2	0.0116	0.0122	0.0136	0.0134	0.0306	0.006 54
110	0.009 34	0.009 64	0.0109	0.0115	0.027	0.015	0.0229	0.006 66
120	0.014 2	0.014 6	0.0163	0.0173	0.0197	0.0268	0.0187	0.010 8
130	0.030 6	0.032 1	0.0372	0.0425	0.0483	0.0459	0.0175	0.017 4
140	0.096 9	0.088 9	0.0919	0.0838	0.0815	0.0396	0.0208	0.075 3
150	0.054 4	0.050 8	0.054	0.0538	0.0537	0.0345	0.0389	0.039 1
160	0.047 5	0.046 3	0.0549	0.0506	0.0516	0.0367	0.0859	0.033 5
170	0.058 2	0.059 6	0.0688	0.0729	0.0797	0.0611	0.116	0.037 5
180	0.256	0.201	0.241	0.212	0.213	0.145	0.103	0.176
μ	0.886	0.887	0.878	0.878	0.873	0.894	0.875	

Table 4.3. Mean values of scattering (attenuation) coefficient in km^{-1} for $\lambda = 0.7$ µm for various cloud types.

Coef- ficient	Cu	Sc	St	Ns	St fr.- Frnb.	As	Ac
ε_1	100	48	43	29	37	25	20
σ_1 wide	69	26	26	37	37	26	24
σ_1 narrow	132	50	50	62	62	50	47
σ_p	89	43	43	38	38	31	31

The integration over r was carried out using Simpson's method, with an interval of 0.2 µm. The accuracy of the result was checked by comparison with the calculations of [7, 9]. The agreement with the data of [7] was everywhere within 1%, while the difference from [9] proved to be somewhat greater, especially in the "tail" part of the scattering function. This was apparently because a low-capacity computer was used for the calculations in [9].

Data on the complex index of refraction $m = i\varkappa$ given in [10] were used in the calculations.

The calculation results are presented in Tables 4.1 and 4.2. The scattering coefficients and scattering functions for $\lambda = 0.71$ and 2.81 µm, which will be used often in the following chapters, are designated as, respectively, σ_1 and σ_2, γ_1 and γ_2 ("narrow" or "wide"). In order to evaluate the degree of correlation of the calculations with the observational data, Table 4.3, taken from [25], compares experimental values of ε_1 for stratiform clouds with calculated values of σ_1.

The first row in the table was taken from Table 2.5, with the addition of other data from [17]; the figures in the second and third rows were obtained with the aid of the formula $\sigma_1 = \tilde{\sigma}_1 w$ for $\tilde{\sigma}_1 = 1300$ and 2500 cm^2/g, respectively, corresponding to a wide or narrow droplet-size distribution. The cloud water content w was selected on the basis of data in Chap. 2.

A comparison of rows 1–3 in Table 4.3 indicates that a wide distribution provides an unsatisfactory description only in the case of St, Sc clouds, while a narrow distribution fits these clouds well and in addition does not give a worse representation of Cu clouds than a wide distribution does. The fourth row gives values calculated using the following formula from [14]:

$$\sigma_p = w / \rho r_{av} \qquad (4.3)$$

Here $\rho = 1$ g/cm^3 is the density of water; the values of r_{av} and w were taken from Chap. 2. The calculations carried out with this formula showed a satisfactory fit with measurement data.

4.2. An approximate calculation method

The described approximate representation of parameter σ is applicable and valid only for visible radiation.

Because of the complexity of the calculations using the Mie theory, it is convenient to have for any wavelength simpler methods for approximating the optical parameters. In [27] formulas were obtained for the volume coefficients of attenuation ε and scattering σ of radiation by clouds, and these were compared with the data of "precise" calculations. The spectrum of cloud droplets was described by a gamma distribution (see above and Chap. 2); the effectiveness factor of the attenuation is given by a formula from [2], and the absorption effectiveness factor by an interpolation formula from [28].

Strictly speaking, the expression for K_ε in [2] was obtained assuming "soft" particles, that is, when the index of refraction $m \to 1$. However, as pointed out in [2], even for $m = 1.5$ the approximate formula is quite close to the accurate expression obtained with the aid of the Mie theory.

Under these conditions in [27] the following expression was obtained for ε:

$$\varepsilon = 2\pi N_0 r_{av}^2 \left\{ \frac{n+2}{n+1} - \frac{2\cos z}{(n+1)\,\delta\beta^{(n+2)/2}} \times \right.$$
$$\times \sin\left[(n+2)\arctg\frac{\delta}{n+\gamma+1} - z\right] + 2\frac{\cos^2 z}{\delta^2}\left[\cos 2z - \frac{1}{\beta^{(n+2)/2}} \times \right.$$
$$\left.\left.\times \cos\left[(n+1)\arctg\frac{\delta}{n+\gamma+1} - 2z\right]\right]\right\}. \qquad (4.4)$$

Here $\quad z = \arctg\dfrac{\varkappa}{m-1};\qquad \delta = \dfrac{4\pi r_{av}(m-1)}{\lambda};\qquad \gamma = \dfrac{4\pi\varkappa r_{av}}{\lambda};$

$$\beta^2 = \left(1 - \frac{\gamma}{n+1}\right) + \left(\frac{\delta}{n+1}\right)^2; \quad \text{and } m - i\varkappa \text{ is the complex}$$

index of refraction.

Similarly, for the absorption coefficient we have

$$\alpha = \pi N_0 r_{av}^2 \frac{n+2}{n+1}\left[1 - \left(1 + \frac{2\gamma}{n+1}\right)^{-(n+3)}\right]. \qquad (4.5)$$

Expressions (4.4) and (4.5) can be used to calculate the coefficients of attenuation, scattering, and absorption for all wavelengths. For long-wave radiation all the terms in (4.4) are of comparable importance. For short-wave radiation, on the other hand, for $\lambda \leq 4$ µm estimates in [27] showed that, with an error of 1.5% for narrow droplet spectra and an error of 0.5% for wide spectra, formulas (4.4) and (4.5) can be simplified. Then, after dividing by the water content $w = {}^{4}/_{3}\pi r_{av}^{3}\rho\,\dfrac{(n+2)(n+3)}{(n+1)^2}\,N_0$, we obtain the mass coefficients of attenuation, scattering, and absorption:

$$\tilde{\varepsilon} = \frac{3}{2\rho r_{av}}\frac{n+1}{n+3}\left[1 + \frac{n+1}{n+2}\cdot\frac{\lambda^2}{4\pi^2 r_{av}^2}\cdot\frac{[(m-1)^2 - \varkappa^2]}{[(m-1)^2 + \varkappa^2]}\right], \quad (4.6)$$

$$\tilde{\alpha} = \frac{3}{4\rho r_{av}}\frac{n+1}{n+3}\left[1 - \left(1 + \frac{8\pi\varkappa r_{av}}{\lambda(n+1)}\right)^{-(n+3)}\right] \quad (4.7)$$

where $\tilde{\sigma} = \tilde{\varepsilon} - \tilde{\alpha}$. Here $\rho = 1$ g/cm^3. It is evident from (4.6) that coefficient $\tilde{\varepsilon}$ has an anomalous variance, a result which is in accordance with [12].

A comparison of the calculations using formula (4.6) and (4.7) with the calculations based on the Mie theory (see Table 4.1) indicates that the error in computing $\tilde{\sigma}$ does not exceed 9%, and for $\lambda < 2.4$ µm it is less than 5%. The accuracy in calculating α is also acceptable for a number of problems.

4.3. Optical parameters of an inhomogeneous cloud for a bimodal droplet-size distribution

Chapter 2 included data on the height distribution of droplet sizes for St, Sc clouds. There it was also mentioned that a great number of tiny particles, condensation nuclei surrounded by water, are present in stratiform clouds.

The optical parameters of a bimodal system of particles were calculated according to these data using the formulas of the Mie theory, for the following model:

$$N(r) = \sum_{i=1}^{2} A_i r^{n_i} e^{-\alpha_i r}. \quad (4.8)$$

The required values of parameters n_i and α_i were evaluated using the known relations (see Chap. 2):

$$\frac{v_i}{r_{av\,i}} = \frac{1}{\sqrt{n_i + 1}}; \quad \alpha_i = \frac{n_i + 3}{r_{mod\,i}}, \quad (4.9)$$

where v is the rms deviation of the radii, and r_{av} and r_{mod} are, respectively, the mean and mode radii of the droplets. For small particles ($i = 2$) it was assumed that: $v_2/r_{av\,2} = 0.3$; $r_{mod\,2} = 0.3$ µm; the total number of particles $N_2 = 1000$ cm^{-3}; the refractive index takes the value of the aerosol model in Chap. 5.

The mode radius for large particles varied with height

RADIATION IN A CLOUDY ATMOSPHERE

Table 4.4. Spectral and height dependences of parameters ε, σ (km^{-1}), and ω.

λ, μm	z_{1b}			$z_{1b} + 200$ m			$z_{1b} + 400$ m			z_{ub}			$z_{ub}, v/r_{av} = 0.3$		
	ε	σ	ω	ε	σ	ω	ε	σ	ω	ε	σ	ω	ε	σ	ω
0.40	3.33	3.28	0.985	11.0	11.0	1.000	13.9	13.8	0.997	16.0	16.0	1.000	26.7	26.6	0.998
0.60	3.43	3.41	0.995	11.3	11.3	1.000	14.2	14.2	1.000	16.5	16.5	1.000	27.4	27.4	1.000
0.80	3.40	3.39	0.997	11.3	11.3	1.000	14.2	14.2	1.000	16.4	16.4	1.000	27.3	27.3	1.000
1.00	3.42	3.41	0.998	11.4	11.4	1.000	14.3	14.3	1.000	16.6	16.6	1.000	27.7	27.7	1.000
1.20	3.45	3.44	0.998	11.5	11.5	1.000	14.4	14.4	1.000	16.7	16.7	1.000	27.8	27.8	1.000
1.40	3.49	3.48	0.998	11.7	11.5	0.995	14.6	14.5	0.994	16.8	16.7	0.994	28.8	27.8	0.994
1.60	3.51	3.50	0.998	11.9	11.9	1.000	14.9	14.8	0.997	17.2	17.1	0.997	28.6	28.5	0.997
1.80	3.48	3.48	1.000	12.2	12.1	0.997	15.2	15.1	0.997	17.5	17.4	0.996	28.7	28.5	0.996
2.00	3.39	3.35	0.989	12.5	12.2	0.977	15.5	15.1	0.974	17.8	17.3	0.971	28.5	27.6	0.969
2.40	2.99	2.91	0.993	13.2	13.1	0.987	16.4	16.2	0.985	18.8	1.85	0.984	30.1	29.5	0.982
2.80	2.14	1.17	0.549	11.5	6.34	0.553	14.8	8.08	0.547	17.3	9.37	0.542	30.5	16.5	0.542
3.20	3.23	2.11	0.655	12.6	7.09	0.556	15.8	8.57	0.543	18.2	9.77	0.535	30.0	15.0	0.506
3.60	2.90	2.80	0.965	13.8	13.0	0.939	17.2	16.0	0.929	19.7	18.2	0.923	32.0	29.3	0.917
4.00	2.25	2.17	0.968	14.0	13.4	0.956	17.8	16.9	0.949	20.6	19.5	0.945	37.2	35.2	0.947

according to the law

$$r_{\text{mod}}(z) = r_{\text{mod}}(0)\left[1 + 2(z)^{1/3}\right]. \qquad (4.10)$$

Here z is reckoned from the lower boundary of a cloud layer 0.6 km thick. Parameter v/r_{av} = 0.3 for a narrow droplet-size distribution or 0.6 for a wide distribution. The index of refraction was taken from [37] (see also Table 4.1).

Table 4.4 presents calculated values of the attenuation coefficient ε_λ, the scattering coefficient σ_λ, and the albedo of a unit volume ω_λ of the bimodal system being considered. The calculations were mostly made for v/r_{av} = 0.6; for comparison, the last section of the table gives parameter values for v/r_{av} = 0.3.

Figure 4.3 shows a plot of the attenuation coefficient $\varepsilon_\lambda(z)$ for λ = 0.6 μm, so normalized that $\varepsilon(z_{\text{ub}})/\varepsilon_{\text{av}}$ = 1.5. The agreement between the mean experimental data and the calculations proved to be satisfactory.

4.4. Measurements of attenuation of visible light in clouds

In [17] the results of many years of measurements of the attenuation (extinction) coefficient ε for clouds of various types were generalized. The equipment used (recorder, Central Aerological Observatory) provided measurements of ε in the wavelength interval from 0.5 to 0.7 μm, with an error no greater than 20%, provided that ε lies between 2.5 and 250 km^{-1}. The main results can be summarized as follows. Parameter ε is an extremely variable cloud characteristic, which varies from point to point. On the average, if we consider stratiform clouds of the middle latitudes and the Arctic, we see that lower-level clouds are denser and upper-level clouds are less dense. Clouds of the tropical Atlantic, on the other hand, have a loose structure and are optically very non-uniform, while upper-level Cs and Ci sp. clouds have, on the average, extinction coefficients comparable to those of more low-lying clouds. In dense regions ε is twice or three times as high as the mean, being comparable to the ε values in clouds of the middle latitudes.

Table 4.5 gives the mean $\bar{\varepsilon}$ and the median ε_{med} values of the extinction coefficient for stratiform clouds. The differences between $\bar{\varepsilon}$ and ε_{med} range from 20 to 40%, and always $\bar{\varepsilon} > \varepsilon_{\text{med}}$, that is, the visibility in the clouds is greater than the mean in more than half of the cases. Table 4.5 also gives the reciprocals $l = (\bar{\varepsilon})^{-1}$ and $l_{\text{med}} = (\varepsilon_{\text{med}})^{-1}$, characterizing the mean free paths of photons in clouds [23]. Figure 4.1 shows plots of distributions $f(\varepsilon)$ and $\varphi(l)$ for clouds of various types.

In convective clouds of the middle latitudes ε increases linearly with an increase in the cloud depth H. As H varies from 0.5 to 4 km, this relation is described satisfactorily by the formula

$$\varepsilon = 50H + 70, \qquad\qquad (4.11)$$

where H is in km and ε is in km^{-1}.

Table 4.5. Mean $\bar\varepsilon$ and median ε_{med} values of extinction coefficient in km^{-1} and mean free paths $\bar{l}$ and l_{med} in meters, for stratiform clouds in various geographical regions.

Cloud type	European USSR				Arctic				Eastern tropical Atlantic			
	$\bar\varepsilon$	ε_{med}	$\bar{l}$	l_{med}	$\bar\varepsilon$	ε_{med}	$\bar{l}$	l_{med}	$\bar\varepsilon$	ε_{med}	$\bar{l}$	l_{med}
Sc	48	40	21	25	25	27	40	37	13	6	77	170
St	43	31	23	32	42	36	21	28	—	—	—	—
Ns	29	21	34	48	32	25	31	40	—	—	—	—
Ac	20	17	50	59	19	19	53	53	8	4	125	250
As	25	18	40	56	4	2.5	250	400	7	6	140	170
Cs,Ci sp	<2.5	<2.5	>400	>400	>2.5	>2.5	—	—	8	4.6	125	220

Note. Measurements made more recently than [23] gave considerably lower (about half as great) values of ε_{med} and ε for clouds of the middle latitudes.

Clouds in the eastern part of the tropical Atlantic are, on the average, considerably less optically dense [23], the densest

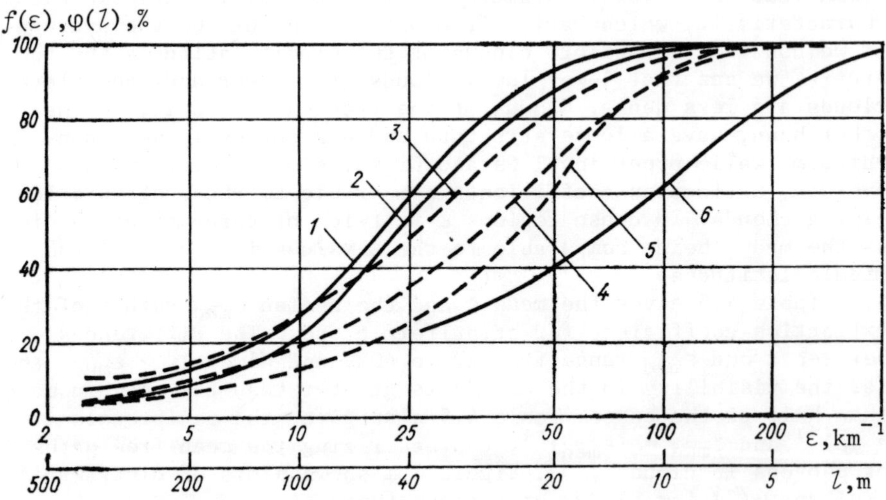

Fig. 4.1. Curves of cumulative frequency of $f(\varepsilon)$ and $\varphi(l)$. 1) Ac, 2) As, 3) Ns, 4) St, 5) Sc, 6) Cu.

kind there being Cu med., for which $\bar\varepsilon \approx 40$ km^{-1} ($\varepsilon_{med} \leq 20$ km^{-1}).

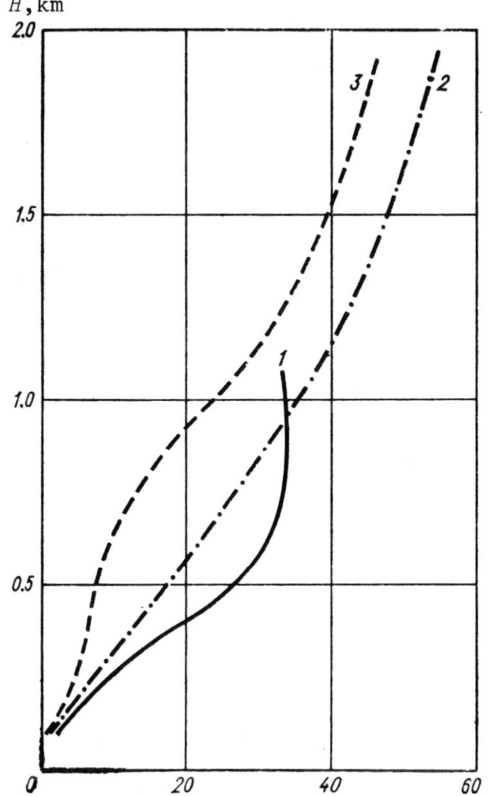

Fig. 4.2. Optical thickness
of clouds as function of
cloud depth.
1) St, Sc;
2) Ns, St fr.-Frnb;
3) As, Ac.

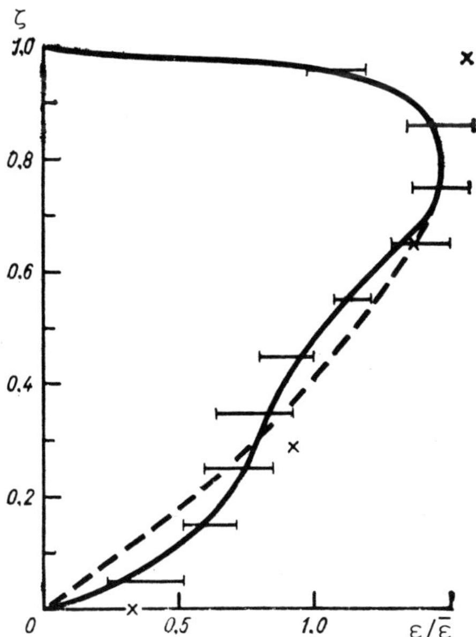

Fig. 4.3. Ratio $\varepsilon/\bar{\varepsilon}$ as function of relative height ζ
for stratiform clouds.
Dashed curve shows approximation (4.12) and crosses
show calculations (see Section 4.3).

The values of $\bar{\varepsilon}$ and ε_{med} for Cu hum. and Cu cong. are only 2/3 to 1/2 as great. A comparison of these values with formula (4.11) shows that in tropical clouds the extinction coefficients are several times less than in the middle latitudes.

The optical thicknesses τ_0 of stratiform clouds are, on the average, quite intimately related to the geometrical thicknesses. For clouds of the middle latitudes, the deviations of τ_0 from the mean curves in Fig. 4.2 are in each specific case no greater than 50%, in 80 to 90% of the cases.

In order to ascertain the mean vertical stratification of the extinction coefficient for stratiform clouds, the results of aircraft soundings were used to plot the sections in $\zeta = z/H$ and $\varepsilon/\bar{\varepsilon}$ coordinates. On the average, the normalized profiles so constructed approached a universal form for all clouds. Figure 4.3 portrays the universal mean curve of $\varepsilon/\bar{\varepsilon}$. The horizontal line segments indicate the maximum deviations from the universal curve of the mean profiles of the attenuation factor for stratiform clouds. As a rough approximation, the universal curve can be described by the expression (dashed curve in Fig. 4.3)

$$\varepsilon/\bar{\varepsilon} = 2.8\zeta\,(1-\zeta)^{1/4}. \qquad (4.12)$$

4.5. Optical characteristics of ice clouds

The microphysical properties of ice-crystal clouds differ from those of droplet clouds. First of all, the particles of these clouds consist of ice with a hexagonal crystal lattice, the optical properties of which differ from the properties of liquid water [37, 43]. Secondly, the crystals can have a variety of shapes and their mean size is larger than the size of a droplet. As a consequence of all these factors, some optical characteristics of ice-crystal clouds differ from those of droplet clouds [7].

The published quantitative data on the optical properties of natural ice-crystal clouds are very scanty. Theoretical and laboratory studies have focused on finding methods of calculating the optical characteristics of crystalline media and on determining the effect of the particle shape, size, and orientation on these characteristics.

4.5.1. Attenuation coefficient

In natural clouds and fogs measurements are made mainly for visible radiation. Table 4.6 gives mean values of the coefficient of attenuation (or scattering) according to data of [23, 26, 40] (see also Section 4.4). Values of $\sigma \geq 7$ km^{-1} for Sc and Ac pertain to clouds of mixed phase [23]. The spectral transmission,

measured in the atmosphere with crystals present [26] and in Ci
clouds [30, 41], turned out to be almost neutral in the range
λ = 0.5–12 µm. Therefore, Table 4.6 gives an idea of values of
the attenuation coefficient in the infrared range as well.

Table 4.6. Attenuation coefficients for visible radiation in ice-
-crystal clouds and fogs (km^{-1}).

Cloud type	Eastern trop- ical Atlantic	Middle latitudes	Arctic
Sc	Droplet	Droplet	25
Ac	7. . . 20	25	4
Cs,Ci sp.	8.3	2.5	2.5
Fogs	—	0.5. . . 1.2	2.3. . . 32

4.5.2. *Scattering function*

Measurements of the light scattering function $\gamma(\varphi)$ have been
made only for artificially created crystalline fogs [3, 4, 8, 13,
15, 19, 29, 35, 36, 39], and mainly for visible radiation, where
wavelength dependences are not observed. Figure 4.4 shows normalized scattering functions, obtained experimentally in [4, 19] for
φ = 10–170° and in [4, 13, 19) for φ = 2° and φ = 180°. The
maxima for $\varphi \approx 22°$ and 46° are known as the small and large halos.
With an error of ±30% the values of $\gamma(\varphi)$ characterize the angular
distribution of scattered light in the range of φ from 10 to 180°
for crystals of various shapes and sizes; for φ = 2° this error
was not evaluated. In order to determine $\gamma(\varphi)$ for $\varphi < 10°$, it is
necessary to know the sizes and shapes of the cloud crystals, since
diffraction contributes significantly to the scattering at small
angles.
Models of ice spheres and cylinders have been used to calculate the scattering function of an ice-crystal cloud [34, 38].
The experimentally obtained function best fits the model function
for hexagonal ice prisms (see Fig. 4.4), which in [4] was calculated using analytical expressions. If the large axes of the
prisms are oriented in a single plane, then the distribution of
the scattered radiation in space will depend on the angle of in-
cicence of the radiation onto this plane. For perpendicular incidence (θ = 0°), the values of $\gamma(\varphi)$ in different planes will be
the same. For $\theta \neq 0°$ the distribution of scattered radiation is
anisotropic. The measurements and calculations show that for
θ = 90° in the interval of φ from 10° to 180° the anisotropy of
the light scattering can be neglected, as a first approximation
[3, 4]. For $1 < \varphi < 10°$ the asymmetry of the scattering depends

on the prism shape b/d (where b is the length and d is the diameter). The ratio of the $\gamma(\varphi)$ values for scattering planes perpendicular to and parallel to the plane of prism orientation is approximately $36/\pi^2 \left(\frac{b}{d}\right)^2$. In a real cystal medium, in which only some of the crystals may assume the preferred orientation, the anisotropy of the scattering should be less.

For the infrared and ultraviolet ranges both measurements and calculations are lacking, except for $\lambda = 10.6$ μm, where the relative scattering function has been measured for $\varphi = 10-60°$ [15]. Thus it can be said that so far the optical properties of ice-crystal clouds have not yet been studied at all.

The following conclusions can be drawn from the published data:

1) ice-crystal clouds are ten times as transparent to visible radiation as droplet clouds, and the attenuation in the range λ from 0.5 to 12 μm is almost neutral;

2) the angular distribution of scattered radiation in the visible spectrum is independent of λ and differs from the distribution for droplet clouds over almost the entire range of angles. For scattering angles from 10° to 180°, with an error of ±30% this distribution can be described by the experimental scattering function plotted in Fig. 4.4 (curve 3).

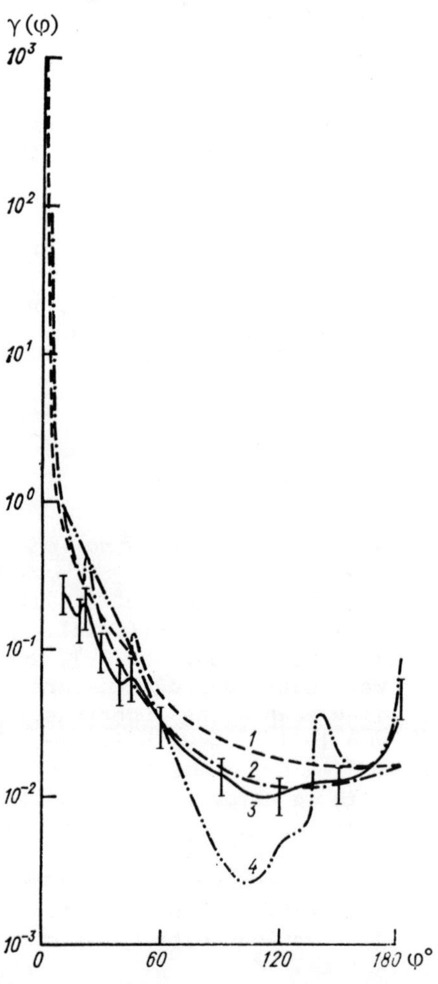

Fig. 4.4. Scattering functions $\gamma(\varphi)$.
1) and 2) for ice prisms oriented in a single plane (1 in the plane of prism orientation containing the beam of incident radiation, 2 in an orthogonal scattering plane) [4]; 3) for crystals of arbitrary shape [4]; 4) for water droplets $\gamma_2(\varphi)$ (see Table 4.2).

4.6. An evaluation of aerosol absorption in clouds

In the visible part of the spectrum water barely absorbs any radiation. According to the theory of multiple scattering, the "visible" albedo of a sufficiently thick scattering layer in the absence of absorption should be close to unity. Yet numerous measurements have revealed that under actual conditions, however thick a cloud layer may be, its albedo practically never is greater than 0.8, whereas the transmission is often only a few percent. Direct measurements (for instance, [32, 42]) indicate that the absorption in clouds may be as high as 30%, which is greater than the possible absorption by atmospheric gases (see Chap. 9).

The absorption in clouds can be evaluated using experimental data on the spectral variation of the luminosity R of solar radiation reflected from clouds. The necessary relations between the optical and microphysical parameters of clouds, on the one hand, and their luminosity, on the other, were obtained in [18, 20] (see also Section 4.8), for the following assumptions: relatively low specific absorption, great optical thickness, horizontal homogeneity, and great extent of the cloud layer. These relations are used to find the optical thickness of the layer from the brightness of the light reflected from it in a "transmission window" of gas absorption for a known albedo of the underlying surface. Then the spectral variation of the specific absorption $\beta = \alpha/\sigma$ is determined. This problem was solved in [16] on the basis of a large number of measurements of cloud luminosities made with satellites *Cosmos 149* and *Cosmos 320*; telephotometers operating in specific regions of the visible and near-infrared parts of the spectrum, up to $\lambda = 1.03$ μm, were used for this. Only dense homogeneous clouds of great horizontal extent were considered. Simultaneous temperature measurements made it possible to select just liquid-water clouds.

If, on the other hand, in some spectral interval the absorption by water vapor is negligibly small, as is the case, for example, for $\lambda = 0.74$ μm, then the optical thickness of the cloud τ_0 can be found using the relation $R = f(\tau_0, \gamma, \beta = 0)$ from [18]. The explicit form of this relation includes angular distribution functions for reflected and transmitted light in the case of clouds of great optical thickness, functions depending on γ (see Section 4.8). These functions were calculated for the clouds being considered too, and it was found that the dependence on γ is insignificant for cloud scattering functions. After τ_0 had been determined, function $R = f(\tau_0, \gamma, \beta)$ was applied to the spectral intervals in the presence of absorption to find β. The results of these calculations are given in Table 4.7, which shows that the specific absorption of clouds is of the order of 10^{-3} for $\lambda < 1$ μm and 10^{-2} for $\lambda \approx 1$ μm. These values are two orders of magnitude higher than those obtained from laboratory measurements of absorption, in these same spectral regions, by water droplets a few microns in radius. The spectral variation of this absorption for

Table 4.7. Specific absorption β and optical thickness τ_0, calculated from measurements of brightness coefficients R for clouds
with upper-boundary heights z_{ub} for zenith distance of Sun ζ;
R_0 is brightness coefficient in transmission interval, obtained
by linear extrapolation.

$\mu_0 = \cos\zeta$	z_{ub}, km	τ_0	R_0	$\lambda=0.723$ μm		$\lambda=0.738$ μm		$\lambda=0.744$ μm		$\lambda=1.03$ μm	
				R	β	R	β	R	β	R	β
0.450	5.3	12	0.39	0.36	0.004	0.33	0.007	0.33	0.007	0.28	0.015
0.454	4.0	12	0.40	0.38	0.004	0.36	0.007	0.36	0.007	0.30	0.011
0.458	4.7	7	0.25	0.22	0.001	0.19	0.002 1	0.21	0.014	0.19	0.021
0.463	4.0	11.4	0.38	0.33	0.006	0.28	0.001 3	0.28	0.013	0.25	0.020
0.572	10.0		0.95	0.85	0.0)0 13	0.75	0.000 76				
0.618	6.8	71	0.82	0.76	0.006	0.69	0.001 6				
0.656	7.0	117	0.89	0.82	0.000 4	0.75	0.001 0				
0.960	6.8	11.3	0.32	0.29	0.005 0	0.26	0.001 1				

0.7 μm $\leq \lambda \leq 1.03$ μm is qualitatively close to the spectrum of
pure water. Consequently, it appears that clouds contain a substance which differs from pure water or ice but which has the same
spectral variation of radiation absorption.

Similar results were obtained by American investigators [31],
on the basis of measurements of cloud albedos. The approximate
relations between the albedo and the absorption in the layer offered
in [31] give the same value $\beta \approx 10^{-3}$ in the visible part of the
spectrum. The reasons for the absorption of solar radiation by
clouds in the visible range have been discussed by a number of
authors [21, 44]. Various hypotheses have been put forth: regarding condensation nuclei, regarding the presence of water-coated
particles with interference phenomena in their water envelopes,
and regarding absorption by finely divided haze. Although the
causes of the observed effect have not yet been definitely ascertained, still it is an undoubted fact that the strong absorption
of radiation in a cloud exceeds the absorption by pure water or
ice.

4.7. Spectral transmission of thin clouds

Results of ground-based measurements of the spectral transmission of thin clouds relative to the Sun, illustrating their
marked variability in time, were first presented in [33]. That
paper contains a large volume of material for clouds of various
types. The equipment consisted of a filter radiometer and a
spectrometer, used to measure the intensity of the solar radiation
I_λ passing through a cloud and the radiation I_λ^0 in the intervening
spaces between clouds, in spectral regions corresponding to $\lambda =$
0.477; 0.614; 1.02, 1.25; 2.2; 3.8; 4.7; 9.3; 10.0; 27.4 μm.

The optical thicknesses of clouds $\tau_\lambda = \ln I_\lambda / I_\lambda^0$ were compared with the value of τ_λ for $\lambda = 0.477$ μm. This comparison of the ratios $k_\lambda = \tau_\lambda / \tau_{0.477}$ indicated that for $\lambda \leq 2.2$ μm there are only slight differences between the values of k_λ for clouds of different types. The greatest differences are observed close to $\lambda \approx 10$ μm.

It was shown in [33] that for $\lambda \approx 10$ μm ratio $k < 1$ for droplet clouds and $k = 1$ for ice clouds.

Results of similar measurements were presented in [5]; these synchronous measurements for $\lambda = 0.5$ and 12.2 μm agreed with the data of [33] for cumulus clouds ($k = \tau_{12.2} / \tau_{0.5} \approx 0.6$). The results differed considerably for cirrus: according to the data of [5], k ranges from 1.7 to 2.

Figure 4.5 presents new data for cirrus and cumulus clouds, obtained with the aid of five synchronously operating spectrometers tuned to $\lambda = 0.31, 0.33, 0.63, 1.1,$ and 12.2 μm. The main difference between this setup and the instruments used in [5, 33] is the possibility of selecting for synchronous measurements five spectral regions practically anywhere from 0.3 to 13 μm. This provides access to transmission windows of the atmosphere as well as to absorption bands of atmospheric gases (ozone, oxygen, etc).

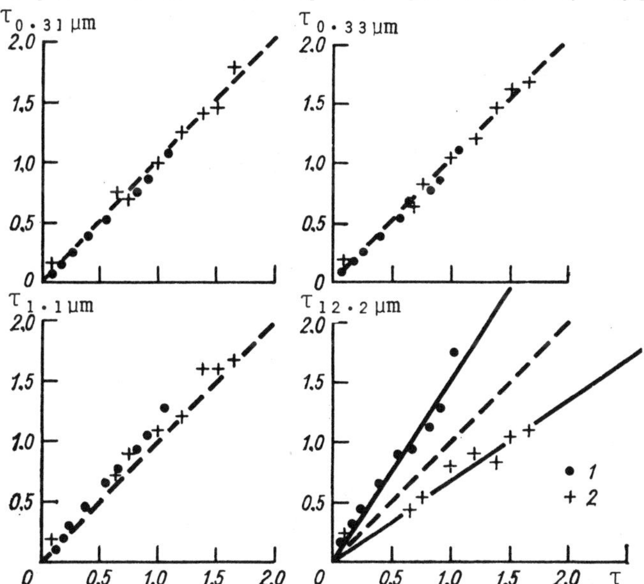

Fig. 4.5. Optical thickness τ_λ as a function of τ for $\lambda = 0.63$ μm for cirrus (1) and cumulus (2).

Inspection of Fig. 4.5 shows that ratio $k_\lambda = \tau_\lambda / \tau_{0.63}$ for $\lambda = 0.31$ and 0.33 μm is practically constant and close to unity for both cumulus and cirrus clouds. This result is significant with regard to measuring the ozone in terms of the amount of ultraviolet radiation passing through thin clouds. The values of k for cirrus and cumulus also differ slightly for $\lambda = 1.1$ μm, which agrees with [33]. A marked difference in k values is observed only for λ = 12.2 μm, where k = 0.7 for **cumulus and** k = 1.5 for **cirrus** This result agrees with the data of [5].

The data plotted in Fig. 4.5 are typical for cumulus and cirrus clouds, but a detailed, reliable analysis of the variations in transmission by various kinds of clouds must await the gathering of new data pertaining to other regions of the spectrum.

4.8. Variability of optical thickness for stratiform clouds

The transmission and reflection of short-wave radiation by an optically thick cloud layer depend not only on the mean parameters of the layer but also on the variability of these parameters. Of primary importance is the variability of the optical thickness, due to the irregularity of the upper and lower boundaries and also due to the spatial inhomogeneity of the water content and microstructure of the cloud. Here only inhomogeneities with a scale considerably greater than the mean free path of photons in the cloud are significant.

The effect of small-scale inhomogeneities of clouds is not great, because of the smoothing and averaging action of multiple scattering. Consequently, the reflection and transmission of radiation by an inhomogeneous cloud layer can be described adequately in terms of asymptotic solutions of the transfer equation for thick layers [20, 22]. On the basis of [20], the transmission coefficient of a layer T and the reflection coefficient A can be expressed, respectively, as

$$T\,(\tau_0) = \frac{\mu_0 c_1}{\pi\,[1 + c_2 \tau_0]}. \tag{4.13}$$

and

$$A\,(\tau_0) = \frac{\mu_0}{\pi}\,h - (1 - A_u)\,T, \tag{4.14}$$

where μ_0 = cos ζ, angle ζ being the zenith distance of the Sun, μ = cos θ, angle θ being the zenith distance of the direction of

observations; $c_1 = \dfrac{\pi}{\mu_0}\,g(\mu_0)$ for flow at the boundary of the cloud

layer or $c_1 = g(\mu_0)\left[\dfrac{1 + 2A_u}{3} + (1 - A_u)\mu\right]$ for the intensity;

$g(\mu_0) = \dfrac{1}{3} + \mu_0$; $c_2 = \dfrac{1 - A_u}{l}$; A_u is the albedo of the underlying

surface; h is a function close to unity, which is independent of the optical thickness; and l is a parameter characterizing the degree of elongation of the scattering function.

Let $\bar{\tau}_0$ and τ be the mean optical thickness of the layer and its deviation from the mean. Using the method of moments and retaining just the first terms of the expansion in (4.13) and (4.14), we obtain the statistics of the coefficients of transmission and reflection [18, 24]:

for the mean values

$$T\left(\bar{\tau}_0 + \tau\right) = T\left(\bar{\tau}_0\right)\left[1 + c^2\sigma_\tau^2\right], \qquad (4.15)$$

$$A\left(\bar{\tau}_0 + \tau\right) = \frac{\mu_0}{\pi}\,h - (1 - A_u)\,T\left(\bar{\tau}_0 + \tau\right); \qquad (4.16)$$

for the variances

$$\sigma_T^2 = c^2\sigma_\tau^2\left(1 + 8c^2\sigma_\tau^2\right)\left[T\left(\bar{\tau}_0\right)^2\right], \qquad (4.17)$$

$$\sigma_A^2 = \left(1 - A_u\right)^2\sigma_T^2; \qquad (4.18)$$

and for the correlation functions

$$r_\tau(t) = r_A(t) = \frac{\left(1 + 6c\sigma_\tau^2\right)r_\tau(t) + 2c^2\sigma_\tau^2 r_\tau^2(t)}{1 + 8c^2\sigma_\tau^2}. \qquad (4.19)$$

Here $c = \dfrac{c_2}{1 + c_2\bar{\tau}_0}$, σ_τ^2 is the variance of the optical thickness, and $r_\tau(t)$ is the normalized autocorrelation function of the optical thickness.

It is evident from formulas (4.15)-(4.19) that the mean transmission is always greater, and the mean reflection always less, than the values given by formulas (4.13) and (4.14) for a homogeneous layer for the mean optical thickness. Consequently, the transmission of radiation by stratus and stratocumulus clouds is enhanced by 5 to 15%, in comparison with a homogeneous layer of mean optical thickness $\bar{\tau}_0$, because of the inhomogeneity of the clouds [24]. The autocorrelation function of the transmission and reflection coefficients decreases with a time lag more rapidly than does the autocorrelation function of the optical thickness.

With the aid of relations (4.15)-(4.18), the parameters of the optical thickness (mean value $\bar{\tau}_0$ and variance σ_τ^2) can be determined from the measured statistical characteristics of the luminosity of the cloud layer [16, 18, 24]:

$$\bar{\tau}_0 = \frac{\overline{F}^{\downarrow}_{ub}\, \lg(\mu_0)\,(1 + y^2)}{\overline{F}^{\downarrow}_{lb}} - l, \qquad (4.20)$$

$$\sigma_{\tau}^2 = y\left(l + \bar{\tau}_0\right), \qquad (4.21)$$

where y is found from the equation

$$\frac{y\sqrt{1 + 8y^2}}{1 + y^2} = V_F; \qquad (4.22)$$

$\overline{F}^{\downarrow}_{ub}$ and $\overline{F}^{\downarrow}_{lb}$ represent the downward flow at the upper and lower

boundaries of the cloud layer; and $V_F = \sigma_F/\overline{F}^{\downarrow}_{lb}$ is the coefficient

of variation of the downward flow at the lower boundary of the
cloud layer.

4.9. Spatial inhomogeneity of optical parameters of clouds

A cloud can be represented as a very inhomogeneous disperse
medium. Various cloud characteristics (including optical characteristics) associated with the dispersed phase (that is, with the
droplets or crystals) may vary considerably from point to point
within the cloud space. Thus the following question arises:
How can the spatial variability of cloud parameters be described?
It would be most natural to measure this variability in terms of
some spatial scale L, which we will call the scale of spatial
homogeneity (SSH). The SSH is defined as the size of the horizontal
region within which the quantity in question differs from the value
at the beginning of the region by no more than a specified factor.
In [17] this factor was taken to be two (3 decibels). Therefore,
at a distance L the quantity being studied varies by no more than
a factor of two. If the field of the measured quantity is statistically uniform,*) then the correlation scale L_0 of the quantity
can be taken as the number characterizing the scale of inhomogeneity; the correlation scale is defined as the distance at which
the autocorrelation function decreases by a factor of e.
Scale L_0, like scale L, is a random quantity, and is measured
in clouds over a wide range of distances, from tens of meters to
kilometers. Fluctuations of L inside a cloud can be described in

---

*) Uniform fields are the spatial analog of steady-state
random processes.

terms of a variability coefficient $\delta_L = \sigma_L/\overline{L}$, where σ_L is the square root of the variance. In clouds δ_L usually ranges from 0.5 to 0.8. However, both more homogeneous (with regard to the variability L) clouds, for which $\delta_L < 0.3$, and less homogeneous clouds, for which $\delta_L > 1$, are encountered. Therefore, let us introduce L_{av}, the mean value of the SSH in an individual cloud, and also $\overline{L}_{av}$ and $\overline{L}_0$, the mean values of L_{av} and L_0 for clouds of the given type. Tables 4.8 to 4.10 give the 25, 50, 75, and 90%

Table 4.8. Portions of frequency of SSH in meters for various cloud types (portions of $P(L)$ distribution).

Frequen-cy %	St	Sc	St fr.-Frnb	Ns	As	Ac	Cu
25	100	120	140	150	100	100	—
50	300	320	260	340	240	220	50
75	1050	1000	600	980	650	700	130
90	3000	3050	1500	3250	1900	1750	360

Table 4.9. Portions of mean SSH values over the cloud in meters for various cloud types (portions of $P(L_{av})$ distribution).

Frequen-cy %	St	Sc	St fr.-Frnb	Ns	As	Ac	Cu
25	420	480	380	460	480		60
50	800	840	480	790	660	540	100
75	1900	1850	720	1450	960	920	140
90	3600	3050	1300	2300	1900	1800	260

Table 4.10. Portions of correlation scales in meters for various cloud types ($P(L_0)$ distribution).

Frequen-cy %	St	Sc	St fr.-Frnb	Ns	As	Ac	Cu
25	500	380	240	580	400	280	60
50	1000	780	440	1100	880	680	100
75	1800	1850	660	2250	1800	1400	180
90	2500	3050	860	3650	2800	2150	290

integral frequencies P of quantities L, L_{av}, and L_0 for various types of clouds [17]. Here the SSH is evaluated on the basis of measurements of the attenuation factor of visible light in clouds.

Table 4.11 gives the mean values of $\overline{L}_{av}$ and $\overline{L}_0$ corresponding to the distributions in Tables 4.8 to 4.10.

Measurements carried out in recent years at the Central Aerological Observatory indicate that the values given in

Table 4.11. Mean values of characteristic scales $\overline{L}_{av}$ and $\overline{L}_0$ in kilometers of spatial uniformity of attenuation factor for various cloud types.

Uniformity scale, km	Cu	Sc	St	St fr. — Frnb	Ns	As	Ac
$\overline{L}_{av}$	0.15	1.47	1.62	0.68	1.16	0.89	0.82
$\overline{L}_0$	0.14	1.40	1.41	0.52	1.62	1.15	0.83

Tables 4.8 to 4.10 may sometimes differ from those obtained in other series of measurements by more than a factor of two.

For instance, we see that clouds are not as often optically homogeneous in regions longer than 1 km as is indicated by Tables 4.8 to 4.10. In Sc and Ac clouds this is the case no more frequently than 10–15% of the cases, while in St, Ns, and As clouds no more frequently than 25% of the cases. At the same time, in more than 50% of the cases L is less than 0.2–0.25 km.

The characteristic scales of homogeneity (median and mean values) for the spatial variability of the water content have about the same values. The differences do not exceed some tens of percents.

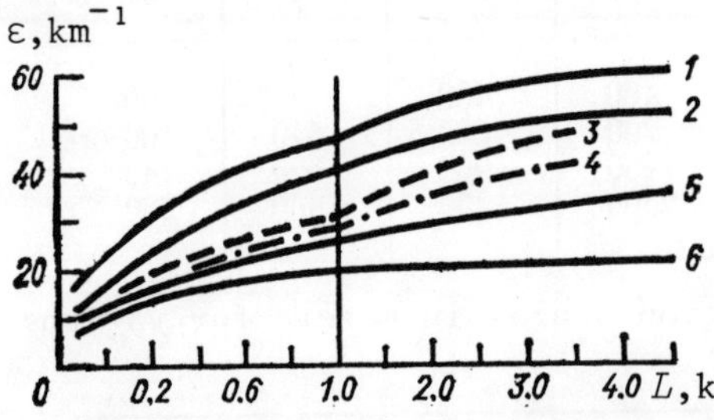

Fig. 4.6. Mean attenuation factor over homogeneous region as function of length of this region. 1) Sc, 2) St fr. –Frnb, 3) St, 4) As, 5) Ns, 6) Ac.

The denser a cloud, the more uniform it is, that is, the longer, on the average, its homogeneous regions are. This is illustrated well by Fig. 4.6, which shows the mean attenuation factor ε over a homogeneous region as a function of the length of this region L. In individual cases the values of ε differed from the curves plotted in Fig. 4.6 by 5 to 15 km^{-1}.

CHAPTER 5

AN OPTICAL MODEL OF THE ATMOSPHERE

5.1. A model of a cloudless atmosphere

When calculating fluxes of ultraviolet and visible radiation, we can arrive at an optical model by specifying the quantities

$$\sigma_\lambda(z) = \sigma_{R\lambda}(z) + \sigma_{a\lambda}(z), \tag{5.1}$$

$$\gamma_\lambda(z,\,\varphi) = \frac{\sigma_{R\lambda}(z)\,\gamma_{R\lambda}(\cos\varphi) + \sigma_{a\lambda}(z)\,\gamma_{a\lambda}(\cos\varphi)}{\sigma_{R\lambda}(z) + \sigma_{a\lambda}(z)}, \tag{5.2}$$

and also the coefficient of ozone absorption $\alpha_{oz\lambda}$ and the aerosol absorption $\alpha_{a\lambda}$.

Table 5.1 gives the spectral optical thicknesses of molecular (Rayleigh) scattering $\tau_{R\lambda} = \int_0^\infty \sigma_{R\lambda}(z)\,dz$ and total scattering $\tau_\lambda = \int_0^\infty [\sigma_{R\lambda}(z) + \sigma_{a\lambda}(z)]\,dz$, taken from [11]. Data on the spectral variation of the coefficient of ozone absorption $\alpha_{oz\lambda}$ from [33] (see also [26]) are also included in the table.

Table 5.1. Total τ_λ and Rayleigh $\tau_{R\lambda}$ optical thicknesses of standard atmosphere, coefficients of ozone absorption $\alpha_{oz\lambda}$.

Character-istic	λ, µm						
	0.3	0.32	0.34	0.36	0.38	0.40	0.45
$\tau_{R\lambda}$	1.22	0.93	0.72	0.56	0.45	0.36	0.22
τ_λ	4.97	1.55	1.05	0.87	0.74	0.62	0.46
$\alpha_{oz\lambda}$, cm^{-1}	4.39	0.39	0.024				2.33

Character-istic	λ, µm					
	0.50	0.55	0.60	0.65	0.70	0.80
$\tau_{R\lambda}$	0.14	0.098	0.069	0.050	0.036	0.022
τ_λ	0.37	0.33	0.30	0.25	0.22	0.19
$\alpha_{oz\lambda}$, cm^{-1}	14.5		58.0	28.6	10.7	

Table 5.2. Coefficients of molecular ($\sigma_{R\lambda}$) and aerosol ($\sigma_{a\lambda}$) scattering in individual atmospheric layers for λ = 0.514 μm.

Δz km	$\sigma_{R\lambda}$ km^{-1}	$\sigma_{R\lambda}$ km^{-1}		Δz, km	$\sigma_{R\lambda}$ km^{-1}	$\sigma_{R\lambda}$ km^{-1}	
		S_M=25km	S_M=5 km			S_M=25km	S_M=5 km
0	0.01500	0.1680	0.8200	7... 8	0.0689	0.0036	0.0036
0... 1	0.0143	0.1120	0.496	8... 9	0.0062	0.0035	0.0035
1... 2	0.0129	0.0486	0.181	9... 10	0.0055	0.0034	0.0034
2... 3	0.0116	0.0207	0.0663	10... 11	0.0049	0.0033	0.0033
3... 4	0.0105	0.0098	0.0242	11... 12	0.0044	0.0032	0.0032
4... 5	0.0951	0.0062	0.0088	12... 13	0.0038	0.0032	0.0032
5... 6	0.0856	0.0045	0.0045	13... 14	0.0033	0.0030	0.0030
6... 7	0.0768	0.0036	0.0036	14... 15	0.0028	0.0029	0.0029

Table 5.2 portrays the height distributions $\sigma_{R\lambda}(z)$ and $\sigma_{a\lambda}(z)$ for λ = 0.514 μm, according to [28]. Values of coefficient $\sigma_{a\lambda}(z)$ are given for a clear (S_M = 25 km) and a turbid (S_M = 5 km) atmosphere (where S_M is the horizontal visibility in the ground layer). In the former case $\sigma_{a\lambda}(z)$ is so selected as to agree with the model for λ = 0.55 μm [22, 23]. These same values of $\sigma_{a\lambda}(z)$ are assumed for a turbid atmosphere at height $z \geq 5$ km. Further down $\sigma_{a\lambda}(z)$ is extrapolated to the value $\sigma_{a\lambda}(0)$ = 4.1/S_M, where S_M = 5 km.

Table 5.3 gives the aerosol scattering function (1) for $\lambda \approx 0.5$ μm, averaged over the depth of the atmosphere and over data from numerous measurements made in the steppes of Kazakhstan.*) The table also includes values of another form of the scattering function (2), suggested in [6] as being typical for the continental aerosol: haze model C for λ = 0.45 μm.

Table 5.3. Aerosol scattering function.

φ^o	$\gamma(\varphi)$		φ^o	$\gamma(\varphi)$		φ^o	$\gamma(\varphi)$	
	1	2		1	2		1	2
0	—	7.9	15	0.440	—	90	0.0309	0.020
2	12.5	—	20	0.311	—	100	0.0278	0.015
4	3.92	—	30	0.188	0.5	120	0.0238	0.010
5	2.91	1.7	40	0.119	0.17	140	0.0214	0.013
6	1.90	—	60	0.0585	0.063	160	0.0200	0.025
8	1.11	—	80	0.0358	0.028	180	—	0.039
10	0.757	0.95						

*) The authors are grateful to G. Livshits, who supplied these data.

5.2. An aerosol model of the atmosphere

Synchronous measurements of the microphysical parameters of the high-altitude aerosol were carried out using different instruments mounted on aerostats at heights of up to 30 km and aboard Il-18 aircraft; measurements in 1975 at the aerological test site of the Central Aerological Observatory (Ryl'sk) were followed by measurements in 1976 at the aerostatic sounding site of the University of Wyoming (Laramie).*)

The complex of measuring equipment, the instrument characteristics, the experimental method, and the results obtained are described in detail in monograph [1]. These studies provided quite reliable data on the height variation of the size spectrum of aerosol particles, as well as of their concentration and chemical composition. After statistical processing, this material formed the basis for constructing a numerical model of the optical characteristics of aerosols up to a height of 30 km.

Table 5.4. Coefficients of attenuation ε_a and scattering σ_a of aerosol in height distribution.

z km	$\lambda = 0.55$ μm		$\lambda = 1.06$ μm		$\lambda = 2.36$ μm	
	$\varepsilon_a \cdot 10^{-2}\,\mathrm{km}^{-1}$	$\sigma_a \cdot 10^{-2}\,\mathrm{km}^{-1}$	$\varepsilon_a \cdot 10^{-2}\,\mathrm{км}^{-1}$	$\sigma_a \cdot 10^{-2}\,\mathrm{км}^{-1}$	$\varepsilon_a \cdot 10^{-2}\,\mathrm{км}^{-1}$	$\sigma_a \cdot 10^{-2}\,\mathrm{км}^{-1}$
0	25.5	25.4	13.8	13.7	3.85	3.45
0.5	9.83	9.76	4.79	4.76	2.54	2.41
1.0	9.42	9.38	4.71	4.64	2.44	2.39
1.5	7.56	7.36	4.55	4.44	2.40	2.33
2.0	7.24	6.88	4.37	4.19	2.37	2.11
2.5	5.28	4.99	3.49	3.32	2.24	2.02
3.0	5.04	4.86	3.38	3.20	2.10	1.79
3.5	5.24	4.45	2.24	1.85	1.68	1.57
4.0	1.87	1.64	1.76	1.65	1.03	0.87
4.5	0.435	0.408	0.325	0.310	0.234	0.224
5.0	0.328	0.318	0.216	0.212	0.112	0.106
6.0	0.277	0.268	0.204	0.200	0.103	0.099
7.0	0.207	0.198	0.154	0.150	0.086	0.083
8.0	0.192	0.181	0.147	0.143	0.093	0.090
10.0	0.292	0.280	0.208	0.203	0.135	0.115
12.0	0.290	0.284	0.194	0.192	0.125	0.116
14.0	0.249	0.244	0.157	0.155	0.089	0.087
16.0	0.302	0.246	0.154	0.152	0.0179	0.0171
18.0	0.256	0.248	0.145	0.143	0.0164	0.0156
20.0	0.176	0.1720	0.0730	0.0728	0.0078	0.0074
25.0	0.010	0.0100	0.0033	0.0030	0.0003	0.0003
30.0	0.002	0.0018	0.0004	0.0004	0.0000	0.0000

---

*) These studies were carried out within the framework of a Soviet-American agreement on cooperation with regard to conservation of the environment.

up to a height of 30 km.

The chemical composition of particles in the troposphere and lower stratosphere turned out to be close to the composition of a previously used [15] "synthetic" model of the aerosol. Consequently, data [7] on the refractive index were employed to calculate the coefficients of scattering and attenuation.

Table 5.4 gives the calculated values of ε_a and σ_a for three wavelengths (λ = 0.55, 1.06, and 2.36 μm).

An analysis of more detailed spectral relationships indicated that Angström's formula can be used as an approximate description: $\varepsilon_a = A\lambda^{-n}$, with n = 1.0 to 1.2 for the lower troposphere and n = 1.2 to 1.5 for the stratosphere. The formula is applicable for $\lambda \leq 10$ μm.

5.3. Transmission functions of solar radiation in water-vapor absorption bands. Integral transmission function

In [19] transmission functions for water vapor $P_{\Delta\lambda}$ for individual absorption bands and also an integral transmission function (ITF) P_{Σ} ($m_1,\ldots,m_n$) were suggested for calculating the spectral and integral fluxes and influxes of solar radiation. The integral function takes into account absorption by cloud particles as well as by gaseous absorbing agents.

The spectral transmission functions for water vapor were obtained on the basis of experimental data [3] using the formula

$$P_{\Delta\lambda}(m_v) = \frac{\sum\limits_{i=1}^{n} I_{0i} \exp\left[-\beta_i (m_{ef\ i})^{k_i}\right]}{\sum\limits_{i=1}^{n} I_{0i}}. \qquad (5.3)$$

Here i denotes the spectral interval $\Delta\nu_i \approx 20$ cm^{-1} corresponding to the band $\Delta\lambda$. Parameters β_i and k_i were taken from [3]; the effective masses of water vapor $m_{ef\ i}$ were found using the formula

$$m_{ef\ i} = m_v \left(\frac{p}{p_0}\right)^{n_i}, \qquad (5.4)$$

where m_v is the water-vapor mass along the radiation path in a homogeneous layer, p is the air pressure in this layer, and $p_0 = 10^5$ pascals. Values of parameter n_i were also given in [3]. When calculating the ITF, it was assumed that n_i = 1 for all i and all $\Delta\lambda$.

For a plane-stratified atmosphere and for a layer (z_1, z_2), assuming an inclined direction of the radiation, formula (5.4) becomes

$$m_{\text{ef } i}(z_1,\ z_2) = \sec\theta \int_{z_1}^{z_2} \rho_v(z)\left(\frac{p(z)}{p_0}\right)^{n_i} dz. \qquad (5.5)$$

Here θ is the angle a light ray makes with the vertical.

Table 5.5 gives the transmission functions for water-vapor absorption bands (a; 0.8; $\rho\sigma\tau$; φ; ψ; Ω; $\varkappa$; 3.2) in the spectral range from 0.7 to 3.6 μm. These bands are assigned numbers 2, 4, 6, 8, 10, 12, 14, 15; number 1 denotes the range of visible radiation 0.4 μm $\leq \lambda \leq$ 0.712 μm; numbers 3, 5, 7, 9, 11, 13 denote the "transmission windows" of water vapor, in which liquid water is able to absorb (see Table 4.1 and Chap. 9).

The integral transmission function of solar radiation $P_\Sigma(m_1,\ldots,m_n)$ is calculated using the formula

$$P_\Sigma(m_1,\ \ldots,\ m_n) = \frac{\displaystyle\int_{\lambda_0}^{\lambda_1} I_{0\lambda} \prod_{j=1}^{n} P_{j\lambda}(m_j)\, d\lambda}{\displaystyle\int_{\lambda_0}^{\lambda_1} I_{0\lambda}\, d\lambda}. \qquad (5.6)$$

Here m_j is the content of the jth absorbing substance along the radiation path, which is found with the aid of a formula of type (5.5) (with $n_j = 0$ for water); $P_{j\lambda}(m_j)$ is the corresponding spectral transmission function; $\lambda_0 \approx 0.70$ μm is the left-hand limit of the infrared range, and $\lambda_1 \approx 4$ μm is the right-hand limit.

For the troposphere in the near infrared range it is sufficient to take into account three absorbing substances: water vapor ($j = 1$, $\rho_1 = \rho_v$, $m_1 = m_v$), carbon dioxide ($j = 2$, $m_2 = m_{CO_2}$ air col.), and water droplets of clouds ($j = 3$, $\rho_3 = w$, $m_3 = m_w$).

The integral transmission function of water vapor has been used for a long time [16, 27]. More recent tables of function $P(m_1, m_2)$ have been published [8, 35], based on spectral data [24]. Finally, in [19] an integral transmission function for water vapor and water droplets was obtained. Absorption by CO_2 was neglected in [19], which according to estimates in [8] introduces errors of the order of a few percent in determinations of the integral function for $\lambda < 4$ μm. As in the case of formula (5.3), the data on absorption by water vapor were taken from [3], while the absorption coefficients for water droplets α_w were found with the aid of Table 13 in [9]. The spectral transmission functions for water droplets were calculated using the formula

$$P_{\Delta\lambda}(m_w) = \exp\left(-\alpha_{w,\,\Delta\lambda} m_w\right). \qquad (5.7)$$

The integral transmission functions for water vapor given in [8, 19, 35] agree well with one another (see [19]). In order to convert from directional radiation to diffuse radiation, a

Table 5.5. Transmission function of water vapor in absorption bands $\Delta\lambda$ (from - to) µm.

m_v g/cm^2	0.712 0.765	0.810 0.865	0.890 1.00	1.085 1.21	1.28 1.535	1.66 2.08	2.25 3.0	3.0 3.6
	2	4	6	8	10	12	14	15
0	1	1	1	1	1	1	1	1
0.000 01	0.999	0.999	0.998	0.996	0.985	0.986	0.949	0.992
0.000 04	0.999	0.999	0.996	0.992	0.971	0.972	0.905	0.983
0.001	0.998	0.998	0.994	0.986	0.955	0.945	0.864	0.974
0.005	0.995	0.997	0.986	0.971	0.906	0.913	0.766	0.944
0.01	0.993	0.994	0.980	0.958	0.872	0.882	0.714	0.922
0.05	0.985	0.988	0.956	0.911	0.762	0.785	0.614	0.838
0.1	0.974	0.983	0.939	0.878	0.701	0.734	0.577	0.783
0.2	0.970	0.975	0.916	0.834	0.634	0.682	0.543	0.714
0.4	0.957	0.965	0.883	0.779	0.568	0.634	0.512	0.636
0.6	0.948	0.958	0.870	0.742	0.542	0.626	0.506	0.620
0.8	0.940	0.951	0.844	0.711	0.501	0.591	0.482	0.548
1.0	0.934	0.946	0.829	0.687	0.480	0.578	0.472	0.519
2	0.903	0.920	0.765	0.598	0.410	0.536	0.441	0.423
4	0.874	0.895	0.707	0.517	0.352	0.501	0.415	0.347
6	0.846	0.871	0.659	0.460	0.315	0.476	0.397	0.299
10	0.811	0.841	0.606	0.402	0.279	0.448	0.380	0.255
20	0.745	0.782	0.519	0.317	0.227	0.404	0.351	0.196

diffusivity coefficient r = 1.67 has to be introduced in formula (5.6) as a multiplier of m_j.

For many purposes it is useful to have an analytical expression for function $P(m_v, m_w)$. The following representation [10] is natural and convenient for use in the theory of radiation transfer:

$$P(m_v, m_w) = P(m_v) P^*(m_v, m_w), \qquad (5.8)$$

where $P(m_v)$ is the integral transmission function for water vapor. Function $P^*(m_v, m_w)$ is given by the formula

$$P^*(m_v, m_w) = \frac{\int_{\lambda_0}^{\lambda_1} I_{0\lambda} P_\lambda(m_v) e^{-a_{w\lambda} m_w} d\lambda}{\int_{\lambda_0}^{\lambda_1} I_{0\lambda} P_\lambda(m_v) d\lambda} \qquad (5.9)$$

the dependence on m_v being slight.

Table 5.6. Coefficients of formula (5.10) in range $0 \le m_v \le 20$ g/cm^2.

N	Coefficient							
	a_1	a_2	a_3	α_1	α_2	α_3	ε_{min} %	ε_{max} %
2	0.174	0.826	—	6.27	0.0276	—	0.06	3.4
3	0.0772	0.145	0.778	58.3	1.45	0.019	0.01	1.2

Table 5.7. Coefficients of formula (5.11) for two intervals of m_v.

Mass range, g/cm^2	N	Coefficient				
		a_1	a_2	a_3	a_4	b_1
$1 \leqslant m_v \leqslant 10$	2	0.464	0.536	—	—	2.182
	3	0.207	0.320	0.474	—	14.21
	4	0.116	0.182	0.260	0.442	55.68
$2 \cdot 10^{-4} \leqslant m_v \leqslant 1$	2	0.522	0.477	—	—	2.819
	3	0.247	0.342	0.412	—	20.03
	4	0.143	0.193	0.279	0.385	57.53

Mass range, g/cm^2	N	Coefficient			
		b_2	b_3	b_4	ε %
$1 \leqslant m_v \leqslant 10$	2	0.0163	—	—	5
	3	0.606	0.002 46	—	3
	4	2.779	0.372	0.0033	2
$2 \cdot 10^{-4} \leqslant m_v \leqslant 1$	2	0.0199	—	—	7
	3	0.704	0.008 45	—	5
	4	3.791	0.006	0.004 37	5

It is reasonable to express function $P(m_v)$ in the form of a series:

$$P(m_v) = \sum_{i=1}^{N} a_i e^{-\alpha_i m_v}. \tag{5.10}$$

Table 5.6 gives the coefficients of formula (5.10) for $N = 2$ and $N = 3$. The minimum and maximum approximation errors $\varepsilon_{\min}(m_v)$ and $\varepsilon_{\max}(m_v)$ are also given.

Function $P^*(m_v, m_w)$ can also be represented as

$$P^*(m_v, m_w) = \sum_{i=1}^{N} a_i e^{-b_i m_w}. \tag{5.11}$$

Table 5.7 gives the parameters of this formula and the mean approximation error. In [10] approximations (5.11) are made more precise.

5.4. Transmission function of thermal radiation

At real temperatures in the troposphere (200–300 K) the thermal radiation is practically negligible for wavelengths less than 3–4 μm and greater than 40–60 μm. The transmission function for thermal radiation can be written as

$$P_{\Delta\lambda}(m_1, \ldots, m_k) = \frac{\int_{\lambda_1}^{\lambda_2} B_\lambda(T) \prod_{j=1}^{k} P_{j,\lambda}(m_j, \lambda)\, d\lambda}{\int_{\lambda_1}^{\lambda_2} B_\lambda(T)\, d\lambda} \qquad (5.12)$$

For λ_1 = 3–4 μm and $\lambda_2 \approx$ 50 μm the denominator of expression (5.12) is equal to σT^4, with an error of less than 3%. Masses $m_{i\lambda}$, just as in the case of the transfer of solar radiation, are given by formulas (5.4) and (5.5). In the expression for the effective absorbing mass as a function of pressure, exponent n is equal to unity for strong lines and equal to zero for weak lines. The choice of the mean value over the spectrum depends on the temperature and pressure of the atmospheric layer transmitting the radiation. For the main atmospheric gases absorbing thermal radiation (water vapor, carbon dioxide, and ozone), it can be assumed that n_{H_2O} = 0.5–1, n_{CO_2} = 0.8, and n_{O_3} = 0.4, regardless of λ [18, 28].

The transmission function of thermal radiation depends on the temperature in a more complicated manner than does the transmission function of solar radiation. In addition to the dependence of the absorption coefficients on the temperature of the medium, which is the same for both types of radiation, for thermal radiation there also exists a dependence of the weighting function (Planck function) on the temperature of the radiating layer or surface. The effect of the temperature on the absorbing capacity of the atmosphere, just as for solar radiation, will not be taken into account below, but when comparing different transmission functions it will be specified to which temperature the function corresponds.

The monochromatic functions $P_{j\lambda}(m_j)$ have the form of (5.7). Upon transition from monochromatic radiation to individual absorption bands, functions $P(m)$ will assume different forms depending on the band model chosen.

For water vapor a statistical model of the absorption band [5] is generally employed, and for carbon dioxide a model of a regular absorption band. Accordingly, the integral transmission functions take the form of a combination of functions such as:

$$P(m) = A \exp\left(-am/\sqrt{c+m}\right), \qquad (5.13)$$

$$P(m) = B[1 - \mathrm{erf}(dm)].\qquad(5.14)$$

Coefficients A, a, c, B, d are to be determined during the course of the approximation. The integral of error probability is unsuitable for practical calculations, but it can be replaced by the natural logarithm. The two functions agree well within a sufficiently large range of the independent variables, but the logarithm does not behave in a correct asymptotic manner for very low absorbing masses; at zero mass the transmission function has to equal unity.

Expressions (5.13) and (5.14), together with analogous expressions, were used in [21, 24, 32] to describe the transmission by the atmosphere in the range from 25 to 2975 cm^{-1} for individual spectral intervals 25 cm^{-1} in width. Tables of the integral transmission functions were compiled [13] and an approximation worked out [4, 18], on the basis of data of [21, 24, 32, 34]. An approximation was also worked out [14] on the basis of somewhat different initial data [29–31].

Let us consider the analytical forms used to represent the integral transmission functions for water vapor. The most reasonable form for the ITF is expression (5.10), where the weighting factors a_i are equal to the fraction of the blackbody radiation $\beta_\lambda(T)$ transferable in all the spectral intervals in which $\alpha_\lambda = \alpha_i$. However, to attain accuracies of a few percent, we have to use (5.10), with values of N from 4 to 5 (see below). The diffusive ITF (12.1) is represented as

$$D(m) = 0.46 \exp\left(-0.674\sqrt{m}\right) + 0.54 \exp\left(-7.75\sqrt{m}\right).\qquad(5.15)$$

In [14] a new variable $x = 10m/\sqrt{10^{-4} + 10m}$ is introduced, and the diffusive ITF becomes

$$D(m) = 0.74 \exp(-0.28x) + 0.26 \exp(-5.5x);\qquad(5.16)$$

for low m expression (5.16) behaves like (5.10), and for high m it behaves like (5.15).

In [25] a simple, but less physically valid, approximation of the form

$$D(m) = [1 + a(rm)^b]^{-1}\qquad(5.17)$$

was proposed, where $a = 1.75$, $b = 0.42$, and $r = 1.67$ is the diffusion coefficient. A similar approximation was offered in [12], on the basis of data of [13], for $a = 1.72$ and $b = 0.41$. In [29–31] the emissivity of the atmosphere $\varepsilon(m)$ is calculated, where $D(m) = 1 - \varepsilon(m)$.

For comparison, Table 5.8 shows ITF values obtained by various investigators. All the functions were calculated for a uniform radiation path, that is, for fixed pressure and temperature.

Table 5.8. Diffusive integral transmission functions for water vapor according to data of various investigators.

m_v,g/cm^2	T = 273 K [20]	T = 273 K [13]	T = 300 K [14]	T = 290 K [29]	[25]
10^{-5}	0.985	0.994	0.985	0.986	0.983
10^{-4}	0.956	0.968	0.954	0.950	0.957
10^{-3}	0.873	0.888	0.889	0.869	0.894
10^{-2}	0.679	0.749	0.721	0.724	0.762
0.1	0.418	0.572	0.555	0.556	0.548
1.0	0.235	0.349	0.306	0.318	0.364
2.5	0.158	0.243	0.178	0.209	0.239
6.3	0.084	0.132	0.077	0.098	0.175
10	0.055	0.082	0.043	0.040	0.149

Inspection of Table 5.8 shows that, even for a uniform path, the difference between transmission functions obtained on the basis of different spectral data is great for $m_v \approx 1$ g/cm^2 and considerably exceeds the errors in calculating and approximating the indicated functions. As yet, however, there is no good reason to prefer one set of experimental data over another, and further studies are needed.

The absorption of thermal radiation by carbon dioxide is taken into account most simply, albeit roughly, by multiplying together the ITF for water vapor and CO_2. However, strictly speaking, this is valid only for monochromatic radiation or else when at least one of the absorbing gases can be considered "gray", that is, its absorption coefficient is independent of the wavelength.

In [4, 18] the ITF for CO_2 and water vapor are described by the expression

$$P(m_1, m_2) = P_1(m_1) + P_2(m_1) P_3(m_2) \qquad (5.18)$$
$$\left(m_1 = m_v, \quad m_2 = m_{CO_2} \right),$$

where P_2 and P_3 correspond to the CO_2 band for λ = 15 μm and P_1 pertains to the rest of the spectrum. All the P_i in (5.18) have form (5.10). The parameters of P_i are given in Table 5.9 (m_{CO_2} in air column).

Table 5.9. Parameters of formula (5.10) for function P_i.

P_i	a_1	a_2	a_3	α_1	α_2	α_3
$P_1, m_v < 1$g/cm^2	0.19	0.21	0.35	470	11	0.30
$P_1, m_v < 10$g/cm^2	0.19	0.26	0.29	430	10	0.15
P_2	0.05	0.19	0.00	2.9	0.23	0.00
P_3	0.32	0.40	0.28	0.55	0.012	0.000 15

The ITF for CO_2 and water vapor can be put into a form similar to the ITF for solar radiation (5.8), that is, $P(m_1, m_2) = P_1(m_1)P_2(m_1, m_2)$, where P_1 and P_2 have form (5.10), but where for P_2 as a function of m_{CO_2} coefficients a_i and b_i depend on m_v, as in (5.11). The mean values of a_i and b_i over the mass m_v for P_2, and the corresponding coefficients for P_1, are given in Table 5.10. This table also shows the approximation errors for various N (calculation and method described in [17]). It should be noted that the introduction of the mean a_i and b_i values over m_1 for P_2 actually reduces (5.8) to $P(m_v, m_{CO_2}) = P_1(m_v)P_2(m_{CO_2})$, which as already mentioned may lead to a sizable increase in the error of $P(m_v, m_{CO_2})$.

Table 5.10. Approximation parameters for functions $P_1(m_v)$ and $P_2(m_{CO_2})$ according to (5.11).

P_i	N	a_1	a_2	a_3	a_4	b_1
P_1	2	0.558	0.442			21.87
	3	0.342	0.306	0.352		172.6
	4	0.201	0.254	0.225	0.320	769.4
P_2	2	0.187	0.813			0.254
	3	0.102	0.109	0.789		0.993
	4	0.058	0.088	0.086	0.768	3.014

P_i	N	b_2	b_3	b_4	ε_{max} %	ε_{av} %
P_1	2	0.181			18.5	
	3	2.12	0.151		6.5	
	4	17.4	1.12	0.138	2.5	
P_2	2	0.148 (−3)			10.0	2.7
	3	0.421 (−1)	0.748 (−4)		10.0	2.2
	4	0.205	0.145 (−1)	0.571 (−4)	9.0	2.2

Note: Numbers in parentheses denote powers of ten.

Whereas in studies of the transfer of solar radiation the optical properties of clouds must be carefully taken into account, in the thermal region the spectrum of absorption by clouds is usually allowed for by introducing the "gray" absorption coefficient, that is, the mean absorption coefficient for water droplets $\bar{\alpha}_w$ over the entire spectral interval (λ_1, λ_2). The values of $\bar{\alpha}_w$ for different kinds of clouds vary from 500 to 2000 cm^2/g. Then the integral transmission function of the atmospheric gases has to be multiplied by $P_3(m_w) = \exp(-\tilde{\alpha}_w m_w)$ [4, 18]. In many problems of the energetics of the atmosphere the so-called "black" approximation is applicable, for which $P_3(m_w) = 0$ or, correspondingly,

$\tilde{\alpha}_w = \infty$ (see Chap. 14). However, if the water reserves of the cloud are low, the "gray" approximation is no longer applicable, and the "black" approximation even less so. Thus it is of interest to find the transmission functions in specific spectral intervals. For water droplets absorption coefficients $\tilde{\alpha}_{w\lambda}$ can be calculated with the aid of the approximate formulas of Section 4.2.

Table 5.11. Transmission functions for cloudless and cloudy atmosphere

Spectral Region	m_v 10^4 g/m^2	Cloudless atmosphere		Cloud m_w g/m^2			Cloudy atmosphere m_w g/m^2	
		1	2	10	30	70	10	70
I	0.1	0.052	0.052	0.021	4.0 (—3)	3.0 (—4)	0.021	3.0 (—4)
	2.0	0.022	0.022				0.009	1.6 (—4)
	5.0	0.015	0.015				0.006	1.2 (—4)
II	0.1	0.166	0.083	0.043	6.0 (—3)	1.0 (—5)	0.030	7.0 (—5)
	2.0	0.099	0.071				0.026	6.0 (—5)
	5.0	0.080	0.058				0.021	5.0 (—5)
III	0.1	0.117	0.117	0.026	1.4 (—3)	4.0 (—6)	0.026	4.3 (—6)
	2.0	0.096	0.096				0.021	3.6 (—6)
	5.0	0.075	0.075				0.017	2.9 (—6)
IV	0.1	0.232	0.063	0.039	1.1 (—3)	7.0 (—7)	0.010	2.5 (—7)
	2.0	0.109	0.031				0.005	9.0 (—8)
	5.0	0.065	0.020				0.003	6.0 (—8)
V	0.1	0.083	0.083	0.015	5.2 (—4)	5.0 (—7)	0.015	5.2 (—7)
	2.0	0.007	0.007				0.001	3.1 (—8)
	5.0	0.002	0.002				0.0003	1.0 (—8)
Σ	0.1	0.599	0.398	0.144	13 (—3)	4.2 (—4)	0.104	3.9 (—8)
	2.0	0.333	0.227				0.072	2.3 (—4)
	5.0	0.237	0.168				0.047	1.7 (—4)

Note. Numbers in parentheses denote powers of ten.

The spectral intervals are so selected as to distinguish between the effects of absorption by water vapor, CO_2, and ozone, and also to be able later to take into account continual absorption in the transmission window and the role of traces of gaseous admixtures. The intervals are: I (3.5–8.5 μm) H_2O absorption band; II (8.5–10.5 μm) short-wave part of transmission window, ozone absorption band; III (10.5–12.5 μm) long-wave part of window; IV (12.5–18 μm) CO_2 absorption band; V (18–40 μm) H_2O absorption bands.

Table 5.11 shows the contributions of each of the intervals to the transmission of the integral thermal radiation for the following cases: a cloudless atmosphere, taking into account absorption by water vapor alone (1) and taking into account absorption by CO_2 and O_3 as well (2); clouds, taking into account only absorption by water droplets, water vapor, CO_2, **and** O_3. All the values of the transmission functions given in Table 5.11 have been normalized by σT^4.

As was to be expected, a standard stratiform cloud with water reserves of the order of 70 g/m^2 (see Section 10.2) will be absolutely opaque to thermal radiation. However, if the cloud water reserves are reduced to 10 g/cm^2, the transmission of the cloud layer remains close to the integral transmission function of a cloudless tropical atmosphere with $m_v = 5 \cdot 10^4$ g/m^2. Clearly, when solving cloud-formation problems, it is important to take into account the spectral and vertical structure of the absorption by a cloud layer.

INTRODUCTION

A flux of luminous energy from a source, the Sun for instance, undergoes a number of transformations when it enters a cloud. As a result of scattering at water droplets or ice crystals, the flux is divided up into rays of different intensities, propagating in different directions and following different trajectories. Thus the light may be partly absorbed by atmospheric gases or by scattering particles. This process can be described conveniently if we represent the luminous flux as an aggregate of photons moving along ray tubes [18].

The distribution of photon trajectories in a cloud is determined by its scattering and absorbing properties. Absorbing substances are generally distributed nonuniformly in a cloud, so that a different number of photons will be absorbed over different trajectories of equal length. If in addition the light is nonmonochromatic and Bouguer's law for absorption is not obeyed, then this process cannot be described by the transport equation. The only way to calculate the radiation properties of clouds in this general case is to the apply the Monte-Carlo method, making it possible to model the photon trajectories and to allow for absorption along each of these. To do this, the transmission function along each of the trajectories has to be determined.

However, certain simplifying approximations can be made. If, for instance, the absorption in a given spectral interval is sufficiently low, then the scattering and absorption processes can be assumed to be independent. In [6] the effect of the imaginary part of the refractive index $\varkappa$ of water droplets on the scattering process was estimated. These estimates indicated that for a real part of the refractive index $n = 1.3$ and for optical thicknesses of clouds τ_0 from 1 to 10^3 the influence of absorption on the scattering process can be neglected if $\varkappa \leq 3 \cdot 10^{-2}$–$10^{-3}$. If it is also assumed that the absorbing substances in a cloud are distributed uniformly, then different trajectories of equal length will be equivalent. Let $J(x, y, z, r, l)$ be the light intensity, reduced to a source of unit power, at point x, y, z in direction r, produced by photons traversing a path l from the moment they enter the cloud until they reach point x, y, z. For a purely scattering cloud with scattering characteristics corresponding to a spectral interval $\Delta\nu$, let us call this function the photon distribution over the paths l (photon-path distribution). In addition, let $I_0, \Delta\nu$ be the source strength, and let $P_{\Delta\nu}(m)$ be the transmission function (m is the amount of absorbing mass on path l). Assuming independence of the scattering and absorption in a cloud with a uniform distribution of absorbing substance, we can write the light intensity $I_{\Delta\nu}$ as

$$I_{\Delta\nu}(x,\ y,\ z,\ \mathbf{r}) = I_{0,\,\Delta\nu} \int_0^\infty J(x,\ y,\ \dot{z},\ \mathbf{r},\ l)\, P_{\Delta\nu}(m)\, dl. \quad (1)$$

Expression (1) is valid for any form of the transmission function. Since the scattering properties of a cloud depend on the light frequency much less markedly than does the absorption by atmospheric gases, the very same function $J(x,\ y,\ z,\ \mathbf{r},\ l)$ can be used over a quite wide spectral range. The photon-path distribution $J(x,\ y,\ z,\ \mathbf{r},\ l)$ satisfies a transport equation with the same form as the unsteady-state equation which has to be applied when studying the general properties of $J(l)$. The path distribution can be calculated either using this equation or using the Monte-Carlo method. A survey of methods for calculating $J(l)$ is included in [17]. It is important to remember, however, that the transport equation can be obtained directly for the intensity $I_{\Delta\nu}(x,\ y,\ z,\ \mathbf{r})$ only within rather narrow spectral intervals, where Bouguer's law for the transmission function holds true. A combined path distribution $J(l_1,\ l_2,\dots,l_n)$ can also be considered, where l_i are the lengths of paths traversed in different cloud layers [15].

This section presents data on the following characteristics of the photon-path distribution.

The probability density for paths l at point x, y, z in the absence of absorption is

$$p(x,\ y,\ z,\ \mathbf{r},\ l) = \frac{J(x,\ y,\ z,\ \mathbf{r},\ l)}{\overline{I}_0(x,\ y,\ z,\ \mathbf{r})}, \quad (2)$$

where $\overline{I}_0(x,\ y,\ z,\ \mathbf{r}) = \int_0^\infty J(x,\ y,\ z,\ \mathbf{r},\ l)\,dl$ is the light intensity at a given point by the photons traversing any path l in the absence of absorption. The corresponding probability density function for l, sometimes also called the cumulative probability, is

$$P(x,\ y,\ z,\ \mathbf{r},\ l) = \int_0^l p(x,\ y,\ z,\ \mathbf{r},\ l)\, dl. \quad (3)$$

Functions (2) and (3) are of interest not only in order to calculate the radiation properties of clouds in the absorption bands of atmospheric gases; they are also important for estimating the depth of light penetration in a cloud, when determining upper cloud boundaries with the aid of satellites.

A considerably simpler characteristic than the above is the mathematical expectation, that is, the mean value of the photon path length from the entry of the light into the cloud until arrival at point x, y, z in direction $\mathbf{r}$:

$$l_{av}(x,\ y,\ z,\ \mathbf{r})=\int_{0}^{\infty} lp(x,\ y,\ z,\ \mathbf{r},\ l)\,dl. \qquad (4)$$

In formulas (3), (4) it should be kept in mind that $p(l) = 0$ if $l < l_{min}$, where l_{min} is the distance from the source to point x, y, z. The effective path length l_{eff} is a very useful integral characteristic, which depends not only on the photon-path distribution for pure scattering, but also on the transmission function. According to the theorem of the mean, formula (1) can be rewritten as

$$\frac{I_{\Delta\nu}(x,\ y,\ z,\ \mathbf{r})}{I_{0,\Delta\nu}\,\overline{I}_{0}(x,\ y,\ z,\ \mathbf{r})}=P_{\Delta\nu}\,[\alpha,\ l_{eff}(x,\ y,\ z,\ \mathbf{r})]=$$

$$=\int_{0}^{\infty} p(x,\ y,\ z,\ \mathbf{r},\ l)\,P_{\Delta\nu}(\alpha,\ l)\,dl, \qquad (5)$$

where α is the generalized absorption coefficient.

Since any existing physical idea of the transmission function in the vicinity of an arbitrary l can be represented as a linear function αl, it follows that

$$\lim_{\alpha l\to 0} l_{eff}=l_{av}. \qquad (6)$$

When approximate expressions are used for the transmission function, it may well be that formula (6) is not valid. For example, for a transmission function of the form $P(\alpha,\ l) = \exp\left[-(\alpha l)^{k}\right]$, where $k < 1$, formula (6) does not hold true at an upper illuminated cloud boundary, since the derivative $P_l'(0)$ does not exist. However, in any case where l_{av} and l_{eff} are close, it will be possible to use the value of l_{av} determined for pure scattering, instead of l_{eff}, in order to estimate the absorption or the depth of light penetration into the cloud. The mean path length l_{av} can also be calculated in the absorption bands:

$$l_{av}(\alpha)=\int_{0}^{\infty} lp(x,\ y,\ z,\ \mathbf{r},\ l)\,P_{\Delta\nu}(\alpha,\ l)\,dl. \qquad (7)$$

The data in Chaps. 6 and 7 pertain mainly to the above-mentioned quantities, reduced to fluxes of light (integrals over hemispheres) reflected and transmitted by clouds. Sometimes integrals over smaller solid angles are considered.

For reflection the following symbols are used:

$$p^{\uparrow}(l), \quad P^{\uparrow}(l), \quad l^{\uparrow}_{av}, \quad l^{\uparrow}_{eff}, \tag{8}$$

and for transmission:

$$p^{\downarrow}(l), \quad P^{\downarrow}(l), \quad l^{\downarrow}_{av}, \quad l^{\downarrow}_{eff}. \tag{9}$$

For stratiform clouds, which are assumed to be horizontally homogeneous, these quantities pertain to any point on the upper, or correspondingly the lower, boundary of the cloud or cloud system. For cumulus clouds these characteristics are calculated as averages over the entire cloud.

The mathematical expectation of the path lengths of all photons emerging from the cloud in all directions will be written as

$$l_{av} = \int_{\Sigma} dx\, dy\, dz \int_{\Omega} d\Omega \cos(\Omega,\ \mathbf{n}\,[x,\ y,\ z]) \int_{0}^{\infty} l p\,(x,\ y,\ z,\ \mathbf{r},\ l)\, dl, \tag{10}$$

where Σ is the total surface of the cloud boundaries, $(\Omega,\ n)$ is the angle between the direction of photon emergence and the outward normal to the boundary, and Ω is the set of all directions of photon emergence at point x, y, z. This quantity can be used to estimate the fraction of the energy R absorbed by the whole cloud in a weak band,

$$R = 1 - P_{\Delta\nu}\,(\alpha,\ l_{av}). \tag{11}$$

Some values of quantities (8), (9), and (10), calculated using the Monte-Carlo method, are given in Chap. 6. This same method was employed to obtain data on light propagation in clouds with a nonuniform distribution of absorbing substances (also assuming the scattering and absorption to be independent). An experimental study of l_{eff} in a weak absorption band is described in Chap. 7.

CHAPTER 6

CALCULATIONS FOR VARIOUS CLOUD TYPES

6.1. Photon-path distribution, mean and effective path lengths in an isolated stratiform cloud

Here we will present photon-path distributions, and also $l_{av}^{\uparrow\downarrow}$ and $l_{eff}^{\uparrow\downarrow}$, for an isolated stratiform cloud with the vertical stratification of water content w and scattering coefficient σ described in Chaps. 2 and 4.

The mean values of these quantities over the height are taken to be $w_{av} = 0.2$ g/m^3 and $\sigma_{av} = 50$ km^{-1}. The height distribution of water vapor is assumed to be uniform, with a density $\rho_v = 10$ g/m^3. The scattering functions correspond to a "wide" droplet-size distribution (see Chaps. 2 and 4).

Figures 6.1 through 6.3 show, in the form of histograms, the probability densities $p^{\uparrow\downarrow}(l)$ for photon path lengths in the reflected and transmitted fluxes, in the case of pure scattering, as functions of the fundamental parameters $\tau_0 = \sigma_{av} H$ and $\mu_0 = \cos \zeta$, where ζ is the zenith angle of the Sun. The distributions were calculated for two scattering functions $\gamma_1(\varphi)$ and $\gamma_2(\varphi)$, "wide" relative to wavelengths 0.714 and 2.811 µm; these represent extreme cases of scattering functions for the visible and infrared regions of the spectrum (see Section 4.1).

Table 6.1 compares the values of $l_{av}^{\uparrow\downarrow}$ and $l_{eff}^{\uparrow\downarrow}$ for $\tau_0 = 20$ and $\mu_0 = 1$ for a uniform cloud layer and for a cloud with the above-mentioned vertical stratification of the scattering coefficient and the water content w. In the χ and 3.2 µm bands function γ_2 was used, and in the others function γ_1. Path lengths $l_{av}^{\uparrow\downarrow}$ and $l_{eff}^{\uparrow\downarrow}$ were calculated for pure scattering and in the water-vapor absorption bands (the notation of Section 5.3 was used). The values of $l_{av}^{\uparrow\downarrow}$ taking absorption by water droplets into account were also calculated (in Table 6.1 the values of $l_{av}^{\uparrow\downarrow}$ and $l_{eff}^{\uparrow\downarrow}$ in the absence of absorption by water droplets are given in parentheses). In a nonuniform cloud with a nonuniform height distribution of water droplets, formula (7) cannot be used to calculate l_{av}. In this case $l_{av}^{\uparrow\downarrow}$ is calculated directly during the modeling of the photon trajectories taking absorption along the trajectory into account:

$$l_{av} = \frac{1}{N} \sum_{i}^{N} l_i, \qquad (6.1)$$

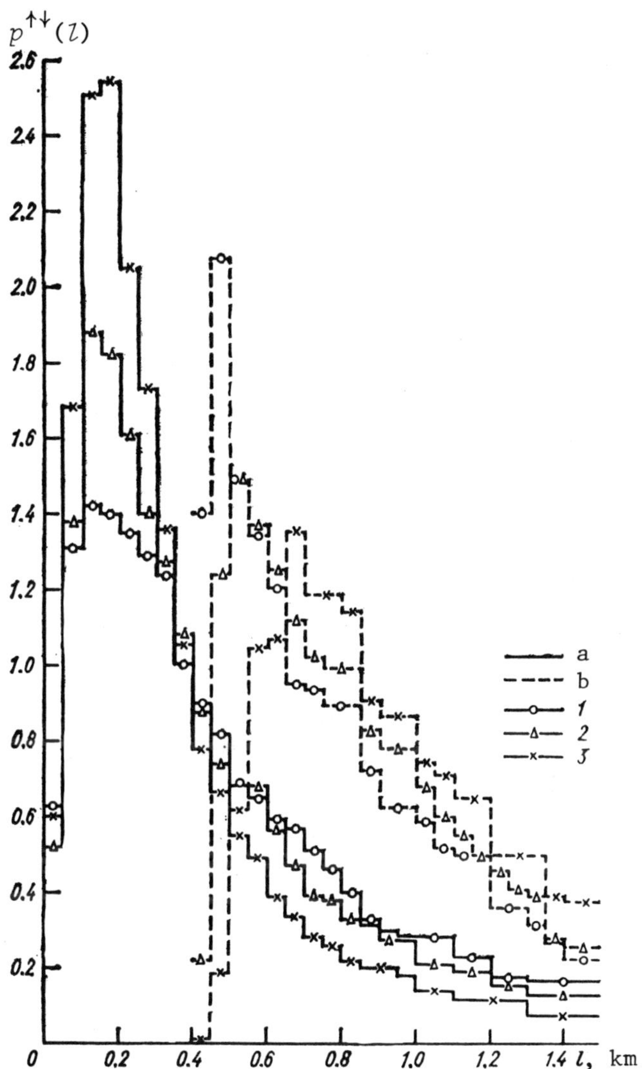

Fig. 6.1. Probability density for photon path lengths in reflected
(a) and transmitted (b) fluxes for nonuniform cloud with scattering
function $\gamma_1(\varphi)$ and $\tau_0 = 20$.
1) $\mu_0 = 1$; 2) $\mu_0 = 0.71$; 3) $\mu_0 = 0.26$.

where N is the number of photons with a history. The concept of
l_{eff} becomes ambiguous in a cloud with nonuniform absorption.
     Inspection of Table 6.1 shows that $l_{av}^{\downarrow}$ and $l_{eff}^{\downarrow}$ do not differ
by more than 10-15% within an absorption band, for practically

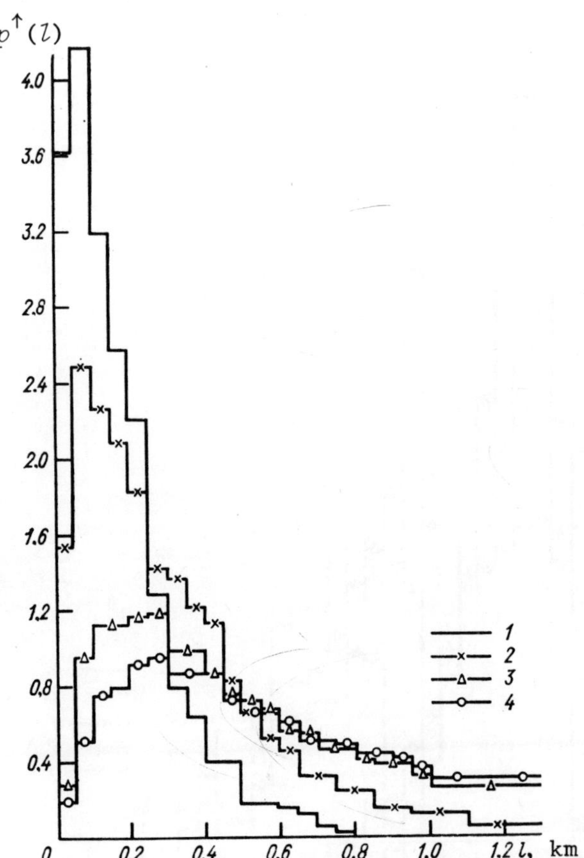

Fig. 6.2. Probability density for photon path lengths in reflection for nonuniform cloud with scattering function $\gamma_1(\varphi)$ and $\mu_0 = 1$.
1) $\tau_0 = 5$; 2) $\tau_0 = 10$; 3) $\tau_0 = 30$; 4) $\tau_0 = 50$.

all bands. The differences between $l_{av}^{\uparrow}$ and $l_{eff}^{\uparrow}$ are much greater, being from 15 to 30% for the various bands. This is because of the nonlinearity of the transmission functions in the water-vapor absorption bands, which is considerably more pronounced at the low values of $\rho_v l$ corresponding to reflection. A comparison of the values of l_{av} for pure scattering and l_{eff} makes it possible to establish the limits of applicability of the linear approximation for the transmission function of water vapor or, what amounts to the same thing (see [15]), to see if it is possible to use the mean path lengths in a purely scattering medium, instead of the actual photon-path distributions, to estimate the albedo and light transmission in the water-vapor absorption bands.

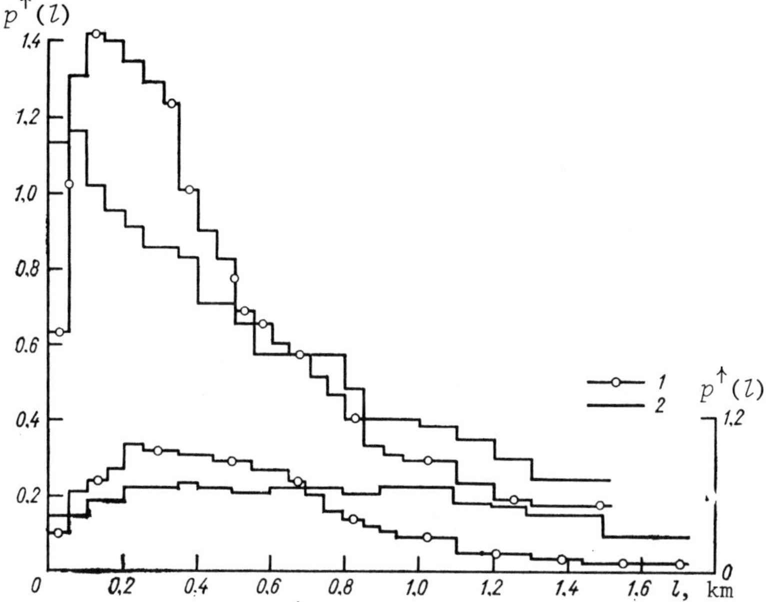

Fig. 6.3. Comparison of $p^\uparrow(l)$ in uniform (1) and nonuniform (2) clouds for $\tau_0 = 20$ and $\mu_0 = 1$. Left-hand scale $\gamma_1(\varphi)$, right-hand scale $\gamma_2(\varphi)$.

Table 6.1. Mean and effective photon path lengths in a cloud for water-vapor absorption bands.

Path lengths, μm	Pure scattering for		Absorption bands							
	$\gamma_1(\varphi)$	$\gamma_2(\varphi)$	a	0.8	$\rho\sigma\tau$	φ	ψ	Ω	χ	3.2
Uniform cloud										
$l^\uparrow_{av}$	0.79	0.90	0.77	0.77	0.74	0.70	0.54 (0.66)	0.53 (0.85)	0.05 (0.86)	0.02 (0.80)
$l^\uparrow_{eff}$	0.79	0.90	0.56	0.64	0.56	0.55	— (0.51)	— (0.61)	— (0.64)	— (0.67)
$l^\downarrow_{av}$	1.02	0.65	1.01	1.01	0.99	0.97	0.87 (0.96)	0.53 (0.63)	0.43 (0.64)	0.42 (0.61)
$l^\downarrow_{eff}$	1.02	0.65	0.90	0.91	0.90	0.89	— (0.84)	— (0.56)	— (0.56)	— (0.55)
Nonuniform cloud										
$l^\uparrow_{av}$	0.62	0.76	0.61	0.61	0.59	0.56	0.44 (0.54)	0.44 (0.72)	0.07 (0.73)	0.05 (0.68)
$l^\uparrow_{eff}$	0.62	0.76	0.49	0.48	0.48	0.45	— (0.39)	— (0.52)	— (0.54)	— (0.55)
$l^\downarrow_{av}$	0.97	0.65	0.96	0.96	0.94	0.93	0.83 (0.92)	0.54 (0.63)	0.44 (0.63)	0.42 (0.61)
$l^\downarrow_{eff}$	0.97	0.65	0.82	0.83	0.86	0.86	— (0.82)	— (0.56)	— (0.55)	— (0.54)

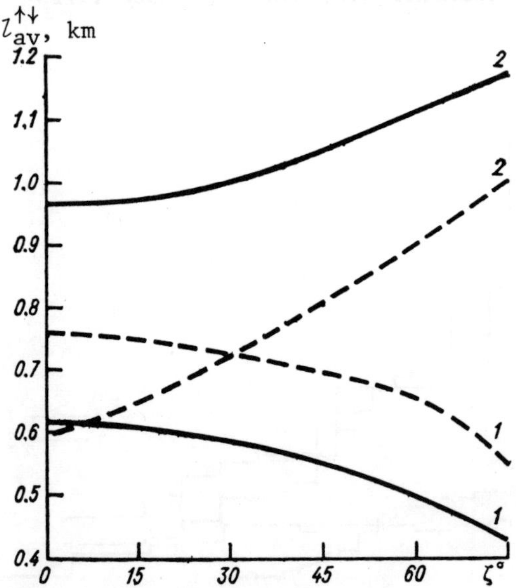

Fig. 6.4. Path lengths l_{av} (1) and l_{av} (2) as functions of ζ for pure scattering in nonuniform cloud. $\tau_0 = 20$.
Solid curves $\gamma_1(\varphi)$, dashed curves $\gamma_2(\varphi)$.

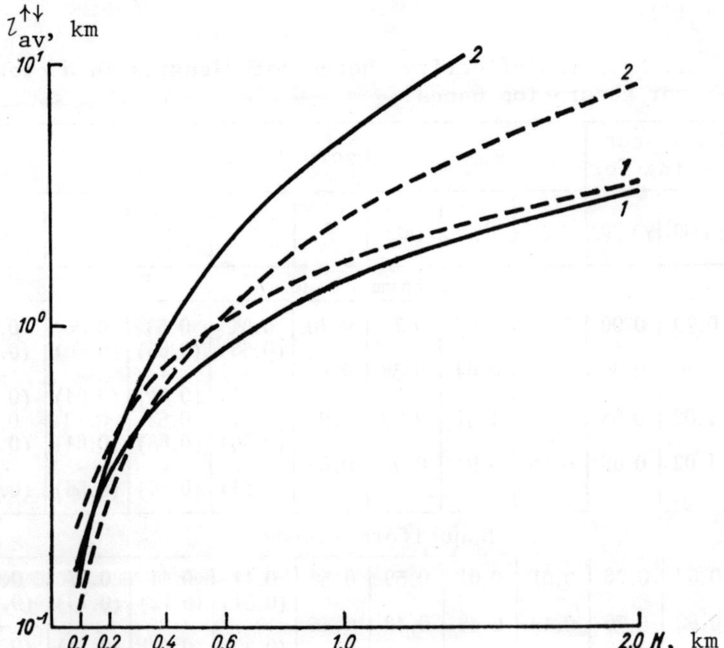

Fig. 6.5. Path lengths l_{av} (1) and l_{av} (2) as functions of H for pure scattering in nonuniform cloud. $\mu_0 = 1$.
Solid curves $\gamma_1(\varphi)$, dashed curves $\gamma_2(\varphi)$.

When the vertical nonuniformity of a cloud is taken into account, the values of $l^{\downarrow}_{av}$ and $l^{\downarrow}_{eff}$ are reduced by no more than 8%, in comparison with a uniform cloud, whereas for some bands $l^{\uparrow}_{av}$ and $l^{\uparrow}_{eff}$ differ from the corresponding values in a uniform cloud by 22%. The mean optical scattering paths $\tau^{\uparrow}_{av}$ for a reflected flux, like the mean number of scatterings, are much less sensitive to the vertical stratification of a cloud than are the geometrical path lengths $l^{\uparrow}_{av}$. Figures 6.4 and 6.5 show $l^{\uparrow\downarrow}_{av}$ for pure scattering as functions of parameters H and ζ. These data indicate a weak dependence of $l^{\uparrow\downarrow}_{av}$ on ζ. The dependence of $l^{\uparrow\downarrow}_{av}$ on the cloud depth H is more marked. In reflected light the mean geometrical path of a photon is 1.5 to 2 times greater than the cloud depth, whereas in transmitted light it is several times greater. The errors in calculating the quantities discussed in this section do not exceed 5%. Similar data for uniform clouds are given in [10].

6.2. Photon-path distribution and mean path lengths in two-layer clouds, taking underlying surface into account

The photon-path distribution and the values of $l^{\uparrow\downarrow}_{av}$ and $l^{\uparrow\downarrow}_{eff}$ are influenced by the size of the cloud system and its macro-structure. Experiments [3, 4] reveal that reflection from the underlying surface has considerable effect on the values of $l^{\uparrow\downarrow}_{eff}$. Accordingly, calculations were made in order to evaluate quantitatively the effect of the albedo of the underlying surface on the effective and mean path lengths in the simplest models of one-layer and two-layer cloud systems [5].

The photon-path distributions $J^{\uparrow}(l)$ and $J^{\downarrow}(l)$ were computed using the Monte-Carlo method for the following situation [5]: two infinitely extensive layers of a scattering medium, each with a fixed thickness of 1 km, are situated over an infinite plane surface reflecting according to Lambert's law with an albedo A_s (Fig. 6.6). The sizes of the intervening spaces are given in fractions of the height of the upper edge of the upper layer $H_{ub} = 10$ km. The scattering coefficient inside the layers was varied from 0.5 to 30 km^{-1}, and in the intervening spaces it was taken to be 0.1 km^{-1}; a value of $\sigma_3 = 0.1$ km^{-1} corresponded to single-layer cloudiness. In all cases (even the "cloudless" intervening spaces) an identical polydisperse "narrow" scattering function $\gamma_1(\varphi)$ was assumed (see Chap. 4). In order to evaluate the effect of the observation conditions, the fluxes between various solid angles (such as θ_1 and θ_2 in Fig. 6.6) were calculated.

The effective photon path length in the cloud model was calculated, as in [5, 10], for various low absorption coefficients α using the expression

$$l^{\alpha}_{eff} = \frac{1}{\alpha} \left| \ln \left[\frac{\int\limits_0^{\infty} J(l) P(\varkappa, l)\, dl}{\int\limits_0^{\infty} J(l)\, dl} \right]^{1/k} \right| , \qquad (6.2)$$

where $P(\alpha, l) = \exp\left[-(\alpha l)^k\right]$ is the transmission function ($k = 1$ for Bouguer's law and $k = 0.5$ for a "square-root" law).*)

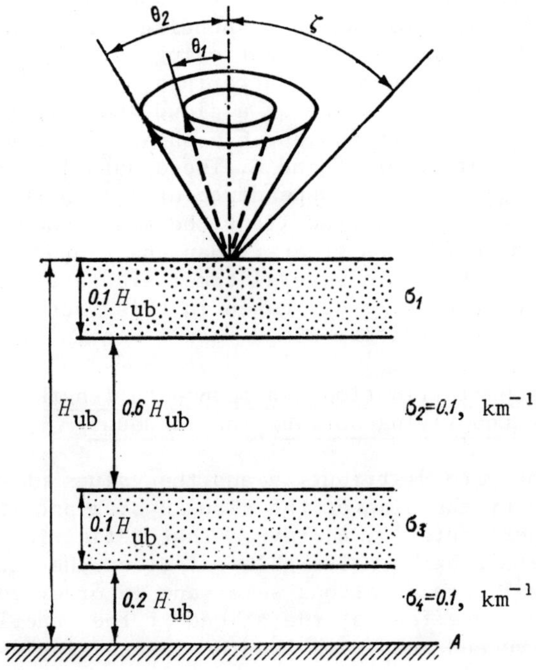

Fig. 6.6. Diagram of calculation model.

For pure scattering the albedo A_Σ and the transmittance T_Σ of the system were calculated, as well as the mean photon path length $l_{av}^{\uparrow\downarrow}$. The latter turned out to be close to l_{eff} for the values of α used. Therefore, in the following only the data for $l_{av}^{\uparrow\downarrow}$ will be considered.

The photon-path distributions were normalized on the basis of the condition

$$A_\Sigma = \int\limits_0^\infty J^\uparrow(l)\, dl = N^\uparrow / N_0, \qquad (6.3)$$

where $N^\uparrow$ is the number of photons emerging from the system into the upper half-space, and N_0 is the total number of photons (statistically determined). Analogously,

---

*) Note that for $k = 0.5$ formula (6) is not valid and $\lim\limits_{\alpha \to 0} l_{eff} \neq l_{av}$. However, l_{eff} remains close to l_{av} for sufficiently low α.

$$T_\Sigma = \int\limits_0^\infty J^\downarrow(l)\,dl = \check{N}^\downarrow/N_0, \qquad\qquad (6.4)$$

where $\check{N}^\downarrow$ is the number of events of photon arrival at the underlying surface. If the albedo of the latter $A_s = 0$, then $\check{N}^\downarrow = N_0 - N^\uparrow$ and thus $A_\Sigma + T_\Sigma = 1$. For $A_s > 0$ it is evident that $N^\downarrow > N_0 - N^\uparrow$ because the trajectory of a given photon may be tangent to the underlying surface in more than one place. In this case

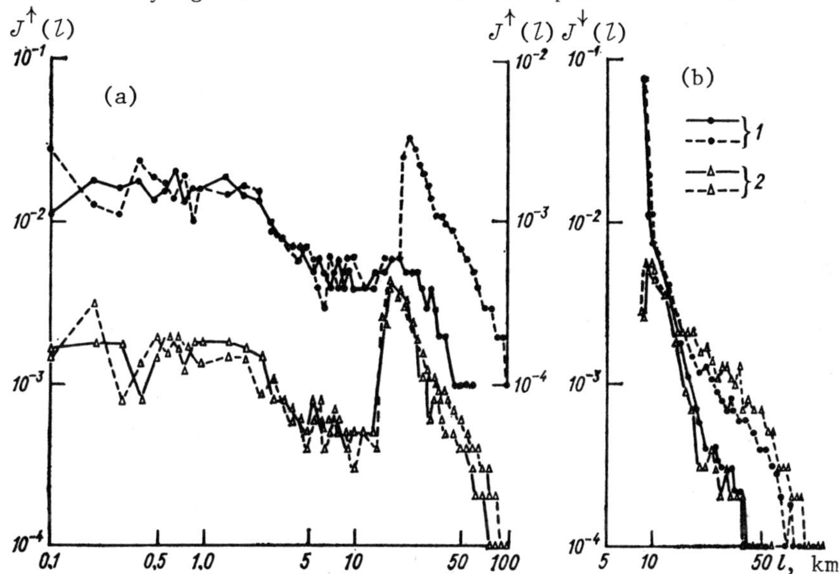

Fig. 6.7. Photon-path distributions $J(l)$ in reflected (a) and transmitted (b) fluxes.
1) $\sigma_1 = 1.5$ km^{-1}, $\sigma_3 = 1.5$ km^{-1}; 2) $\sigma_1 = 1.5$ km^{-1}, $\sigma_3 = 30$ km^{-1}. Solid curves pertain to $A_s = 0$, and dashed curves to $A_s = 0.8$.

$A_\Sigma + T_\Sigma > 1$, and it may even be that in some situations $T_\Sigma > 1$ (an increase in photon density in the space between the lower scattering layer and the underlying surface, see also [14]). In the absence of absorption, A_Σ and T_Σ are independent of the vertical structure of the cloud system. The errors of all the calculations are less than 10%.

Figure 6.7 shows photon-path distributions for the reflected (a) and transmitted (b) fluxes in models of two-layer cloudiness for albedos of the underlying surface $A_s = 0$ and 0.8 and for normal incidence of the illuminating flux. In all these cases properties associated with the presence of weakly scattering intervening spaces are observed. For instance, for reflection in the case of a dense lower layer and a thin upper layer ($\sigma_1 = 1.5$ km^{-1} and $\sigma_3 = 30$ km^{-1}; see Fig. 6.7a, lower pair of curves and left-hand $J^\uparrow(l)$ scale) there is a maximum in distribution $J^\uparrow(l)$, due to the predominant reflection of photons from the lower scattering layer. The diagrams show that in this case the albedo of the underlying surface does not markedly affect the nature of the distribution (the curves for $l > 2H_{ub} = 20$ km differ little). In contrast, if the lower reflecting layer is not dense ($\sigma_1 = 1.5$ km^{-1}, $\sigma_3 = 1.5$ km^{-1}; see Fig. 6.7a, upper pair of curves, right-hand

$J^{\uparrow}(l)$ scale), the poorly defined maximum on the $J^{\uparrow}(l)$ distribution, due to reflection from the lower scattering layer for $A_s = 0$, becomes for $A_s = 0.8$ more pronounced and shifts toward larger path lengths ($l > 20$ km). Clearly, this situation is explained by a predominant reflection of photons from the underlying surface.

Similar patterns are observed in the path distributions corresponding to single-layer cloudiness. Then, reflection from the underlying surface manifests itself clearly even for a dense layer ($\sigma_1 = 30$ km^{-1}).

In the case of transmission (Fig. 6.7b), for single-layer cloudiness reflection from the underlying surface also leads to the appearance of maxima in the path distributions; in two-layer systems this effect is less pronounced. It should also be noted that the path distributions in the case of dense cloud systems with transmission extend up to higher values of l, which clearly illustrates the role of multiple scattering in lengthening the photon paths.

Figure 6.8 shows $l_{av}^{\uparrow\downarrow}/H_{ub}$ as a function of τ_Σ. When analyzing these curves, it is necessary first of all to note the large range of variations of $l_{av}^{\uparrow\downarrow}$, and also the marked dependence of these quantities on the ratio of the optical thicknesses of the cloud layers and on the albedo of the underlying surface. It is also significant that for reflection the mean photon path length $l_{av}^{\uparrow}$ in two-layer cloudiness is considerably greater than in one-layer cloudiness, for the same values of τ_Σ and A_s. This is evident if the families of curves for $\sigma_3 = 1.5$ km^{-1}, $\sigma_3 = 15$ km^{-1}, and $\sigma_3 = 30$ km^{-1} are compared with the curves for $\sigma_3 = 0.1$ km^{-1}. Interestingly enough, if the optical thickness of the system τ_Σ is concentrated mainly in the upper layer ($\sigma_3 = 0.1$ km^{-1}) then $l_{av}^{\uparrow}$ drops rapidly with an increase in τ_Σ, whereas if τ_Σ is concentrated in the lower layer then $l_{av}^{\uparrow}$ depends only slightly on τ_Σ (the effect of the albedo A_s decreasing with an increase in τ_Σ). Consequently, with a lowering of the center of gravity of the system, the reduced role of the photon paths in the intervening space beneath the lower layer is compensated completely by the paths in the space between the cloud layers. For transmission, in most cases the mean path length $l_{av}^{\downarrow}$ for stratification is smaller than for a single-layer model with the same τ_Σ and A_s.

As was to be expected, $l_{av}^{\downarrow}$ increases with an increase in τ_Σ, this increase being more rapid, the higher the albedo A_s. If the thickness τ_Σ is concentrated mainly in the upper layer ($\sigma_1 = 15$ and 30 km^{-1}), then for low A_s the ratio $l_{av}^{\downarrow}/H_{ub}$ will be low and will scarcely increase with an increase in τ_Σ. As Fig. 6.8 shows, path lengths $l_{av}^{\uparrow}$ and $l_{av}^{\downarrow}$ are, as a rule, considerably

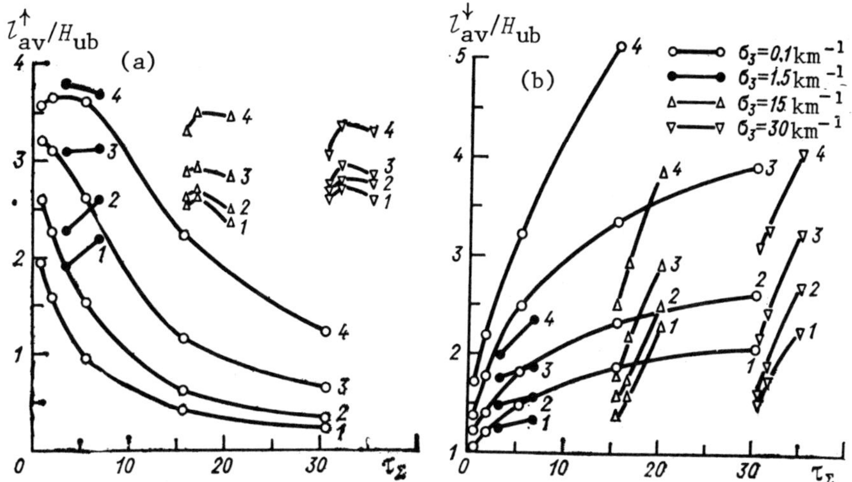

Fig. 6.8. Mean photon path length, reduced to height of upper boundary (l_{av}/H_{ub}), as function of total optical thickness of system τ_Σ for various albedos of underlying surface A_s in reflected (a) and transmitted (b) fluxes.
1) $A_s = 0$; 2) $A_s = 0.2$; 3) $A_s = 0.5$; 4) $A_s = 0.8$.

different for the very same cloud structure.

The values of $l_{av}^{\uparrow\downarrow}$ depend on the observation conditions as well as on the conditions of cloud illumination. Therefore, for some models of a single-layer cloud system ($\sigma_1 = 30$ km^{-1}, $\sigma_3 = 0.1$ km^{-1} for $A_s = 0$ and 0.8; $\sigma_1 = 1.5$ km^{-1}, $\sigma_3 = 0.1$ km^{-1} for $A_s = 0$ and 0.8) path lengths $l_{av}^{\uparrow\downarrow}$ were computed assuming different incidence angles of the solar rays ($\zeta = 0, 30, 60°$) for photons leaving in zenith-angle intervals $\theta = 0-30, 30-60$, and $60-90°$ (see Fig. 6.6). The $l_{av}^{\uparrow}$ values for the total and partial fluxes do not differ too greatly [5]. This justifies the direct comparison of experimental data obtained using a narrow-angle instrument pointed toward the nadir, for zenith angles of the Sun $30° < \zeta < 60°$ [3, 4], with the results of calculations for total fluxes at $\zeta = 0$. In fluxes of transmitted radiation the effect of the angle of observation is quite substantial.

The dependence of the mean paths on the angle of photon incidence ζ is not very pronounced, except in the case where $\theta_1 = 60°$, $\theta_2 = 90°$, and $A_s = 0.8$ for reflection; this can be explained by the fact that the decrease in the number of deeply penetrating photons with an increase in angle ζ is compensated by a rise in the number of gently sloping trajectories.

The calculations carried out indicate that for the possible variations in A_s, τ_Σ, ζ, and θ the values of $l_{av}^{\uparrow}$ exceed the thickness of the cloud system by no more than a factor of 4, while the values of $l_{av}^{\downarrow}$ for high A_s and τ_Σ may be as high as about 10 times the thickness of the system.

6.3. Photon-path distribution and mean path lengths in cumulus clouds

The Monte-Carlo method was used to calculate the distributions, and the numerical characteristics of the photon-path-length distributions, for the paraboloidal model devised in [1] (see Chap. 11). Typical probability density functions for the path lengths in fluxes of reflected and transmitted light are given in [17].

The mean values $l_{av}^{\uparrow\downarrow}$ and the variances $\sigma_{l\downarrow\uparrow}^2$ are plotted in Fig. 6.9. It is interesting that an increase in optical thickness is accompanied by a reduction of $l_{av}^{\uparrow}$ and $l_{av}^{\downarrow}$, but by an increase in the variances (in a stratiform cloud $l_{av}^{\downarrow}$ for the transmitted flux increases with an increase in τ). The mean value l_{av} over the entire cloud (see (10)) varies slightly with a rise in τ, exhibiting a certain tendency to decrease.

Figure 6.10 shows the variation of the mean path lengths of photons scattered by the entire cloud at an angle θ to the direction of the incident flux ($l_{av}(\theta)$ curves). The shortest paths are observed at small scattering angles, corresponding to low multiplicities of photon scattering mainly in thin parts of the clouds. Up to about $\theta = 10°$ the mean path lengths l_{av} increase sharply, by a factor of 3.5 to 4 as θ goes from 0 to $10°$ for $\sigma = 10$ km^{-1} and by a factor of 8 to 10 for $\sigma = 50$ km^{-1}. The longest paths correspond to θ values from 90 to $130°$. With a further increase in the scattering angle, the mean path lengths are seen to be shorter. The effect of the zenith distance of the Sun is not very pronounced: it is enhanced by an increase in the optical density of the cloud.

Figure 6.11 shows some curves of l_{av} found using formula (10) for clouds with arbitrary values of the ratio D/H (where D is the diameter of the cloud base). Inspection of the graphical material indicates that l_{av} depends only slightly on the scattering coefficient; whereas for small and medium zenith distances an increase in σ is accompanied by an increase in the mean path length, for large ζ the opposite is observed. For low cloud heights (up to $H = 0.5$ km) and for moderate zenith distances of the Sun, an increase in the ratio D/H in the region $D/H \geq 3$ practically does not alter the mean path lengths, that is, an asymptotic approach to the case of a plane-parallel layer is observed. For larger zenith distances the region of the asymptote shifts to the right of a D/H value equal to about 5 to 7.

The relative rms errors in determining these mean path lengths $l_{av}^{\uparrow}$ and $l_{av}^{\downarrow}$ do not exceed 5-7%. The data in Fig. 6.11 have accuracies ranging from 10 to 30%, and the errors in calculating $l_{av}(\theta)$ lie in this same range.

Some data on photon-path distributions for statistical models of cumulus clouds are given in [2, 16].

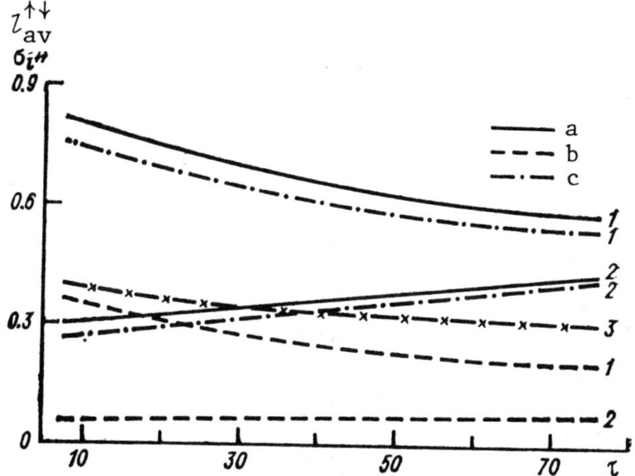

Fig. 6.9. Mean values and variances of photon path lengths as functions of optical thickness for reflection (a), transmission (b), and reflection in nonuniform cloud (c). $\zeta = 60°$.
1) Mean values, 2) variances, 3) l_{av} for entire cloud, km.

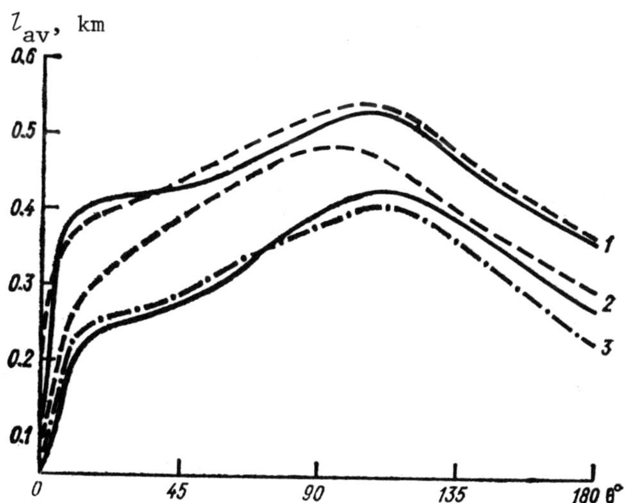

Fig. 6.10. Mean path lengths l_{av} as functions of scattering angle. $H = 0.765$ km.
1) $\tau = 7.65$, 2) $\tau = 38.25$, solid curves $\zeta = 0°$, dashed curves $\zeta = 60°$; 3) nonuniform cloud with $\tau = 38.25$ and $\zeta = 0°$.

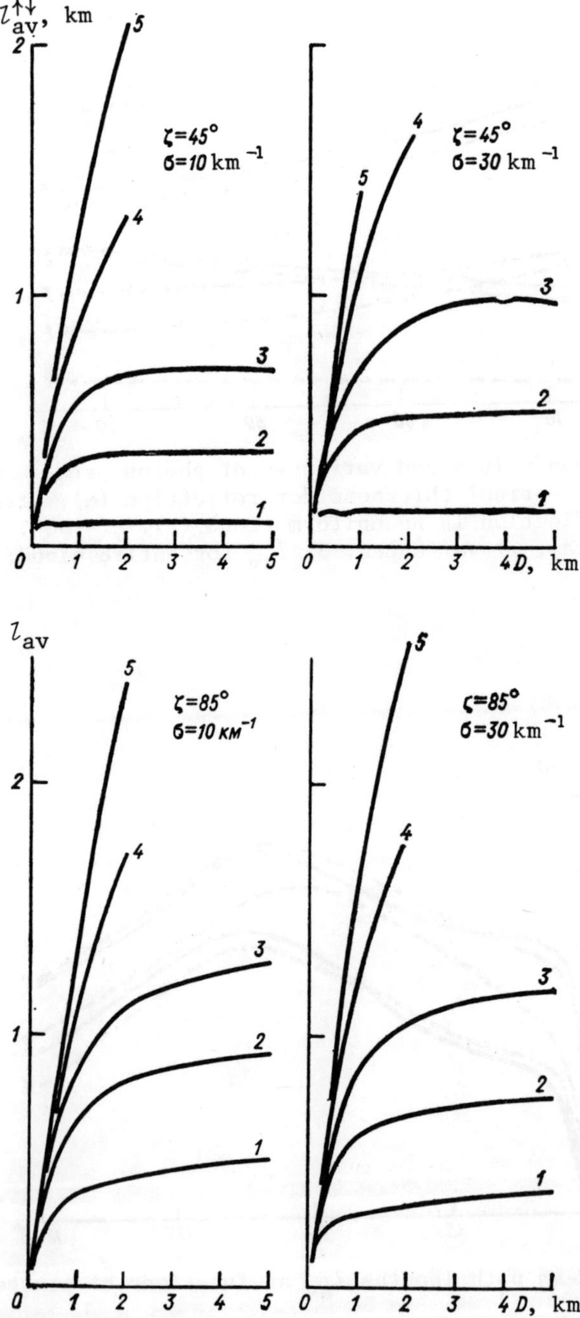

Fig. 6.11. Mean path lengths l_{av} as functions of base diameter of uniform cloud.

6.4. Equivalent trajectories in clouds exhibiting nonuniform absorption

If the absorbing substance is distributed nonuniformly throughout the cloud system, as can be the case, for instance, for oxygen,
water vapor, or water droplets, then photons having trajectories
of equal lengths may not be absorbed to the same extent. Then,
even in the case of quite weak absorption, when the scattering
and absorption can be assumed to be independent, a knowledge of
just the scalar quantities $J(l)$ and l_{av} is insufficient for a
description of the radiation field in a cloud. The scalar quantity
l_{eff} for nonuniform absorption cannot be determined unambiguously
without making additional assumptions. The arrangement of the
trajectories in the medium must be taken into account in order to
solve the problem satisfactorily in this case. For weak absorption
the space network of these trajectories corresponding to the absence of absorption can be used.

Let us consider cloudiness consisting of plane-parallel layers
illuminated by a parallel flux. We replace the set of N real
trajectories by a set of N identical equivalent trajectories for
each of the fluxes, reflected by the cloud system or transmitted
through it [9]. In order to construct the equivalent trajectories
assuming the absence of absorption, let us start by specifying
that the photon density (radiant energy) is conserved in each
elementary cloud region from z_i to $z_i + \Delta z_i$, separately for the
photons constituting the transmitted and reflected fluxes. This is
equivalent to saying that for the two sets of trajectories the sums
of the lengths of all the trajectory segments pertaining to the
given flux are equal in each elementary cloud region. Let us also
require that an equivalent trajectory pass through all regions
without exception, albeit only once. If $S_{ei} = |\Delta l_i / \Delta z_i|$ is a
measure of the length of the segment of the equivalent trajectory
in the ith region, then we can write

$$NS_{ei}\,\Delta z_i = \sum_{j=1}^{N} \sum_{k=1}^{r_{ij}} S_{ijk}\,\Delta z_i$$

or

$$S_{ei} = \frac{1}{N} \sum_{j=1}^{N} \sum_{k=1}^{r_{ij}} S_{ijk}, \qquad (6.5)$$

where N is the number of reflected or transmitted photons, S_{ijk}
is the secant of the inclination angle of the jth real trajectory
in the ith region at its kth intersection, and r_{ij} is the total
number of intersections in the ith region of the jth real trajectory. When $S_{ei} \geq 1$, this quantity can be taken to signify the

secant of the inclination angle of the trajectory in the region, similarly to the quantity S_{ijk}. Values of $S_{ei} < 1$ can occur because some of the photons of the reflected flux do not reach the corresponding layer.

Formula (6.5) implies directly that the total length of such an equivalent trajectory is equal to the mean path length of a photon in the system:

$$L_e = \sum_{i=1}^{M} S_{ei}\,\Delta z_i = \frac{1}{N}\sum_{j=1}^{N}\sum_{i=1}^{M}\sum_{k=1}^{r_{ij}} S_{ijk}\,\Delta z_i = \frac{1}{N}\sum_{j=1}^{N} l_i = l_{av} \tag{6.6}$$

(here M is the total number of elementary regions in the system). Hence, for a system exhibiting uniform absorption, if Bouguer's law is obeyed a calculation of the (weak) absorption along an equivalent trajectory gives the same result as a direct calculation in terms of $J(l)$ with a subsequent use of l_{av} (see Chap. 10), that is, in the uniform case the two approaches are equivalent.

For a homogeneous medium the possibility of using l_{av} to evaluate weak absorption indicates that l_{av} is close to $l_{eff}(\alpha)$. Presumably, the criterion for weak absorption, corresponding to the approximation

$$l_{av} \approx l_{eff}(\alpha), \tag{6.7}$$

can at the same time serve as a criterion for replacement of the real set of trajectories by an equivalent set taking into account absorption in an inhomogeneous medium. A criterion for weak absorption in an inhomogeneous medium can, for instance, be introduced in this way. If we stipulate that $l_{eff} \geq 0.9 l_{av}$, then model calculations for a homogeneous medium [7] show that for the absorption coefficient the condition

$$\alpha l_{av} \leqslant 0.2 \ \text{to} \ 0.5 \tag{6.8}$$

must be satisfied, which corresponds to values of the transmission function

$$P_{\Delta v}(\alpha l_{av}) \geqslant 0.5. \tag{6.9}$$

An equivalent trajectory can be used in an inhomogeneous medium if the transmission function along this trajectory satisfies inequality (6.9). Such absorption is of practical interest.

The values of $S_e^{\uparrow}(z)$ and $S_e^{\downarrow}(z)$ were calculated for the reflected and transmitted fluxes, respectively, using the Monte-Carlo method for 20 models of single-layer and double-layer cloudiness, for various albedos of the underlying surface A_s. Most of the calculations were made for a total depth of the system $z_0 = 3$ km, the system being divided into 30 zones. The intervening space between the layers of the two-layer models was 1 km deep; its position in the system varied. The lower boundary of the system was 1 km above the underlying surface. Calculations were carried out for a "narrow" scattering function $\gamma_1(\varphi)$ in the layers and intervening spaces for a wavelength of 0.714 μm (see Chap. 4); the reflection from the underlying surface was assumed to obey Lambert's law. The scattering coefficient in the intervening space was always 0.1 km^{-1}.

Photons emerging through the upper boundary of the system pertained to the reflected flux A_Σ. The criterion for a photon to belong to the transmitted flux T_Σ was "contact" with the underlying surface. For $A_s = 0$ this corresponds to the traditional definition of T_Σ, when $A_\Sigma + T_\Sigma = 1$. For $A_s \neq 0$ the flux T_Σ^* found in this way characterizes the illumination of the underlying surface, recorded by an instrument placed right on it. Since photons repeatedly contacting the underlying surface (including photons belonging to the reflected flux) also contribute to this illumination, it follows that $T_\Sigma^* > T_\Sigma$ and $T_\Sigma^* + A_\Sigma^* > 1$.

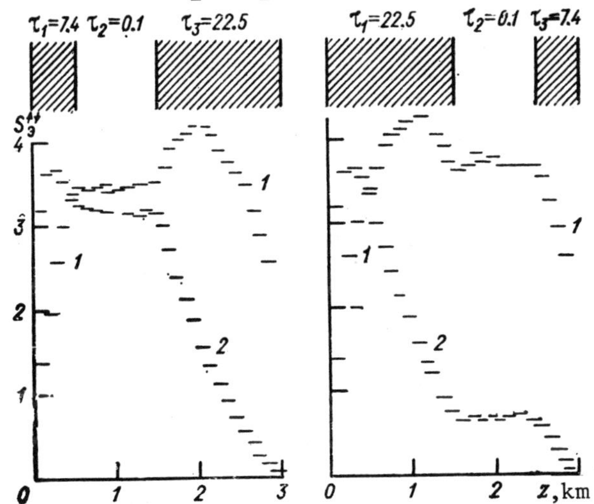

Fig. 6.12. Histograms of S_e for two-layer systems with various positions of the intervening space. $\tau_0 = 30$, $\mu_0 = 1$, $A_s = 0$. Shading indicates dense layers, radiation moves leftward. 1) transmission, 2) reflection.

As an example of the primary histograms, Fig. 6.12 shows the unsmoothed calculated values of $S_e(z)$ for two positions of the interlayer space of systems with the same total optical density τ_0 and $A_s = 0$. Since the random irregularities associated with the Monte-Carlo method are quite small, the curves in the other figures are smoothed. Inspection of Fig. 6.12 also indicates that the values of $S_e(z)$ inside weakly scattering spaces of the system are, as was to be expected, the same as the values which would be observed in an unstratified system with the same optical thickness $\tau(z)$. This means that the relations $S[\tau(z), \tau_0]$, where $\tau(z) =$

$\int_0^z \sigma(z)dz$, are universal, that is, they can be used for systems with an arbitrary number of layers (or, what amounts to the same thing, with an arbitrary distribution $\sigma(z)$, on conditions that the scattering functions in all the layers are the same). Figure 6.13 shows S_e curves plotted in these coordinates for the transmitted and reflected fluxes.

These graphs show that the extreme values of $S_e^{\downarrow}(\tau)$ lie in a region close to the center of the system for low albedos of the underlying surface. As A_s increases, they are shifted downward, until at the limit $A_s = 1$ they reach the level of this surface. A similar tendency for the maximum to shift toward higher τ is observed for the reflected flux (Fig. 6.13b), but it is much less pronounced in this case and the abscissas for the extreme values of $S_e^{\uparrow}(\tau)$ do not go beyond $\tau \approx 10$ (except when $A_s = 1$). Figure 6.13a shows that a variation in the angle of incidence of the radiation $\zeta = \text{arccos } \mu_0$ from 0 to 60° for the transmitted flux affects only the initial parts of the curves ($\tau \leq 10$), after which $S_e^{\downarrow}$ is practically unchanged. Modifications of the reflection conditions change the curves for the reflected flux (see Fig. 6.13b) somewhat more markedly.

The increase in the total density of the radiant energy at the τ level of a nonabsorbing system, in comparison with propagation in vacuo, can for $A_s = 0$ be expressed in terms of S_e as

$$G(\tau) = \mu_0 \left[T_{\Sigma} S_e^{\downarrow}(\tau) + A_{\Sigma} S_e^{\uparrow}(\tau) \right]. \qquad (6.10)$$

The first term in brackets characterizes the contribution of the photons forming the transmitted flux, and the second corresponds to the photons of the reflected flux; T_{Σ} and T_{Σ} are the transmission and albedo of the system, respectively, without taking reflection from the underlying surface into account ($T_{\Sigma} + A_{\Sigma} = 1$).

For $A_s > 0$ relation (6.10) does not hold true, and $G(\tau)$, which is the ratio of the mean secant of the inclination angle of all the photon trajectories in the layer τ, $\tau + \Delta\tau$ to the secant of the angle of incidence of the radiation, was calculated without dividing the flux into a transmitted flux and a reflected flux (Fig. 6.14).

The maximum values of the energy density in a cloud system for low A_s and $\mu_0 = 1$ are, as Fig. 6.14 shows, observed at low optical depths and they shift slightly with an increase in τ_0. A variation in the angle of incidence of the radiation from 0 to 60° shifts the maximum values of G to the surface layers of the system. For high A_s the largest G values correspond to regions adjacent to the underlying surface, the energy concentrations there being considerable ($G = 5$ for $A_s \approx 1$ and $\tau_0 = 30$). This effect is in the main associated with a repeated passage of photons into the space between the lower boundary of the system and the underlying surface.

Equivalent plane photon trajectories for a continuous

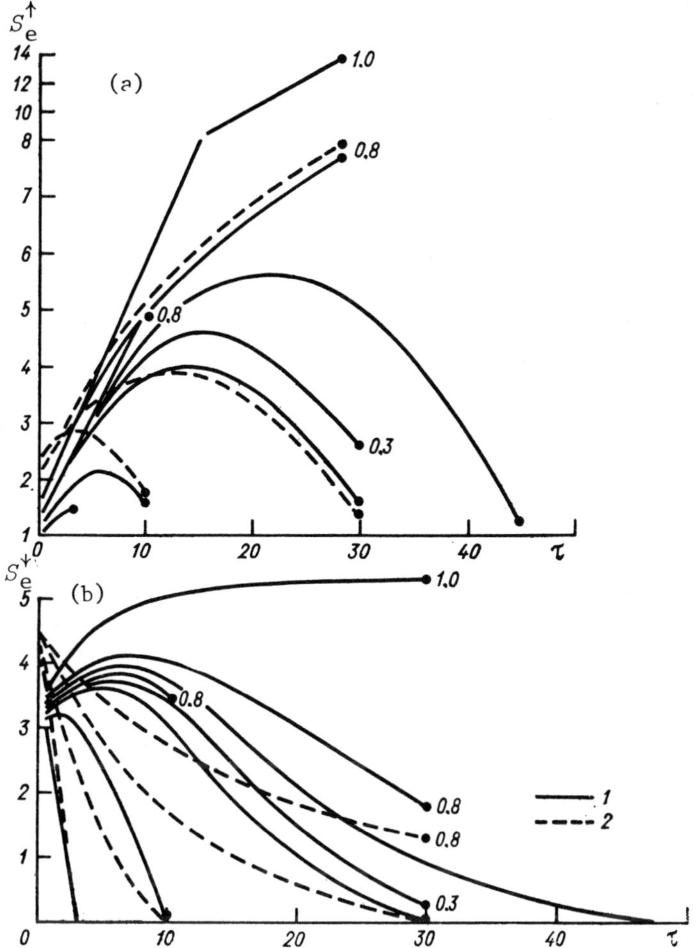

Fig. 6.13. Curves of $S_e(\tau, \tau_0)$ for transmission (a) and reflection (b).
Dots at ends of curves correspond to total optical thicknesses of system τ_0 for each specific case; numbers indicate albedo of underlying surface; for curves without numbers $A_s = 0$.
1) $\mu_0 = 1$; 2) $\mu_0 = 0.5$.

homogeneous layer with $\tau_0 = 10$, calculated on the basis of the data in Fig. 6.13, are shown in Fig. 6.15. In a number of cases the lengths of the equivalent trajectories are seen to be several times greater than the thickness of the cloud system, these lengths increasing substantially with an increase in the angle of incidence of the radiation and the albedo of the underlying surface (see Fig. 6.13). The dot-dash curves in Fig. 6.15 indicate physically unrealizable parts of the equivalent trajectories for the reflected flux, where $S^\uparrow_{ei} < 1$.

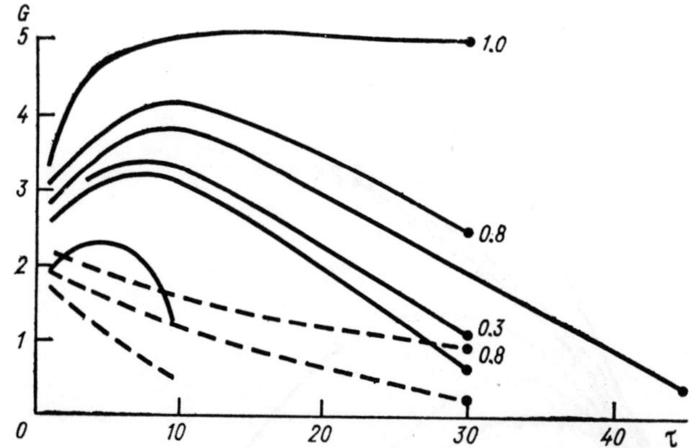

Fig. 6.14. Relative increase in total density of radiant energy
as compared with propagation in vacuo.
Arbitrary notation same as in Fig. 6.13.

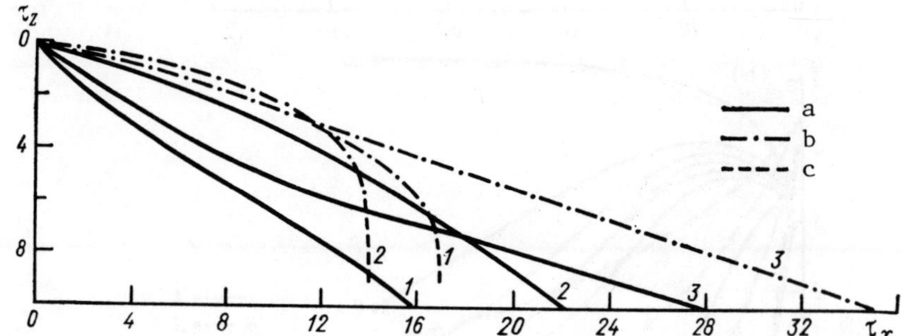

Fig. 6.15. Equivalent trajectories in homogeneous layer for
$\tau_0 = 10$.
τ_z and τ_x are vertical and horizontal optical coordinates.
a) transmission, b) reflection, c) physically unrealizable parts
of trajectories, where $S_e^{\uparrow} < 1$.
1) $\mu_0 = 1$, $A_s = 0$; 2) $\mu_0 = 0.5$, $A_s = 0.3$; 3) $\mu_0 = 1$, $A_s = 0.8$.

CHAPTER 7

EXPERIMENTAL DETERMINATION OF EFFECTIVE PHOTON PATH LENGTHS

7.1. Measuring effective photon path lengths using a weak absorption band

According to formula (5) of the introduction to Part II, the ratio of the total photon flux $I_{\Delta\nu}^{*}$ emerging from a scattering medium with $\alpha \neq 0$ to the flux I_0 for the same medium but without absorption is

$$\frac{I_{\Delta\nu}^{*}}{I_0} = P\left(\alpha,\ l_{\text{eff}}\right), \qquad (7.1)$$

provided the absorbing substance is distributed uniformly throughout the medium. In a cloud system which is homogeneous with regard to absorption, an unambiguous definition of l_{eff} is possible only if a priori information about the shapes of equivalent photon trajectories in the system is available. A consideration of some "extreme" variants of such trajectories indicates that for real lower-level clouds the region of corresponding values of l_{eff} is not too wide, and a satisfactory approximation can be arrived at using the simplest model of an equivalent trajectory, in the form of an inclined line segment ($S_e^{\uparrow\downarrow}$ = const) (see also [4] and below).

Expression (7.1) suggests an experimental determination of the effective photon path length l_{eff} by measuring the fluxes emerging from a scattering system with and without absorption. Such a method can be realized quite simply if spectrum $\alpha(\lambda)$ possesses absorption bands adjacent to windows of almost complete transparency ($\alpha \approx 0$). If the difference between the wavelengths λ_1 and λ_2 corresponding, respectively, to the absorption band and transmission window is sufficiently small, then the scattering characteristics (scattering function, volume scattering coefficient) can be taken to be constant in the interval (λ_1, λ_2). Then the ratio of the luminous fluxes recorded with an instrument $F_{\lambda_1}/F_{\lambda_2} \approx P(\alpha,\ l_{\text{eff}})$, so that for a known form of function $P(\alpha,\ l)$ and a known absorption coefficient α the path length l_{eff} can be found.

This technique was applied to real cloud systems in [3, 4], using as a "sonde" the narrow forbidden band of atmospheric oxygen O_2 with its center around $\lambda_1 = 0.762$ μm, located next to the window of high transparency centered at $\lambda_2 = 0.743$ μm. Because the height distribution of the oxygen concentration is regular and stable, the absorption coefficient of the gaseous medium in the vicinity of a cloud can in this case be determined using universal reference formulas for the optical density of the band as a function of the air mass $D_\odot = f(H, m_\odot)$, where $m_{\odot} = \sec \zeta$, obtained at various heights H [4]. The two-wave "phase" spectrophotometer [13] used in the experiments enabled a direct determination of ratio $F_{\lambda_1}/F_{\lambda_2}$. This automatically excluded any effect **on the measurement** re**s**ults of variations in the absolute values of fluxes F_{λ_1} and F_{λ_2}, due to differences in cloud types and in the conditions of illumination by solar radiation. The instrument measured the spec-

trum $D(\lambda) = P(\alpha) = -1_g \dfrac{F_{\lambda_1}}{F_{\lambda_2}}$ in the range of λ from 0.737 to

0.767 μm for 1–2 min with a resolution of about 2.5 mμm. The viewing angle of the instrument was 1.5 x 0.1°. Figure 7.1 is a diagram of the measurement setup. Reversal of the viewing direction (at the zenith of nadir for measurements in the transmitted and reflected fluxes, respectively) was effected by turning a plane mirror mounted in the cockpit of the aircraft. As an illustration, Fig. 7.2 shows some typical trace recordings. They clearly show the increase in optical density D_{cld} due to the multiple scattering of photons in clouds, as compared with the value $D_\odot$ measured for the same value of $m_\odot$ under cloudless conditions.

The additional absorbing mass Δm in the clouds is determined from the condition

$$D_\odot \left(H_{ub}, m_\odot + \Delta m\right) = D_{cld}\left(H_i, m_\odot\right), \qquad (7.2)$$

and the effective path length is thus

$$l_{eff} = k^{-1} \Delta m H_0 \exp\left(H_{av}/H_0\right), \qquad (7.3)$$

where H_i is the height of the upper cloud boundary H_{ub} for measurements in the reflected flux and the height of the lower boundary H_{1b} in the transmitted flux; $H_{av} = \dfrac{H_{ub} + H_{1b}}{2}$; $H_0 = 8$ km is

the height of the homogeneous atmosphere; and k is a correction factor allowing for the variation of the specific absorption of oxygen with altitude, that is, the inhomogeneity of the cloud with regard to absorption.

As the measurements showed [4], for air masses $m_\odot$ from 1.5 to 4, the mass dependence of the optical density $D_\odot(m_\odot)$ can be approximated satisfactorily by the expression

$$D_\odot = c\left(m_\odot\right)^\eta, \qquad (7.4)$$

where c and η are fitting parameters which depend on the height of the observation point.*)

------

*) In a wider range of $m_\odot$ from 1 to 10, the $D_\odot(m_\odot)$ relation is more complicated.

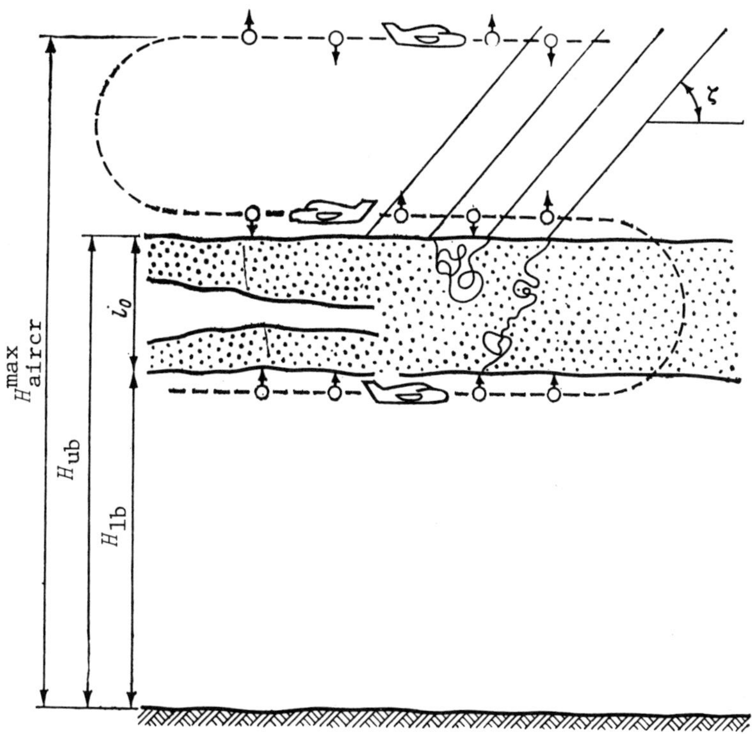

Fig. 7.1. Diagram of measurement setup.
Small arrows along aircraft trajectory show viewing direction.

By finding the difference between the numbers of absorbing
particles on rays beginning at the same observation level but
corresponding to air masses $m_\odot$ and $m_{eff} = m_\odot + \Delta m$ and by
taking (7.4) into account, we can use (7.3) to obtain a working
formula for l_{eff}:

$$l_{eff} = k^{-1} m_\odot H_0 \left[(D_{cld}/D_\odot)^\eta - 1 \right] \exp \left[H_{av}/H_0 \right]. \qquad (7.5)$$

The determination of correction factor k should be considered
specially. In a real cloud system a light ray which has passed
through the layer above the clouds subsequently moves along a
complicated trajectory inside a scattering layer with boundary
pressures corresponding to its upper and lower surfaces. Correction factor k allows for concentration differences and for non-
equivalence of the optical properties of absorbing molecules in
the above-cloud layer and inside the scattering layer of the
cloud. The correction factor was calculated for a ray propagating
at various angles in a two-layer nonscattering atmosphere (Fig.
7.3), using two methods [4]: direct calculation with a computer
of the transmission in the forbidden O_2 absorption band

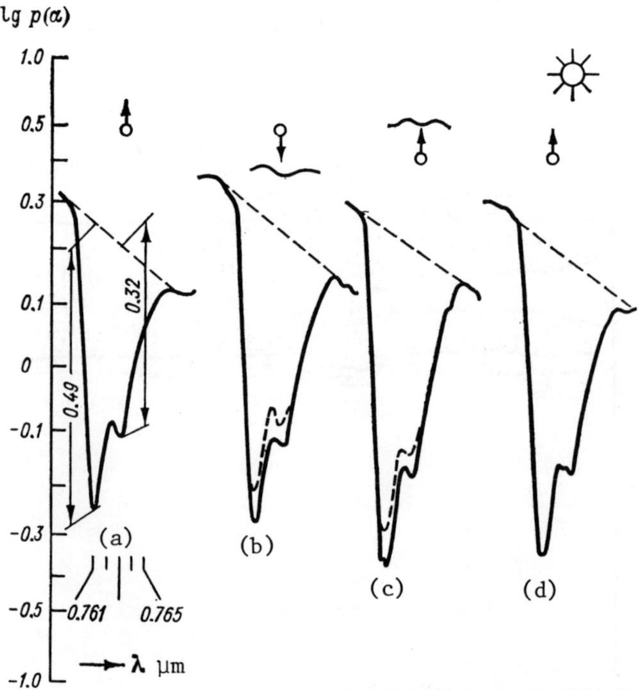

Fig. 7.2. Typical trace recording of effective optical density in O_2 band (0.76 μm) for cloud situation No. 6 (see Table 7.2). a) extinction spectrum for solar radiation in above-cloud layer beginning at upper cloud boundary; b) spectrum of radiation reflected from upper cloud boundary for $H_{aircr} = H_{ub}$; c) spectrum of radiation transmitted by cloud layer for $H_{aircr} = H_{1b}$; d) extinction spectrum in cloudless atmosphere, beginning at level $\dfrac{H_{ub} + H_{1b}}{2}$. Dashed curve on trace b is given for comparison with trace a, and dashed curve on trace c for comparison with trace d.

(see Fig. 7.3, dots), and integration of an element of absorbing mass $\rho(H)^{\beta} \sec \zeta \, dH$, on the basis of an **experimental formula** for the optical density

$$D_{\odot}(H, \zeta) = c\left\{[\rho(H)]^{\beta} \sec \zeta\right\}^{\eta}$$

(see Fig. 7.3, crosses).

During the processing of the experimental data we used the approximate expression $k = 2 - \dfrac{1}{1 + \widetilde{l}_{eff}}$, where $\widetilde{l}_{eff}$ is the effective path length, found with the aid of formula (7.5) for $k = 1$.

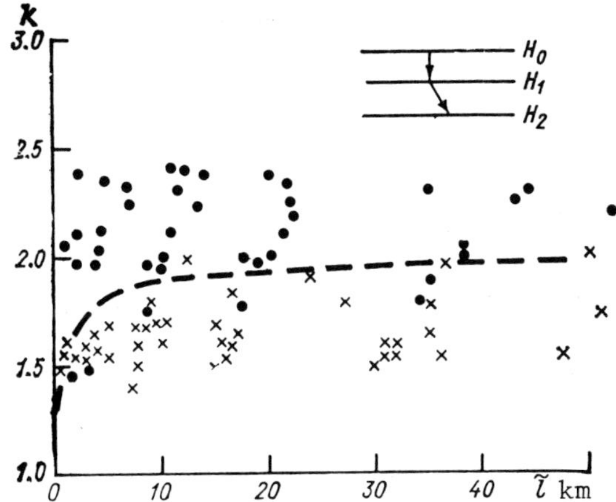

Fig. 7.3. Working model and plot of correlation factor as function of effective path length $\tilde{l}_{eff}$, calculated using approximate method. ● using direct calculation of absorption; × using empirical absorption formula.

7.2. Results of determining effective photon path lengths

The 43 cloud systems studied in [3, 4] included obvious single-layer (Table 7.1), conditionally single-layer (Table 7.2), double-layer (Table 7.3), and multilayer (Table 7.4) cloudiness. The tables give the heights of the upper H_{ub} and lower H_{lb} boundaries of the layers; the layer numbers in Table 7.3 (Roman numerals at left) begin from the upper boundary. Schematic representations of each situation are given for both conditionally single-layer cloudiness and multilayer cloudiness (Tables 7.2 and 7.4).

The results of processing the experimental data with the aid of formula (7.5) for reflection and transmission are plotted on log-log scale in Fig. 7.4. The coordinates are l_{eff} and l_0, where l_0 is the total depth of the cloud system, that is, the difference between the heights of the upper boundary of the upper layer and the lower boundary of the lower layer. The figure also shows calculated curves [7] for a homogeneous scattering layer, without taking into account the atmosphere above the clouds or the albedo of the underlying surface. Symbols of smaller size pertain to individual specific measurements. In order to illustrate the spread of the individual determinations for the corresponding situation, these are given for several cases. Larger symbols denote mean values for each of the situations examined, the numbers with arrows being the numbers N in Tables 7.1-7.4. The vertical dashed segments indicate the rms deviation in terms of l_{eff}. Black dots and x's pertain to multilayer systems, and white dots and +'s to obvious single-layer systems.

Inspection of Fig. 7.4 shows that within the limits of experimental error the effective path lengths in single-layer cloudiness

Table 7.1.

				Obvious single-layer cloudiness	
N	H_{ub} km	H_{lb} km	l_0 km	Note	Diagram
1	1.8	1.1	0.7		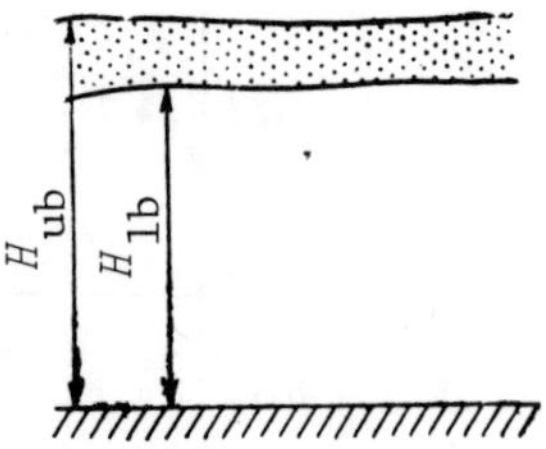
2	2.3	1.4	0.9		
3	5.2	1.6	3.6	Homogeneous loose cloud	
10	2.6	2.2	0.4	Layer located above water sur-face	
11	1.4	1.05	0.35		
12	1.5	1.3	0.2	l_0 varies from	
13	1.6	1.5	0.1	0.15 to 0.05 km	
14	1.6	1.3	0.3		
15	2.0	1.9	0.1		
16	2.2	2.1	0.1		
17	2.4	2.2	0.2		
18	3.0	2.7	0.3		
19	4.9	2.8	2.1		

Table 7.2.

				Conditionally single-layer cloudiness	
N	H_{ub} km	H_{lb} km	l_0 km	Note	Diagram
4	3.5	3.1	0.4	Lower layer cloud cover 1–2	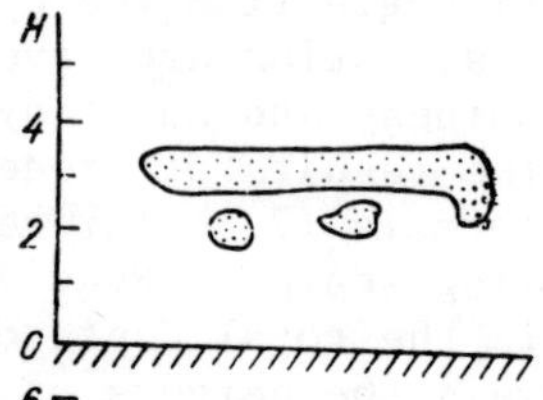
6	2.6	1.3	1.3	Lower layer cloud cover 2–3	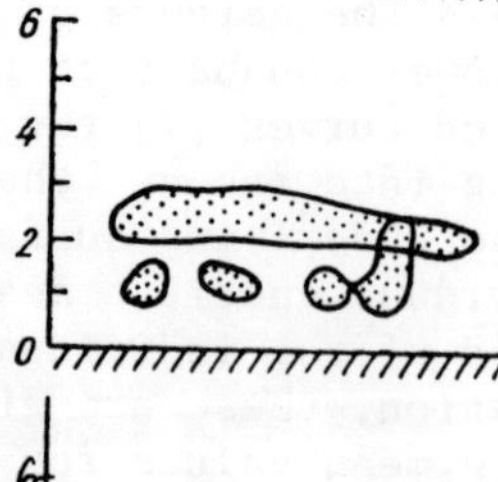
7	1.9	0.6	1.3	Lower layer cloud cover 4–5	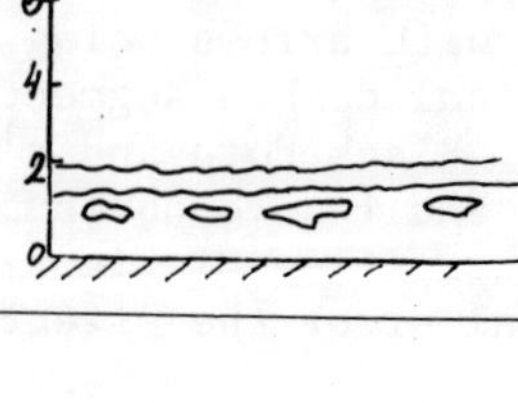

Continually single-layer cloudiness					
N	H_{ub} km	H_{lb} km	Z_0 km	Note	Diagram
20	5.5	0.9	4.6	Lower layer cloud cover 1-2	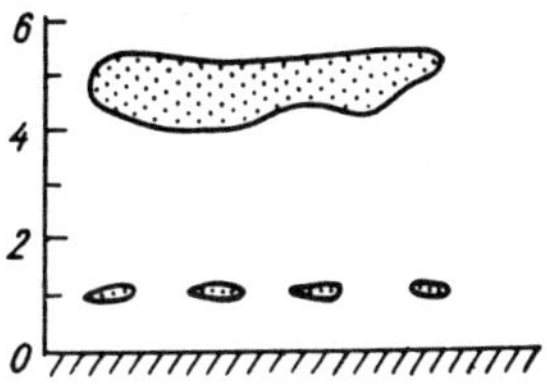
21	4.9	2.4	2.5	Cloudiness has considerable nonuniformity	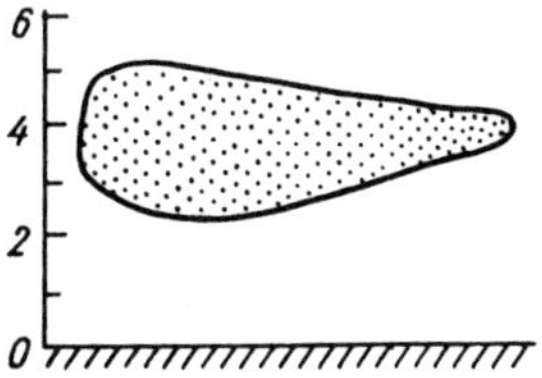
22	3.0	0.3	2.7	No space between layers	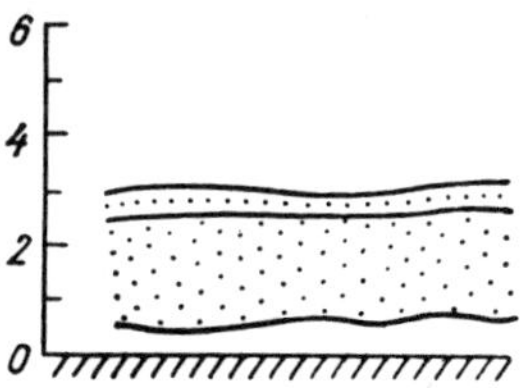
23	4.1	0.8	3.3	Lower layer cloud cover ≈ 3	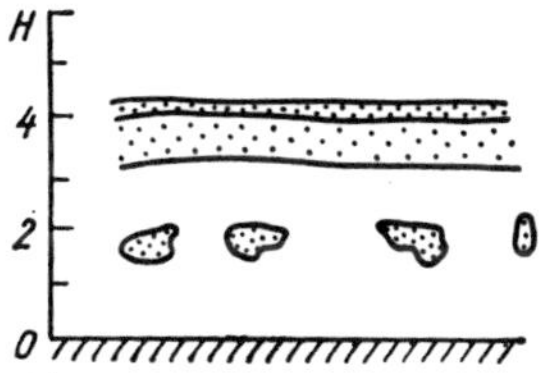
24	2.7	1.5	1.2	Space between layers ≈ 50 m	
25	4.4	1.7	2.7	Lower and middle layers cloud cover 1-2	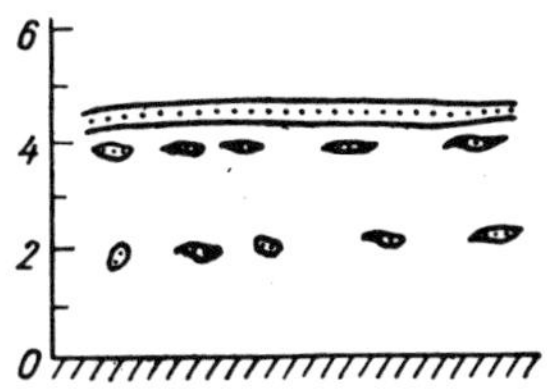

Conditionally single-layer cloudiness					
N	H_{ub} km	H_{1b} km	l_0 km	Note	Diagram
26	4.2	2.0	2.2	Lower and middle layers cloud cover 1–2 Transmission was measured with $H_1 = 4$ km for $l_0 \approx 0.2$ km	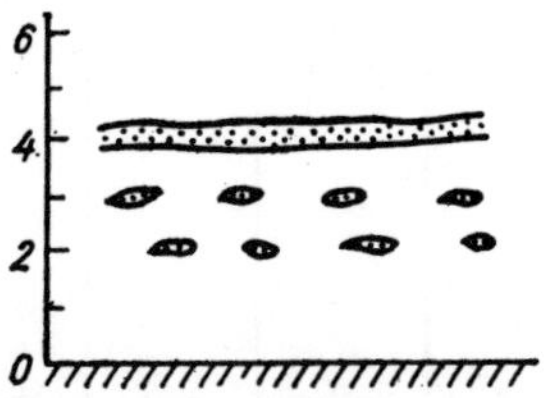
27	5.4	1.3	4.1	Middle layer thin cloud cover about 1 lower layer 2–3	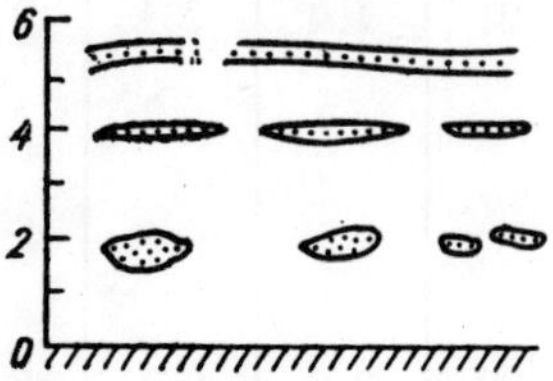
28	4.2	2.4	1.8	Layers in haze Lower layer cloud cover 2–3	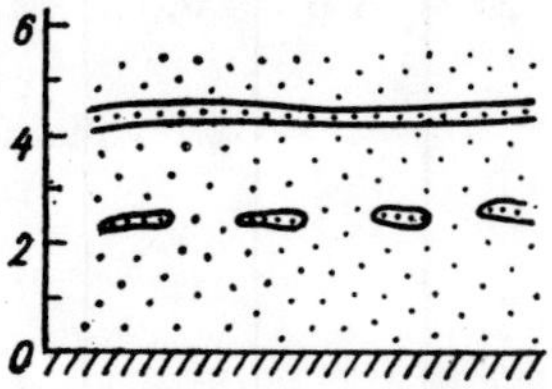
29	3.8	3.65	0.15	At height ≈ 6 km thin haze layer ≈ 0.2 km	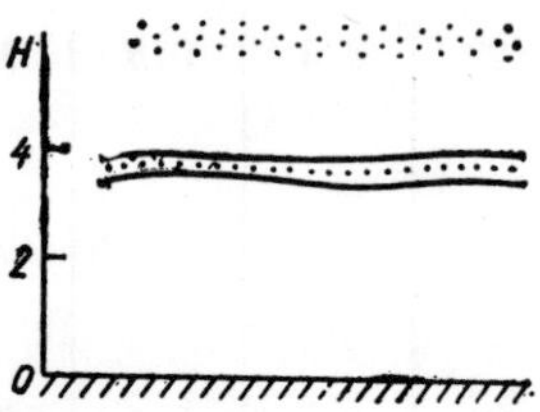
30	5.5	5.3	0.2	Haze under cloud layer	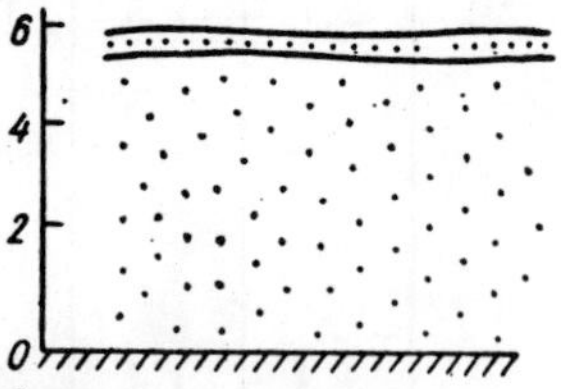
31	1.7	0.8	0.9	Lower layer cloud layer ≈ 2	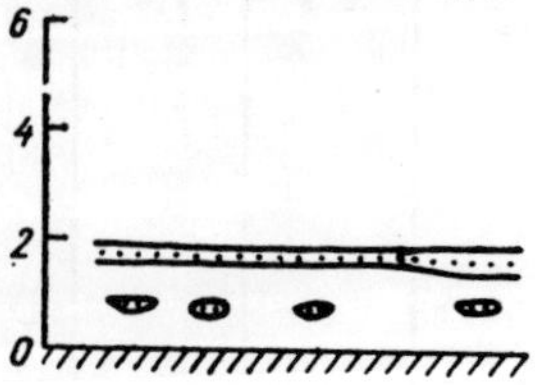

				Conditionally single-layer cloudiness	
N	H_{ub} km	H_{lb} km	l_0 km	Note	Diagram
32	1.6	0.6	1.0	Lower layer cloud cover 2-3	
33	2.5	1.2	1.3	Lower layer cloud cover 2-3	
34	2.3	2.1	0.2	Haze layer at height of 1 km	
35	4.3	1.3	3.0	Lower layer cloud cover 3-5	

Table 7.3.

					Double-layer cloudiness	
N	H_{ub}^{I} km	H_{lb}^{I} km	H_{ub}^{II} km	H_{lb}^{II} km	l_0 km	Note	Diagram
5	3.9	3.7	3.4	1.4	2.5		
36	4.2	3.95	2.5	2.45	1.75		
37	4.6	4.1	2.0	1.95	2.55	Transmission was measured with $H_1^{I} = 4.45$ km for $l_0^{I} = 0.15$ km	
38	4.6	4.4	1.5	1.3	3.3		
39	2.8	2.4	1.3	0.8	2.0		

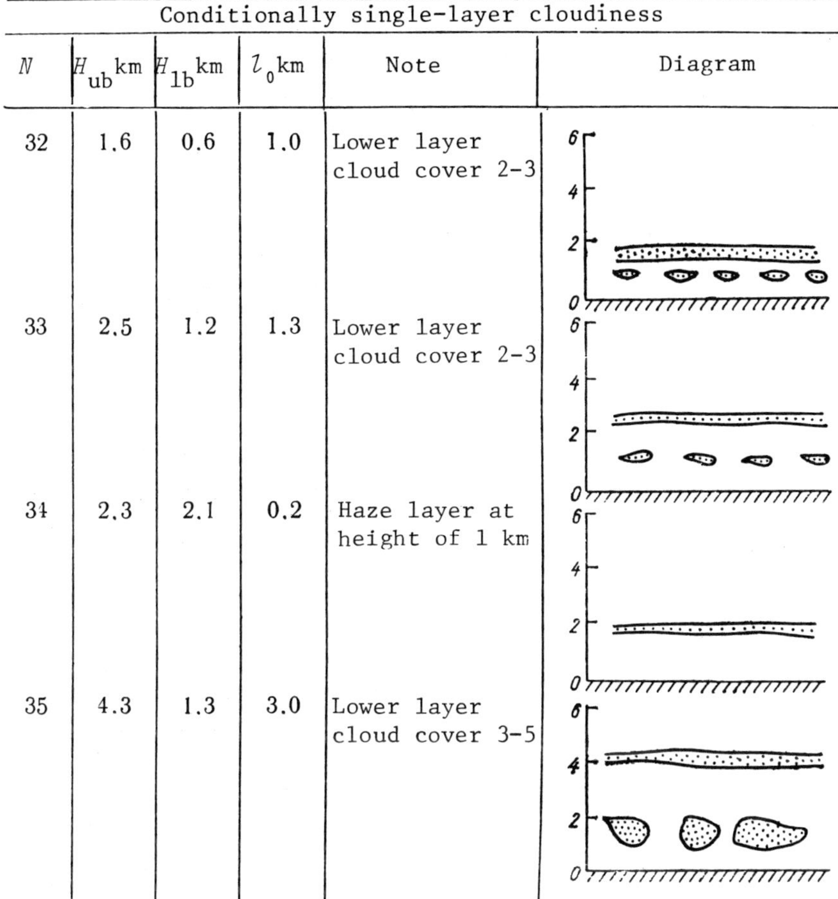

 RADIATION IN A CLOUDY ATMOSPHERE

Table 7.4.

Multilayer cloudiness

N	H_{ub} km	H_{1b} km	l_0 km	Note	Diagram
8	5.4	0.7	4.7	All layers submerged in haze	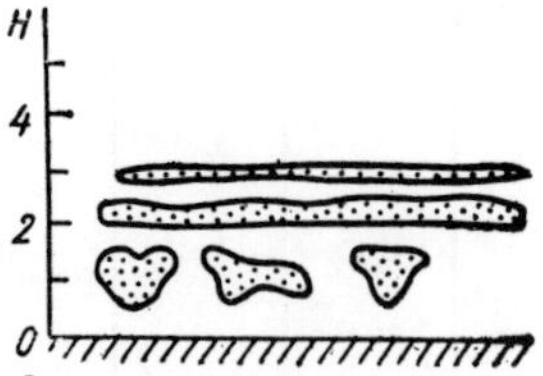
9	4	0.4	3.6	All layers loose clouds, except lower	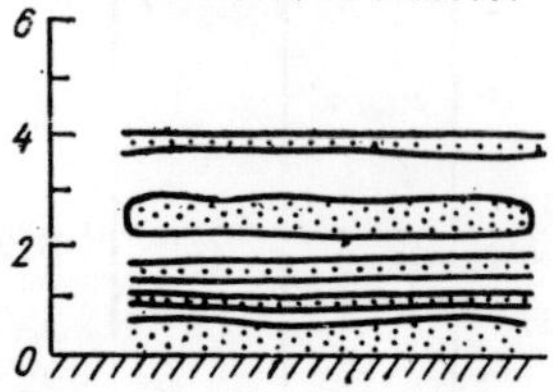
40	3.0	0.8	2.2		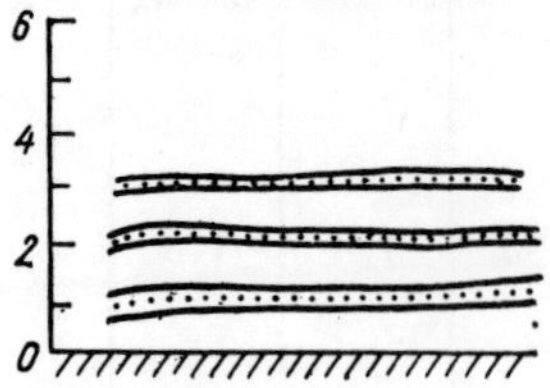
41	4.4	2.0	2.4	Thin haze layer at height of 6.5 km	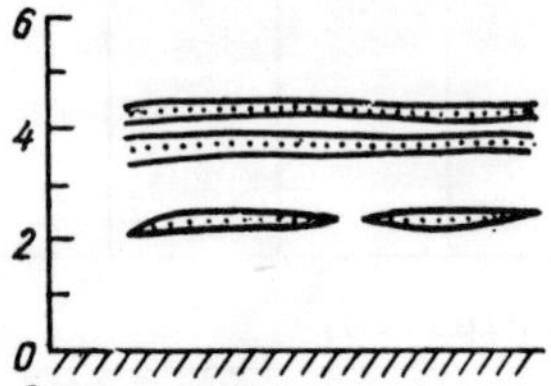
42	2.8	0.7	2.1	Haze layer ≈ 0.5 above clouds	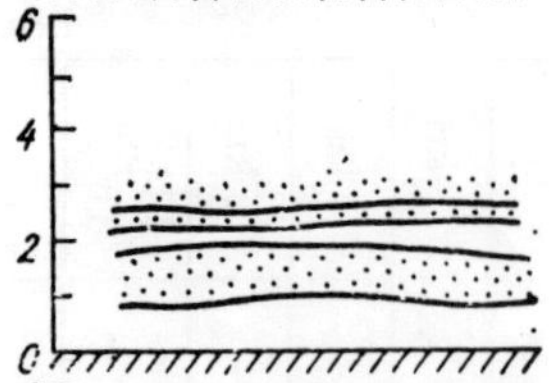
43	5.3	0.4	4.9		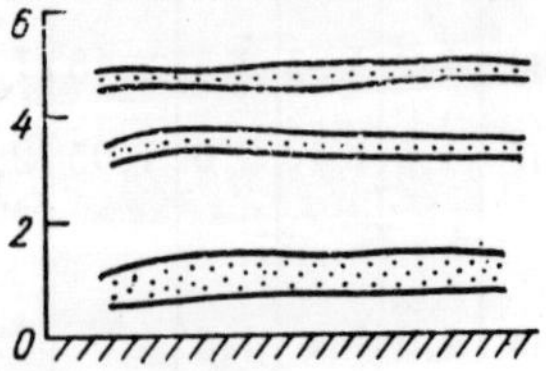

correspond to the calculations for both reflection and transmission
when $l_0 \geq 0.5$ km. For thinner clouds in both situations the experimental values of l_{eff} are considerably higher than the calculated values, this being especially evident for reflection
(Fig. 7.4a). Here a relationship between l_{eff} and the height of
the cloud layer can be distinguished (see Table 7.1 and Fig. 7.4a).
This is because the flux reflected by the underlying surface has
a significant effect. Situation No. 10 in Fig. 7.4a is evidence
of this; it pertains to thin single-layer cloudiness above a calm
water surface (albedo close to zero), where length l_{eff} is close
to the calculated value.

The positions of the dots for multilayer clouds differ considerably in reflected and transmitted light. In the former case
practically all the values of l_{eff} for multilayer clouds are
higher than for single-layer clouds. This is because the flux reflected from the lower layer contributes significantly to the
albedo, producing a perceptible light absorption in the space between the layers, especially if the upper layer is thin. Therefore, for a given geometrical thickness l_0 single-layer clouds
will have the lowest value of l_{eff}, and the calculated curves
in Fig. 7.4a can be taken as a lower limit of these values for
arbitrary cloud systems. This conclusion is valid for the entire
range of cloud depths, including thin clouds, since the effect
of the underlying surface can only be to increase the measured
value of l_{eff}.

The situation is more complicated for the transmitted flux.
As seen from Fig. 7.4b, the effective photon path length for a
thick multilayer system ($l_0 \geq 1$ km) is less than for a single-
layer system with the same value of l_0. This is because l_{eff}
can increase relative to l_0 only via multiple scattering in a
scattering medium. "Cutting out" the part of the scattering
medium inside the cloud system reduces l_{eff}. If just this mechanism is taken into account, then in the limiting case, when the
space between two layers increases, approaching l_0, while the
depths of the scattering layers proper become small in comparison
with this value, it is clear that $l_{eff} \to l_0$.

On the other hand, calculations indicate that for some layer
thicknesses there may be a concurrent multiple reflection between
the layers, leading to an increase in l_{eff} (see Fig. 7 of [8]).
In the case of the thick (deep) systems mentioned above, this
phenomenon was not observed, but this mechanism may well explain
why the measured effective photon path length is appreciably
higher than the calculated values for thin clouds ($l_0 \leq 0.3$ km,
see Fig. 7.4b). Here an underlying surface with a considerable
albedo (dry sand, dry grass) serves as the second reflecting
layer.

With regard to determining the upper cloud boundary H_{ub} using
satellites [12], it is important to remember that l_{eff} varies over
wide limits. This is evident from Fig. 7.5, which shows the effective photon path length in the reflected flux as a function

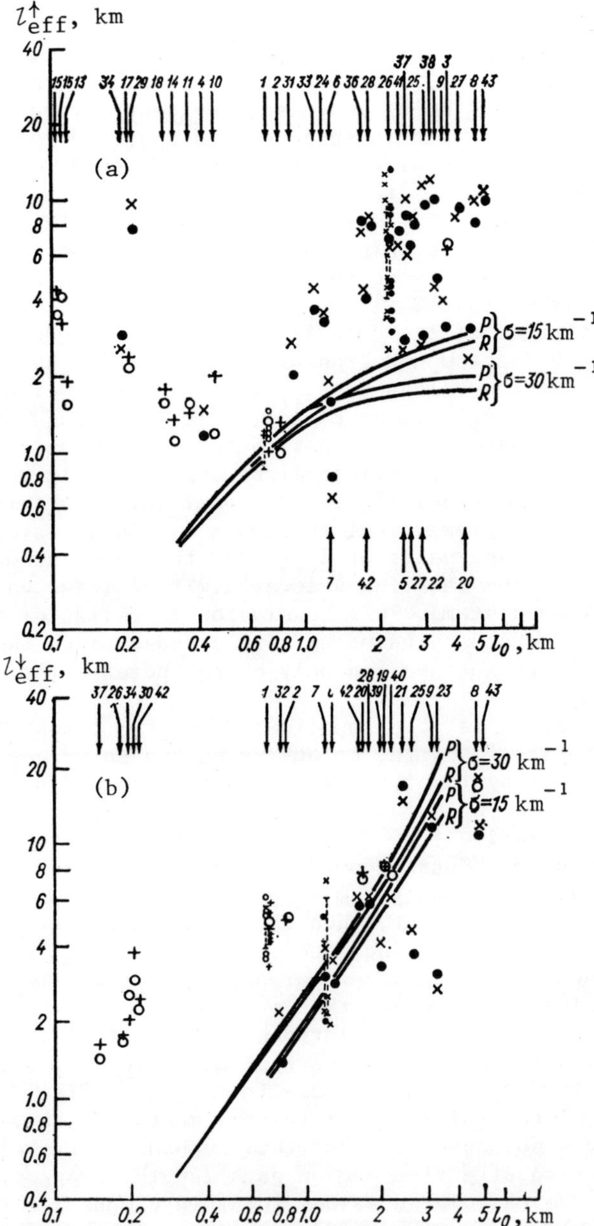

Fig. 7.4. Effective photon path length l_{eff} as function of geometrical depth of cloud system l_0 for reflection (a) and transmission (b) in P and R branches at $\lambda = 0.76$ μm.

of H_{ub} for various cloud systems. Sometimes l_{eff} turns out to be more than twice as great as the true height of the upper cloud boundary (see situation No. 38, where l_{eff} = 11 km for H_{ub} = 4.6 km).

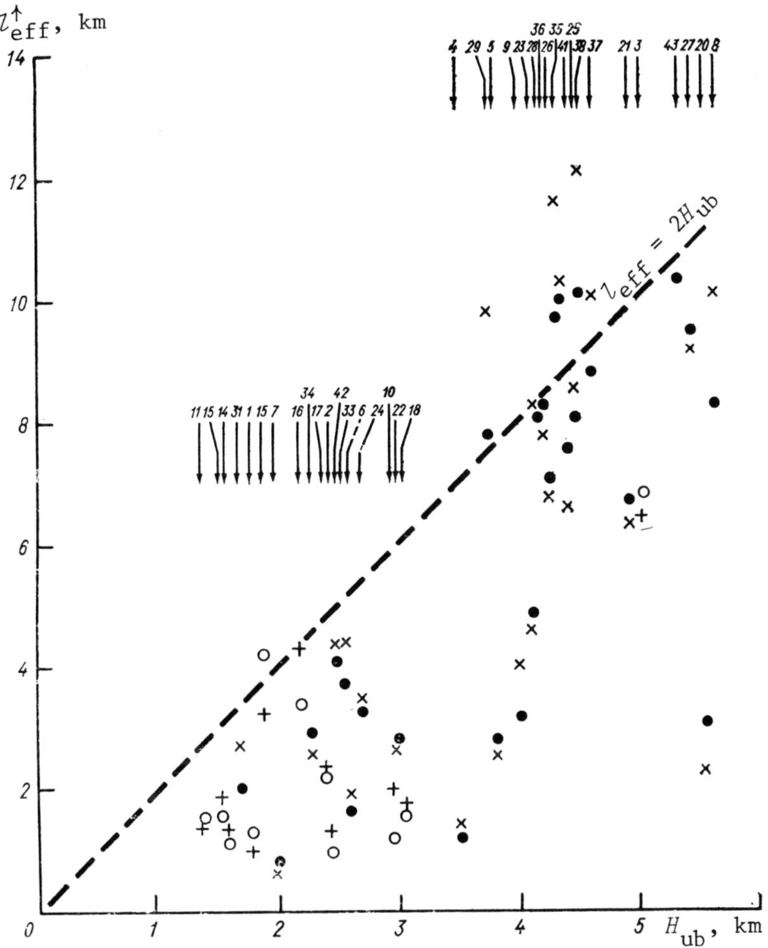

Fig. 7.5. Measured effective photon path length as a function of height of upper boundary of cloud system for reflection.

No definite correlation between the effective path length and H_{ub} is observed, since different types of clouds may have the same upper-boundary heights. Therefore, when finding the upper boundary H_{ub} from photometer readings obtained with satellites, substantial corrections have to be introduced, and these must be calculated for each individual situation.

For the random errors in determining the effective photon path length using the photometric method (without taking into account the accuracy of correction factor k), the following will be true. The total random error in the effective length δ_Σ is made up of the error in determining the reference transmission function δ_D plus the error δ_{cld} associated with the horizontal inhomogeneity of the clouds:

$$\delta_\Sigma^2 (l_{\text{eff}}) = \delta_D^2 + \delta_{cld}^2. \qquad (7.6)$$

Error δ_D is in turn the sum of the photometric instrument error δ_φ, the error in determining the air mass δ_m, and the error δ_a in the reference relation $D_\odot$, associated with the actual instability of the transmission function of the atmosphere:

$$\delta_D^2 = \delta_\varphi^2 + \delta_m^2 + \delta_a^2. \qquad (7.7)$$

The quantity δ_D, which can be evaluated from the scatter of points on the $D_\odot(m_\odot)$ graph, varies over a range $\delta_D = 0.6$–0.8 km for $m_\odot = 1.5$–2. At the same time, it was found independently that in the experiments $\delta_\varphi = 0.4$–0.55 km and $\delta_m \approx 0.15$ km. Thus the effect of variations in the properties of a nonscattering atmosphere on the determination of the effective length can be evaluated: $\delta_a(l_{\text{eff}}) = 0.45$–$0.65$ km. For δ_Σ values from 1.0 to 1.5 km, found directly from the scatter of points for a specific situation (see Fig. 7.4), from formula (7.6) we also get $\delta_{cld}(l_{\text{eff}}) = \sqrt{\delta_\Sigma^2 + \delta_D^2} = 0.6$–$1.2$ km.

PART III. FLUXES OF SOLAR RADIATION

<u>INTRODUCTION</u>

Solar energy is the only external source supplying heat to
the Earth. The amount of energy absorbed by the system made up of
the atmospheric plus the underlying layer determines the global
thermal regime of the system, which is kept in equilibrium by
the departing thermal radiation.

Clouds are the main regulator of the incoming solar energy.
The reflectivities of a cloudy atmosphere and of clouds proper
are almost the same, being an order of magnitude greater than the
reflection under cloudless conditions. The absorption of solar
energy in air is much higher when clouds are present than when
they are not, because of the high absorptivity of water droplets,
and also because photon paths are lengthened due to multiple
scattering in the clouds and reflection in the layers under and
between the clouds.

Chapter 8 begins with a discussion of the methods and results of calculating the radiation parameters of clouds: the
albedo A, the absorptivity Π, and the transmittance $T = 1 - (A + \Pi)$.
The vertical profiles of the solar-radiation fluxes and their
divergences,or the radiation heating, are considered. Various
measurement data are presented, and the measurement results are
compared with calculations.

In order to solve problems in atmospheric energetics, only
the integral values of the radiation parameters (over the wavelengths) have to be known. However, since the conversion of solar
energy takes place differently in different spectral intervals
(ultraviolet $\lambda < 0.4$ μm, visible 0.4 μm $\leq \lambda \leq 0.72$ μm, and infrared $\lambda > 0.72$ μm), these must be considered individually. Never-

$$\text{theless, in view of the inequality} \quad \int_0^{0.4\mu m} I_{0\lambda}\, d\lambda \ll \int_0^{\infty} I_{0\lambda}\, d\lambda = I_0$$

and the closeness of the methodological approaches, determinations
of the ultraviolet and visible radiation fluxes can be carried
out jointly, and only the infrared range of solar radiation need
be considered separately. In each case the radiation properties of
isolated homogeneous cloud layers are examined first. Then the
effects of the inhomogeneity of clouds and a cloudy atmosphere
are taken into account.

Finally, it should be noted that this portion of the book
is based to a large extent on the material of monograph [16]. The
ideology of the latter work is retained here, but the theoretical
part has been shortened and the factual data have been amplified
considerably.

117

VISIBLE AND ULTRAVIOLET RADIATION

Visible radiation and ultraviolet radiation are scattered by air molecules, cloud particles, and other aerosols, being absorbed by ozone but not by an aqueous aerosol. The optical parameters regulating the above processes - the scattering coefficient σ_λ, the absorption coefficient α_λ, and the scattering functions $\gamma_\lambda(\varphi)$ - are continuous functions of the wavelength and can be averaged over spectral intervals of finite width (depending on the accuracy required when calculating A, T, and Π). These parameters were given in Chaps. 4 and 5. A knowledge of these makes it possible to solve the equation of radiation transfer and to calculate the desired radiation characteristics.

8.1. Methods of calculating albedo, transmission, and absorption

In a horizontally homogeneous scattering and absorbing layer the transfer of monochromatic radiation is described by the equation

$$\mu \frac{\partial I(\tau, r)}{\partial \tau} + I(\tau, r) = \frac{\omega}{4\pi} \int_0^{2\pi} \int_{-1}^{1} I(\tau, r') \gamma(\tau, r, r') \, d\mu' \, d\psi', \tag{8.1}$$

$$0 \leqslant \tau \leqslant \tau_0, \quad -1 \leqslant \mu \leqslant 1, \quad 0 \leqslant \psi \leqslant 2\pi.$$

Here $I(\tau, r)$ is the radiation intensity at an optical depth τ in a direction r determined by the vertical angle θ, where $\mu = \cos \theta$, and by the azimuth ψ, reckoned from the vertical of the Sun.

The optical height of thickness τ is expressed in terms of the absorption coefficient $\alpha(z)$ and the scattering coefficient $\sigma(z)$ as

$$\tau(z) = \int_0^z [\alpha(z') + \sigma(z')] \, dz'. \tag{8.2}$$

For $z = H$ the optical thickness of the layer $\tau = \tau_0$. Parameter $\omega(z) = \dfrac{\sigma(z)}{\alpha(z) + \sigma(z)}$ is the probability of quantum survival or albedo of a unit volume. The scattering function $\gamma(r, r') = \gamma(\cos \varphi)$, where φ is the angle between the directions r' and r of the incident and scattered radiations, respectively.

If a parallel beam of solar radiation I_0 is incident upon the upper boundary of the layer, while the lower boundary reflects with a reflection coefficient $R(r, r')$, then the boundary conditions for formula (8.1) are

$$I(\tau_0, r) = I_0 \delta(r - r_0) \quad \text{for} \quad -1 \leqslant \mu \leqslant 0, \quad 0 \leqslant \psi \leqslant 2\pi;$$

$$I(0, r) = \int_0^{2\pi} \int_{-1}^{0} I(0, r') R(r, r') \, d\psi' \, d\mu' \quad \text{for} \quad 0 \leqslant \mu \leqslant 1,$$

$$0 \leqslant \psi \leqslant 2\pi. \qquad (8.3)$$

Here r_0 is the propagation direction of the solar rays, and $\mu_0 = \cos \zeta$, where ζ is the verticle angle of the direction r_0.

The principal energy quantities, the hemispherical radiation fluxes, are calculated for a given intensity $I(\tau, r)$ with the aid of the formulas

$$F^\uparrow(\tau) = \int_0^{2\pi} \int_0^1 I(\tau, \mu, \psi) \, d\mu \, d\psi,$$

$$F^\downarrow(\tau) = \int_0^{2\pi} \int_{-1}^{0} I(\tau, \mu, \psi) \, d\mu \, d\psi. \qquad (8.4)$$

The albedo of the layer and the transmittance are:

$$A(\mu_0) = \frac{F^\uparrow(\tau_0)}{I_0 \mu_0}, \qquad T(\mu_0) = \frac{F^\downarrow(0)}{I_0 \mu_0}. \qquad (8.5)$$

Parameter $A(\mu_0)$ is often called the "plane" albedo of the layer, in contrast to the "spherical" albedo, calculated using the formula

$$A_{\text{sph}} = 2 \int_0^1 A(\mu_0) \, \mu_0 \, d\mu_0. \qquad (8.6)$$

Note that the wavelength symbol has been omitted in formulas (8.1)-(8.6).

Diverse methods have been worked out for solving problem (8.1), (8.3): numerical, asymptotic, and approximate analytical.

The statistical Monte-Carlo method can be used to solve this same problem of radiation transfer without resorting to formula (8.1), using the same entry parameters. The main literature on the subject and a brief description of the methods are given in [16]. The methods currently used in practice are discussed below.

8.1.1. Asymptotic methods

The great optical thicknesses of cloud layers and the degree of expansion of the scattering functions ($\tau_0 \gg 1$ and $\bar{\mu} > 0.7$; see Chap. 4) favor the application of asymptotic methods to solve problem (8.1), (8.3), since these are considerably less laborious than numerical methods [1-4, 8, 11].

The albedo and transmittance of a homogeneous cloud layer are described by the asymptotic formulas

$$
F^{\uparrow}(\tau_0) = I_0\mu_0
\begin{cases}
1 - \dfrac{4u(\mu_0)}{(3 - c_1)\tau_0 + 6\delta} & \text{for } \omega = 1, \\[3mm]
A_s(\mu_0) - \dfrac{MNCe^{-2\nu_0\tau_0}}{1 - N^2 e^{-2\nu_0\tau_0}}\, u(\mu_0), & \omega \neq 1;
\end{cases}
\tag{8.7}
$$

$$
F^{\uparrow}(\tau_0) = I_0\mu_0 u(\mu_0)
\begin{cases}
\dfrac{4}{(3 - c_1)\tau_0 + 6\delta}, & \omega = 1, \\[3mm]
\dfrac{MCe^{-\nu_0\tau_0}}{1 - N^2 e^{-2\nu_0\tau_0}}, & \omega \neq 1.
\end{cases}
$$

The corresponding expressions for fluxes $F^{\uparrow\downarrow}(\tau)$ for $0 < \tau < \tau_0$, together with generalizations to inhomogeneous layers or a multilayer system, are given in [2, 3, 16]. Tables of the auxiliary constants and functions for formulas (8.7) are given in [16].

The accuracy of expressions (8.7) is determined mainly by how accurately the auxiliary quantities are calculated. For the data in [14] for $\tau_0 > 5$, the error in calculating A and T will not exceed 2-3%.

8.1.2. The Monte-Carlo method

This method was improved and applied successfully to radiation-transfer problems at the Computer Center of the Siberian Branch of the Soviet Academy [5, 7]. It is especially suitable for describing infrared solar radiation, taking into account absorption by water vapor (see Chap. 9). Results were also obtained for conservative scattering, that is, in the absence of absorption.

8.1.3. Two-stream approximations

These approximations are described in detail in [16]. They have the advantages of simplicity and the possibility of a direct determination of the hemispherical fluxes in analytical form. Two-stream approximations are also convenient for solving problems of radiation transfer in a stratified inhomogeneous medium: the medium is divided into a number of approximately homogeneous layers, and it is sufficient to specify the conditions of continuity of the hemispherical fluxes at the boundaries of these layers.

One of the most accurate variants of the two-stream equations is the δ-Eddington approximation [19], whereby the problem is reduced to the form:

$$\frac{1}{2}\frac{dX\,[\tau']}{d\tau'} = -(1-\omega')\,Y\,(\tau') + \frac{\omega'}{2}\,I_0 e^{-\dfrac{\tau_0'-\tau'}{\mu_0}},$$

$$\frac{2}{3}\frac{dY\,(\tau')}{d\tau'} = -(1-g'\omega')\,X\,(\tau') - g'\omega'I_0\mu_0 e^{-\dfrac{\tau_0'-\tau'}{\mu_0}}, \quad (8.8)$$

where

$$X\left(\tau'\right) = F^{\uparrow}\left(\tau'\right) - F^{\downarrow}\left(\tau'\right),$$

$$Y\left(\tau'\right) = F^{\uparrow}\left(\tau'\right) + F^{\downarrow}\left(\tau'\right),$$

$$\tau' = (1-\omega f)\,\tau, \quad \omega' = \frac{(1-f)\,\omega}{1-\omega f},$$

$$f = (g')^2, \quad g' = \frac{\bar{\mu}}{1+\bar{\mu}}.$$

Below, in Tables 8.1 and 8.2, the accuracy of the latter approximation is evaluated by comparing it with more precise calculations. Over a wide range of the parameters: $0.1 \leq \tau_0 \leq 100$; $0.5 \leq \omega \leq 1$; $0.2 \leq \mu_0 \leq 1$; $0.6 \leq \bar{\mu} \leq 0.95$, the errors in calculating the albedo and the transmittance scarcely exceed 10%, for a much higher relative error in calculating the absorption, that is, for a small difference $\Pi = 1 - (A + T)$.

8.1.4. An improved transport approximation

A solution of the equation for the source function in the theory of radiation transfer is given in [12] for the case of a homogeneous plane layer. This solution is used in the algorithm for the radiation fluxes described in [9, 10]. This method ensures a high accuracy of the scattering functions, which are series expanded in Legendre polynomials with a number of terms $N \leq 50$. The relative error in the intensity of the departing radiation does not exceed 0.1%, and the error in the fluxes is less than 0.01%. In this case the volume of calculations and the accuracy are independent of τ_0.

The main shortcoming of this method is that the volume of calculations depends on the degree of expansion of the scattering function, that is, on the number N. Thus it is advisable to approximate greatly expanded scattering functions as the sum of a delta function and a smoother scattering function:

$$\gamma(\mu) = 2F\delta(1-\mu) + (1-F)\gamma^*(\mu). \qquad (8.9)$$

The loss of accuracy for this approximation is quite low; from 1 to 5% for the radiation leaving the layer and less than 1% for the fluxes when $\tau_0 \geq 1$, for either a "narrow" or a "wide" distribution of droplet sizes in the wavelength range $\lambda = 0.4-2$ µm. At the same time, the number N of terms in the expansion of scattering function $\gamma^*(\mu)$ does not exceed 25 to 50. In view of approximation (8.9), this method can be called the improved transport approximation. In the classical transport approximation $\gamma^*(\mu) = 1$ in expression (8.9).

Table 8.1 compares A and T values calculated using the δ-Eddington method with data of a precise (1–3% error) numerical solution of the transfer equation [13].

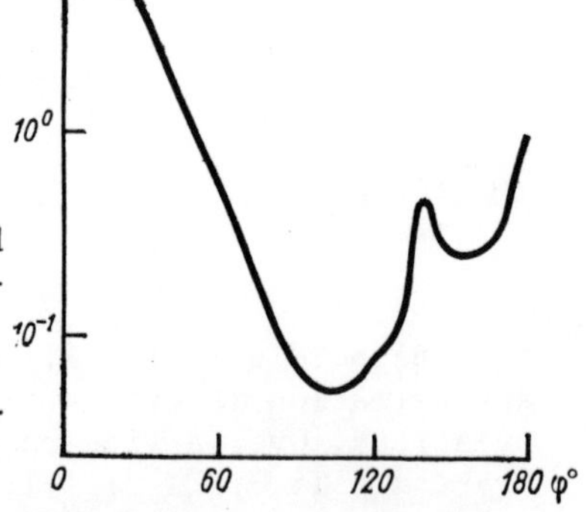

Fig. 8.1. Mean scattering function $\overline{\gamma}^*(\cos \varphi)$.

A model of a cloudless homogeneous atmosphere, for the scattering function represented back in Table 5.4, is considered. The error in the δ-Eddington method rarely reaches 7 to 13% and is generally less than 5%.

Table 8.2 compares the albedos of a homogeneous cloud layer calculated using four different methods, for a "narrow" scattering function γ_1 (see Chap. 4), except in the case of method 8.1.4, for

Table 8.1. Albedo and transmittance of aerosol layer, calculated using δ-Eddington method (A_δ, T_δ) and numerical method (A_T, T_T) [13].

ω	μ_0	$\tau_0 = 0.05$				$\tau_0 = 0.1$				$\tau_0 = 0.5$			
		A_δ	A_T	T_δ	T_T	A_δ	A_T	T_δ	T_T	A_δ	A_T	T_δ	T_T
1	1	0.0066	0.0074			0.0133	0.0149			0.0674	0.0775		
	0.85	0.0093	0.0095			0.0185	0.0190			0.0889	0.0974		
	0.5	0.0216	0.0209			0.0421	0.0415			0.1760	0.1930		
	0.2	0.0636	0.0697			0.1190	0.1310			0.3670	0.4310		
0.9	1	0.0059	0.0066	0.989	0.983	0.0118	0.0130	0.978	0.967	0.0554	0.0619	0.890	0.855
	0.85	0.0083	0.0084	0.986	0.988	0.0164	0.0166	0.972	0.975	0.0736	0.0775	0.864	0.876
	0.5	0.0193	0.0185	0.971	0.985	0.0373	0.0361	0.943	0.971	0.1470	0.1520	0.759	0.826
	0.2	0.0567	0.0613	0.920	0.928	0.1050	0.1130	0.850	0.863	0.3040	0.3390	0.538	0.556
0.5	1	0.0032	0.0034	0.972	0.969	0.0061	0.0065	0.945	0.938	0.0220	0.0232	0.744	0.723
	0.85	0.0045	0.0044	0.966	0.967	0.0085	0.0082	0.934	0.935	0.0301	0.0285	0.704	0.707
	0.5	0.0104	0.0095	0.941	0.941	0.0195	0.0176	0.995	0.899	0.6180	0.0545	0.549	0.573
	0.2	0.0303	0.0313	0.855	0.855	0.0538	0.0542	0.734	0.738	0.1250	0.1210	0.266	0.256

Table 8.2. Albedo of cloud layer, calculated using various methods for $\omega = 1$, 0.95, and 0.9.

τ_0	$\mu_0 = 1$			$\mu_0 = 0.5$			$\mu_0 = 0.3$		
	1	0.95	0.9	1	0.95	0.9	1	0.95	0.9
	1. Asymptotic								
5	0.212	0.118	0.087	0.467	0.235	0.175	0.580	0.320	0.253
10	0.412	0.185	0.112	0.602	0.302	0.200	0.686	0.387	0.278
30	0.708	0.202	0.115	0.802	0.319	0.203	0.844	0.404	0.281
64	0.842	0.202	0.115	0.893	0.319	0.203	0.916	0.404	0.281
	2. Monte-Carlo								
5	0.209			0.43			0.576		
10	0.389			0.574			0.673		
30	0.704			0.793					
100	0.896			0.896					
	3. δ-Eddington								
5	0.235	0.145	0.095	0.430	0.289	0.207	0.531	0.367	0.271
10	0.413	0.189	0.108	0.583	0.321	0.215	0.655	0.391·	0.276
30	0.710	0.201	0.109	0.797	0.328	0.216	0.832	0.396	0.277
64	0.845	0.201	0.109	0.891	0.328	0.216	0.910	0.396	0.277
	4. "Transport" approximation								
5	0.244		0.087	0.464		0.168	0.602		0.263
10	0.418		0.096	0.603		0.172	0.705		0.266
30	0.708		0.100	0.804		0.172	0.855		0.266
64	0.844		0.100	0.895		0.172	0.922		0.266

Table 8.3. Values of parameter F for "wide" (γ_w) and narrow (γ_n) scattering functions.

	λ, μm			
	0.45	0.7	1.19	2.2
γ_w		0.473	0.441	0.409
γ_n	0.458	0.426	0.369	

which the scattering function is represented by formula (8.9), Fig. 8.1, and Table 8.3.

Inspection of Table 8.2 shows that the "precise" methods (see subsections 8.1.1 and 8.1.2) can differ by as much as 5 to 10% (mainly because of differences in representing the scattering function). The δ-Eddington approximation does not go beyond this error range, that is, it is accurate in practice for cloudy conditions. According to the theory behind this method, the accuracy for high τ and $\bar{\mu}$ should be greater than for low values (see Table 8.1). The "transport" approximation, despite the difference of the scattering function, gives quite close values of A.

8.2. Albedo and transmittance of homogeneous cloud layers

The foregoing comparisons indicate that any of the four described methods can be used to calculate parameters A and T. Table 8.4 gives the values obtained using the method described .in subsection 8.1.4 and [9].

The effective scattering function $\gamma^*(\mu)$ in formula (8.9) depends slightly on the type of distribution ("narrow," "wide") and also on the wavelength for $\lambda \leq 2.5$ μm. Therefore, we can introduce the mean scattering function $\overline{\gamma^*(\mu)}$, which in [10] is the smooth part of the "wide" function in Fig. 8.1, calculated using the Mie formulas at $\lambda = 1.19$ μm. The slight dependence of parameter F of the wavelength is portrayed in Table 8.3.

If the scattering function is represented in form (8.9), then (see [9, 12]) the radiation-transfer equation can be reduced to its initial form (8.1) by replacing $\gamma(r, r')$ by $\gamma^*(\mu)$ and by introducing new variables and parameters:

$$\tau^* = (1-\omega F)\tau, \quad \tau_0^* = (1-\omega F)\tau_0, \quad \omega^* = \frac{1-F}{1-\omega F}\omega. \qquad (8.10)$$

The new equation does not contain parameter F in explicit form, and it can be solved for any scattering function of family (8.9), provided it is known for $F = 0$. This makes it possible to construct common tables for the fluxes of transmitted and reflected radiation.

In Table 8.4 the data of [10] are used to calculate the spherical albedo $A_s(\tau^*, \omega^*)$, the plane albedo $A(\zeta, \tau_0^*, \omega^*)$ (first row for each τ_0^*), and the corresponding transmittances $T_s(\tau_0^*, \omega^*)$, $T(\zeta, \tau_0^* \omega^*)$ (second row) for scattering function $\gamma^*(\mu)$.

The calculation error of Table 8.4, if F is taken from Table 8.3 for a "wide" distribution at $\lambda = 1.19$ μm, does not exceed 1%; for a "wide" distribution in the interval $\lambda = 0.7$-2.2 μm and for a "narrow" one at $\lambda = 0.45$-0.7 μm, the error is less than 4%; for a "narrow" distribution at 0.7 to 2.2 μm the error does not exceed 10%.

RADIATION IN A CLOUDY ATMOSPHERE

Table 8.4. Albedo and transmittance (%).

τ_0^*	A_s T_s	$\zeta°$					
		0	30	45	60	75	89
		$\omega^*=1$					
1	200 (—3) 800 (—3)	842 (—4) 916 (—3)	110 (—3) 890 (—3)	155 (—3) 845 (—3)	250 (—3) 750 (—3)	435 (—3) 535 (—3)	690 (—3) 310 (—3)
2	315 (—3) 685 (—3)	172 (—3) 828 (—3)	214 (—3) 786 (—3)	278 (—3) 722 (—3)	388 (—3) 612 (—3)	546 (—3) 454 (—3)	746 (—3) 254 (—3)
3	399 (—3) 601 (—3)	253 (—3) 747 (—3)	302 (—3) 698 (—3)	370 (—3) 630 (—3)	473 (—3) 527 (—3)	609 (—3) 391 (—3)	781 (—3) 219 (—3)
5	515 (—3) 485 (—3)	385 (—3) 616 (—3)	433 (—3) 567 (—3)	495 (—3) 505 (—3)	580 (—3) 420 (—3)	688 (—3) 312 (—3)	825 (—3) 175 (—3)
7	593 (—3) 407 (—3)	481 (—3) 519 (—3)	524 (—3) 476 (—3)	577 (—3) 423 (—3)	648 (—3) 352 (—3)	739 (—3) 261 (—3)	853 (—3) 147 (—3)
10	673 (—3) 327 (—3)	582 (—3) 418 (—3)	617 (—3) 383 (—3)	659 (—3) 341 (—3)	717 (—3) 283 (—3)	790 (—3) 210 (—3)	882 (—3) 118 (—3)
15	753 (—3) 247 (—3)	684 (—3) 316 (—3)	711 (—3) 289 (—3)	743 (—3) 257 (—3)	786 (—3) 214 (—3)	841 (—3) 159 (—3)	911 (—3) 891 (—4)
20	801 (—3) 199 (—3)	746 (—3) 254 (—3)	768 (—3) 232 (—3)	793 (—3) 207 (—3)	828 (—3) 172 (—3)	873 (—3) 127 (—3)	928 (—3) 715 (—4)
25	834 (—3) 166 (—3)	788 (—3) 212 (—3)	806 (—3) 194 (—3)	827 (—3) 173 (—3)	857 (—3) 143 (—3)	894 (—3) 106 (—3)	940 (—3) 598 (—4)
30	857 (—3) 143 (—3)	818 (—3) 182 (—3)	833 (—3) 167 (—3)	852 (—3) 148 (—3)	877 (—3) 123 (—3)	909 (—3) 914 (—4)	949 (—3) ·513 (—4)
35	875 (—3) 125 (—3)	840 (—3) 160 (—3)	851 (—3) 146 (—3)	870 (—3) 130 (—3)	892 (—3) 108 (—3)	920 (—3) 801 (—4)	955 (—3) 450 (—4)
50	909 (—3) 911 (—4)	884 (—3) 116 (—3)	893 (—3) 107 (—3)	905 (—3) 947 (—4)	921 (—3) 787 (—4)	942 (—3) 584 (—4)	967 (—3) 328 (—4)
100	952 (—3) 479 (—4)	939 (—3) 612 (—4)	944 (—3) 560 (—4)	950 (—3) 498 (—4)	959 (—3) 414 (—4)	969 (—3) 307 (—4)	983 (—3) 172 (—4)
		$\omega^*=0.97$					
1	181 (—3) 762 (—3)	771 (—4) 887 (—3)	995 (—4) 858 (—3)	140 (—3) 807 (—3)	225 (—3) 705 (—3)	395 (—3) 516 (—3)	649 (—3) 277 (—3)
2	272 (—3) 618 (—3)	149 (—3) 771 (—3)	184 (—3) 723 (—3)	238 (—3) 652 (—3)	333 (—3) 538 (—3)	480 (—3) 386 (—3)	691 (—3) 212 (—3)
3	331 (—3) 511 (—3)	207 (—3) 662 (—3)	246 (—3) 607 (—3)	302 (—3) 535 (—3)	391 (—3) 433 (—3)	522 (—3) 311 (—3)	713 (—3) 172 (—3)
5	398 (—3) 360 (—3)	287 (—3) 482 (—3)	324 (—3) 432 (—3)	375 (—3) 373 (—3)	451 (—3) 299 (—3)	565 (—3) 215 (—3)	737 (—3) 119 (—3)
7	432 (—3) 259 (—3)	331 (—3) 350 (—3)	365 (—3) 311 (—3)	410 (—3) 267 (—3)	480 (—3) 214 (—3)	585 (—3) 154 (—3)	748 (—3) 852 (—4)

τ_0^*	$\begin{array}{c}A_s\\T_s\end{array}$	$\zeta°$					
		0	30	45	60	75	89
10	455 (−3) 160 (−3)	362 (−3) 217 (−3)	392 (−3) 192 (−3)	434 (−3) 165 (−3)	499 (−3) 132 (−3)	599 (−3) 949 (−4)	756 (−3) 526 (−4)
15	467 (−3) 728 (−4)	377 (−3) 991 (−4)	406 (−3) 877 (−4)	446 (−3) 752 (−4)	508 (−3) 601 (−4)	606 (−3) 432 (−4)	760 (−3) 240 (−4)
20	469 (−3) 334 (−4)	380 (−3) 454 (−4)	409 (−3) 402 (−4)	448 (−3) 344 (−4)	510 (−3) 275 (−4)	607 (−3) 198 (−4)	760 (−3) 110 (−4)

$$\omega^* = 0.95$$

τ_0^*	$\begin{array}{c}A_s\\T_s\end{array}$	0	30	45	60	75	89
1	170 (−3) 738 (−3)	727 (−4) 869 (−3)	933 (−4) 837 (−3)	130 (−3) 781 (−3)	210 (−3) 677 (−3)	371 (−3) 486 (−3)	624 (−3) 257 (−3)
2	248 (−3) 578 (−3)	136 (−3) 736 (−3)	167 (−3) 685 (−3)	215 (−3) 611 (−3)	302 (−3) 495 (−3)	443 (−3) 347 (−3)	658 (−3) 189 (−3)
3	295 (−3) 461 (−3)	183 (−3) 614 (−3)	217 (−3) 557 (−3)	267 (−3) 483 (−3)	348 (−3) 382 (−3)	475 (−3) 268 (−3)	675 (−3) 147 (−3)
5	343 (−3) 301 (−3)	242 (−3) 417 (−3)	274 (−3) 367 (−3)	318 (−3) 311 (−3)	390 (−3) 243 (−3)	504 (−3) 171 (−3)	691 (−3) 938 (−4)
7	364 (−3) 199 (−3)	269 (−3) 279 (−3)	299 (−3) 243 (−3)	340 (−3) 204 (−3)	407 (−3) 159 (−3)	516 (−3) 112 (−3)	698 (−3) 617 (−4)
10	375 (−3) 108 (−3)	285 (−3) 152 (−3)	313 (−3) 132 (−3)	352 (−3) 111 (−3)	416 (−3) 862 (−4)	522 (−3) 607 (−4)	701 (−3) 333 (−4)
15	379 (−3) 390 (−4)	291 (−3) 552 (−4)	318 (−3) 478 (−4)	356 (−3) 400 (−4)	419 (−3) 312 (−4)	525 (−3) 220 (−4)	703 (−3) 121 (−4)

$$\omega^* = 0.9$$

τ_0^*	$\begin{array}{c}A_s\\T_s\end{array}$	0	30	45	60	75	89
1	145 (−3) 682 (−3)	631 (−4) 826 (−3)	798 (−4) 789 (−3)	110 (−3) 729 (−3)	177 (−3) 613 (−3)	318 (−3) 418 (−3)	565 (−3) 214 (−3)
2	200 (−3) 494 (−3)	109 (−3) 659 (−3)	132 (−3) 603 (−3)	170 (−3) 523 (−3)	240 (−3) 406 (−3)	365 (−3) 269 (−3)	587 (−3) 142 (−3)
3	227 (−3) 364 (−3)	139 (−3) 514 (−3)	163 (−3) 454 (−3)	200 (−3) 379 (−3)	266 (−3) 285 (−3)	382 (−3) 189 (−3)	596 (−3) 101 (−3)
5	249 (−3) 201 (−3)	168 (−3) 300 (−3)	190 (−3) 254 (−3)	221 (−3) 205 (−3)	285 (−3) 151 (−3)	395 (−3) 101 (−3)	602 (−3) 542 (−4)
7	256 (−3) 112 (−3)	177 (−3) 170 (−3)	199 (−3) 142 (−3)	231 (−3) 113 (−3)	290 (−3) 832 (−4)	398 (−3) 555 (−4)	604 (−3) 299 (−4)
10	259 (−3) 465 (−4)	181 (−3) 717 (−4)	202 (−3) 592 (−4)	234 (−3) 470 (−4)	292 (−3) 345 (−4)	399 (−3) 230 (−4)	605 (−3) 124 (−4)

$$\omega^* = 0.85$$

τ_0^*	$\begin{array}{c}A_s\\T_s\end{array}$	0	30	45	60	75	89
1	124 (−3) 633 (−3)	549 (−4) 786 (−3)	685 (−4) 745 (−3)	935 (−4) 680 (−3)	150 (−3) 557 (−3)	274 (−3) 362 (−3)	513 (−3) 178 (−3)
2	163 (−3) 427 (−3)	890 (−4) 593 (−3)	107 (−3) 534 (−3)	136 (−3) 452 (−3)	194 (−3) 336 (−3)	306 (−3) 210 (−3)	527 (−3) 108 (−3)

τ_0^*	A_s	ζ°					
	T_s	0	30	45	60	75	89
3	180 (—3)	108 (—3)	126 (—3)	154 (—3)	209 (—3)	315 (—3)	531 (—3)
	292 (—3)	436 (—3)	376 (—3)	304 (—3)	216 (—3)	136 (—3)	708 (—4)
5	191 (—3)	123 (—3)	140 (—3)	166 (—3)	218 (—3)	320 (—3)	534 (—3)
	140 (—3)	224 (—3)	183 (—3)	141 (—3)	982 (—4)	622 (—4)	326 (—4)
7	193 (—3)	127 (—3)	143 (—3)	169 (—3)	220 (—3)	322 (—3)	535 (—3)
	673 (—4)	111 (—3)	886 (—4)	673 (—4)	466 (—4)	296 (—4)	155 (—4)

$$\omega^* = 0.80$$

τ_0^*	A_s	ζ°					
	T_s	0	30	45	60	75	89
1	107 (—3)	479 (—4)	590 (—4)	796 (—4)	127 (—3)	237 (—3)	465 (—3)
	589 (—3)	749 (—3)	704 (—3)	635 (—3)	508 (—3)	313 (—3)	148 (—3)
2	135 (—3)	736 (—4)	870 (—4)	110 (—3)	158 (—3)	258 (—3)	474 (—3)
	371 (—3)	536 (—3)	475 (—3)	393 (—3)	280 (—3)	165 (—3)	828 (—4)
3	145 (—3)	860 (—4)	994 (—4)	122 (—3)	167 (—3)	263 (—3)	477 (—3)
	238 (—3)	373 (—3)	314 (—3)	246 (—3)	166 (—3)	990 (—4)	503 (—4)
5	151 (—3)	945 (—4)	107 (—3)	128 (—3)	172 (—3)	266 (—3)	478 (—3)
	100 (—3)	171 (—3)	135 (—3)	998 (—4)	656 (—4)	394 (—4)	202 (—4)
7	152 (—3)	961 (—4)	108 (—3)	129 (—3)	172 (—3)	266 (—3)	478 (—3)
	425 (—4)	752 (—4)	577 (—4)	417 (—4)	273 (—4)	164 (—4)	844 (—5)

Note: Numbers in parentheses denote powers of ten.

8.3. Radiation properties of ice clouds

Ice clouds differ from liquid-water clouds with regard to depth, scattering power, and the form of the scattering function (see Chaps. 1, 2, and 4). Although the particles of ice-crystal clouds are much larger than those of liquid-water clouds, the concentration of ice particles is much lower. Thus the net effect is that the scattering coefficient for an ice cloud is only one tenth of that for a water cloud. The optical thicknesses of ice clouds and water clouds have ranges of 0.25–15 and 5–50, respectively, for actual depths of these clouds (see Chap. 4).

For visible radiation the scattering functions of droplet clouds and ice-crystal clouds differ over the entire range of angles from 0 to 180° (see Fig. 4.4). Ice crystals scatter more radiation at lateral scattering angles ($\varphi = 60$–$120°$) than droplets do, because the scattering at $\varphi < 60°$ is less. The mean cosines differ correspondingly: $\bar{\mu}_{ice} < \bar{\mu}_{wat}$. Since the scatter of the values of the scattering function for visible radiation, which is determined by the combined effect of the shape, size, and orientation of the ice particles, is ± 30°, therefore, $\bar{\mu}_{ice}$ also depends on the microstructure of the crystalline medium.

Table 8.5. Albedo and transmittance of ice clouds and water clouds for $\mu_0 = 1$.

| | Ice cloud | | Water cloud | |
| | $\bar{\mu} = 0.8$ | | $\bar{\mu} = 0.89$ | |
τ_0	A	T	A	T
		$\omega = 1$		
0.2	0.0123	0.988	0.00622	0.994
0.5	0.0315	0.969	0.0159	0.984
5	0.310	0.690	0.178	0.822
10	0.503	0.497	0.332	0.668
30	0.773	0.227	0.641	0.359
		$\omega = 0.99$		
0.2	0.0121	0.986	0.00613	0.992
0.5	0.0308	0.964	0.0156	0.919
5	0.280	0.642	0.160	0.772
10	0.421	0.401	0.275	0.571
30	0.526	0.0788	0.413	0.169
		$\omega = 0.9$		
0.2	0.0106	0.969	0.005 36	0.974
0.5	0.0256	0.522	0.0129	0.936
5	0.131	0.354	0.0697	0.451
10	0.144	0.0979	0.0803	0.171
30	0.145	0.000 342	0.0817	0.002
		$\omega = 0.8$		
0.2	0.009 03	0.951	0.004 56	0.956
0.5	0.020 6	0.878	0.010 3	0.891
5	0.067 8	0.199	0.033 8	0.261
10	0.069 2	0.0284	0.034 7	0.0545
		$\omega = 0.5$		
0.2	0.005 00	0.898	0.002 51	0.901
0.5	0.009 73	0.762	0.004 81	0.770
5	0.016 0	0.0471	0.007 39	0.0598

To be more precise, let us assume that $\bar{\mu}_{ice} = 0.80$, which corresponds to the mean value $\bar{\mu}_{ice}$ derived from experimental data (Fig. 4.4), neglecting the scattering at $\varphi < 2°$. For a liquid-water cloud at $\lambda = 0.7$ μm we assume $\bar{\mu} = 0.89$, on the basis of the data in Table 4.1.

Table 8.5 compares values of A and T calculated using the δ-Eddington method for ice clouds and water clouds, differing only in the values assumed for μ. The albedo of an ice-crystal cloud is seen to be 20 to 65% (relative to the mean value) higher than that of a water cloud, solely due to the difference in $\bar{\mu}$.

The albedos of water clouds and ice clouds, calculated assuming spherical particles with $r_{mod} = 12$ μm (water cloud) and $r_{mod} = 50$ μm (ice cloud), are described in [20]. For $\omega = 1$ it turns out that A_{ice} is about 20% higher than A_{wat} for $\tau_0 = 10$ and $\lambda = 0.7$ μm. Let us recall that τ_{0ice} is, as a rule, lower than τ_{0wat}. Then, comparing the albedos for typical optical thicknesses ($\tau_0 = 5$ for ice and $\tau_0 = 30$ for water clouds) in the absence of absorption, we see from Table 8.5 that A_{wat} is 70% higher than A_{ice}. For considerable absorption ($\omega \leq 0.9$) A_{ice} again becomes higher than A_{wat} by 50–70% for the above-indicated typical optical thicknesses.

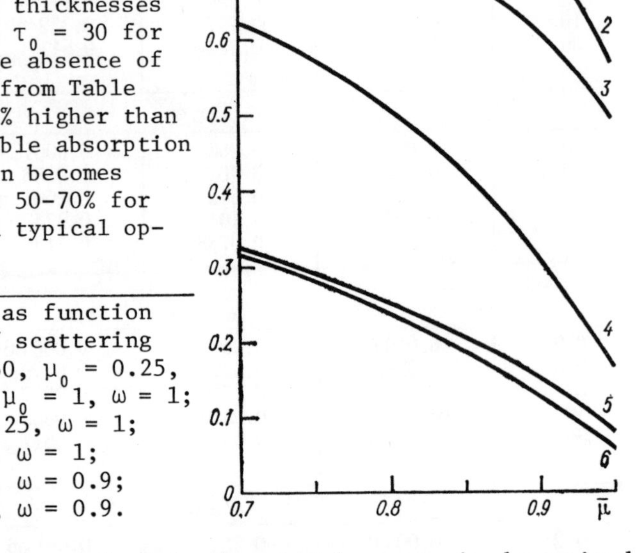

Fig. 8.2. Albedo A as function of mean cosine $\bar{\mu}$ of scattering function. 1) $\tau_0 = 50$, $\mu_0 = 0.25$, $\omega = 1$; 2) $\tau_0 = 50$, $\mu_0 = 1$, $\omega = 1$; 3) $\tau_0 = 10$, $\mu_0 = 0.25$, $\omega = 1$; 4) $\tau_0 = 10$, $\mu_0 = 1$, $\omega = 1$; 5) $\tau_0 = 50$, $\mu_0 = 1$, $\omega = 0.9$; 6) $\tau_0 = 10$, $\mu_0 = 1$, $\omega = 0.9$.

The absorption in a cloud in the visible range is determined by the aerosol: $\omega = (\sigma_{cld})/(\alpha_a + \sigma_{cld})$, where σ_{cld} is the coefficient of scattering by cloud particles, and α_a is the coefficient of absorption by the aerosol located in the cloud. An industrial ("soot") aerosol can contribute significantly to the absorption of visible light in a cloud; for a "turbid" cloudless atmosphere it can be assumed that $\alpha_a = 0.25$ km^{-1} [21]. Assuming the content of this aerosol in the cloud to be the same as outside it, for $\sigma_{cld} = 1$ and 10 km^{-1}, respectively, for ice clouds and droplet clouds, we get $\omega = 0.8$ and 0.97. Now, if we compare the albedos of an ice cloud for $\tau = 5$, $\omega = 0.8$, and $\bar{\mu} = 0.8$ and a water cloud for $\tau = \overline{30}$, $\omega = 0.97$, and $\bar{\mu} = 0.89$, then we see that A_{wat} is quite a bit (140%) higher than A_{ice}, that is, the presence of a common absorbing aerosol increases the difference between the reflectivities of water and ice clouds.

In conclusion, keeping in mind the above-mentioned difficulty in determining $\bar{\mu}_{\text{ice}}$, we have plotted A as a function of $\bar{\mu}$ in Fig. 8.2. Although this relationship is important, it is apparently more vital to take into account the differences in A_{ice} and A_{wat} due to the unlike scattering powers and due to the fact the different reactions may thus correspond to the same absorption.

8.4. Effect of aerosol on radiation characteristics of clouds

Evidence of aerosol absorption in clouds was presented in Chap. 4, and results of direct experiments will be given in Chap. 9. Some qualitative considerations and data of quantitative experiments are offered in the present section.

Aerosol particles may serve as condensation nuclei for cloud droplets, that is, activated particles, and at the same time there may be nonactivated particles present inside the cloud but outside the droplets. An aerosol may be affected by the high humidity in the cloud space (hydrophilic or hygroscopic particles) or it may not react at all to the humidity (hydrophobic particles).

The contribution of an aerosol to producing the albedo and absorbing power of clouds will be different in all these cases. Let us consider the first case, using the data of [22]. On the assumption that the cloud water content w is determined by dynamic processes and that the number of droplets N depends on the number C of condensation nuclei, the author of [22] studied the effect on the albedo of increasing C with $w = $ const. The droplet size r then decreases, while the number of droplets increases.

Let us assume that $N \approx C^{\alpha}$ for $\alpha \approx 0.8$ in clean air. The optical depth of a cloud $\tau_0 = k_{\varepsilon} N$, where k_{ε} decreases with r. Consequently, in general τ_0 can either increase or decrease with C. The author of [22] neglected the latter possibility, apparently because cloud droplets are in any case comparatively large ($r > 1$ μm), so that k_{ε} cannot diminish appreciably. Thus τ_0 increases with increasing C.

Now we further assume that all aerosol particles have the same absorbing power, so that with a rise in C the probability ω that quanta will survive decreases.

Making the valid assumption that τ_0 and ω are the main parameters determining A, the author of [22] considers the derivative

$$\frac{dA}{dC} = \frac{dA}{d\tau_0}\frac{d\tau_0}{dC} + \frac{dA}{d\omega}\frac{d\omega}{dC}, \qquad (8.11)$$

where $dA/d\tau_0$, $d\tau_0/dC$, and $dA/d\omega > 0$, while $d\omega/dC < 0$.

For low τ_0 the influence of the first term is predominant, and for high τ_0 that of the second term (see Table 8.4). Figure 8.3 (taken from [22]) portrays schematically the resultant effect of the two terms.

Let us note, however, that the concentration nuclei are mainly soluble particles: NaCl and sulfur compounds, which do not absorb visible radiation. Moreover, the optical properties of a sufficiently large droplet with an absorbing nucleus are close to those of a droplet of pure water [6], that is, it does not absorb visible light either.

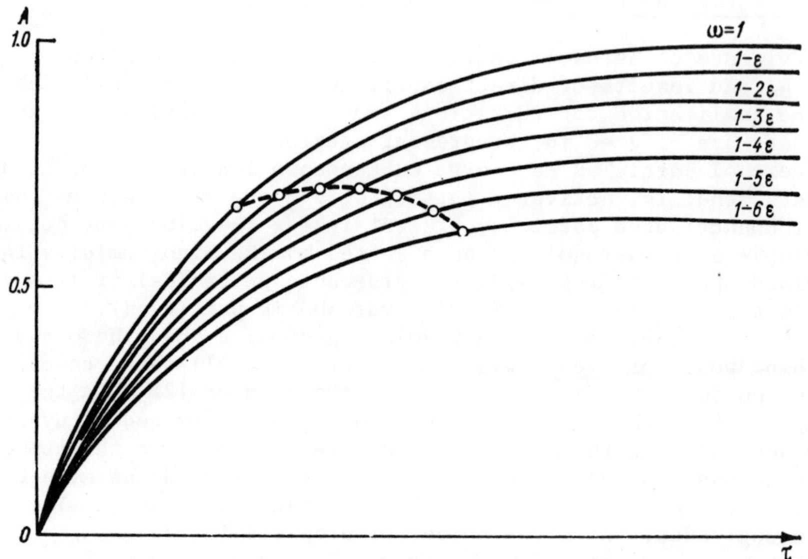

Fig. 8.3. Effect of aerosol pollution of clouds on albedo (qualitative picture, according to [22]).
Dashed curve shows trajectory of albedo variations with increasing pollution; parameter $(1 - n\varepsilon)$ with $n = 1$ to 6 represents decreasing ω.

Consequently, for $d\tau/dc > 0$ it may be that $d\omega/dc = 0$ and the absorbing power of the cloud does not increase.

The absorption is apparently affected appreciably by the aerosol particles outside of droplets but still within the cloud space. Since the particles of this aerosol are small, it does not affect the scattering.

The most important case is probably that of a cloud with pure water droplets or slightly absorbing droplets, which contains a nonactivated highly absorbing aerosol, comprising mainly hydrocarbon particles.

A similar case was considered in [17] for a two-layer system consisting of a cloud and an underlying layer at $\lambda = 0.5$ μm. An aerosol, which is nonabsorbing for $n = 1.55-0.00i$ (model 1) or highly absorbing for $n = 1.55-0.046i$ (model 2*)), is assumed to be distributed uniformly throughout the cloud and under it. The scattering coefficient in the cloud $\sigma = 70.5$ km^{-1}, while outside of the cloud $\sigma = 0.6$ km^{-1}.

The particles of an absorbing or nonabsorbing aerosol may be:
a) hydrophobic;
b) completely soluble;
c) partly soluble.

In the last two cases aerosol particles in a cloud grow in size, due to the high humidity, and their refractive index changes, in comparison with the size and refractive index of particles under the cloud. For the cloud these parameters are calculated as weighted-mean values over the volumes of the particles and droplets.

Figure 8.4 shows the albedo of the two-layer system (ratio of A_S to the albedo A_0 in the absence of aerosol particles in the cloud) as a function of the depth of the cloud layer H. The soluble nonabsorbing particles 4 in the cloud enhance the scattering, and thus increase the albedo. In the other extreme case, highly absorbing insoluble particles 1, the albedo is always reduced. In the intermediate cases of partly and completely soluble, highly absorbing particles the albedo increases for low cloud depths, the effect of absorption

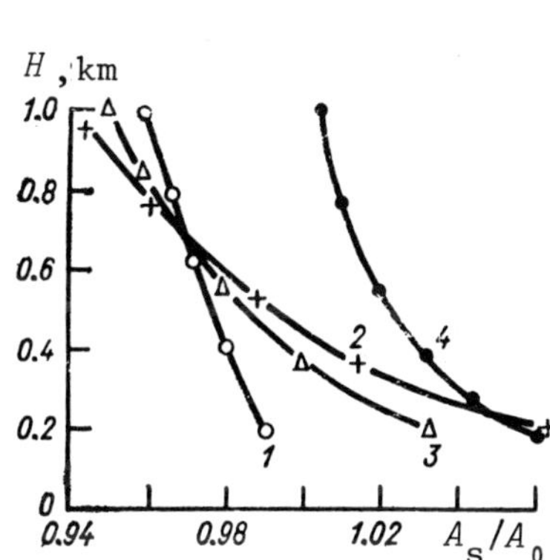

Fig. 8.4. Ratio A_S/A_0 as function of H. 1-3) absorbing aerosol, 4) nonabsorbing; 1) insoluble, 3) partly soluble, 2 and 4) completely soluble in cloud particles.

for thin clouds being insignificant in comparison with the effect of the increasing scattering. With an increase in cloud depth, the effect of absorption becomes appreciable and A_S decreases. In [17], by the way, the effect of absorption was underestimated, since the value assumed for $\sigma = 70$ km^{-1} in a cloud was too high (cf. Tables 4.3 and 4.4).

The data presented demonstrate the significant effect of an aerosol on the radiation regime of clouds and a cloudy atmosphere in the visible range, which is the main range affecting the albedo of the atmosphere. Until quite recently, the absorption of visible light by an aerosol, which is sometimes comparable to gas absorption in the near infrared, was completely ignored in problems of atmospheric energetics. It is easy to allow for this absorption for specified τ_0 and ω. In particular, various calculation methods

*) Model 3 according to [17].

are suggested above and detailed results are given in Table 8.4.
At the same time, it is extremely difficult to determine parameters τ_0 and ω, due to the diversity of the quantities and chemical
compositions of aerosols.

The following simple means of determining these parameters is
proposed, on the basis of [14, 15].

The total optical depth of cloud and aerosol attenuation
τ_Σ is:

$$\tau_\Sigma = \tau + \tau_a \quad \text{for} \quad \tau \leq 1.0$$
$$\tau_\Sigma = \tau \quad \text{for} \quad \tau > 1.0 \tag{8.12}$$

Here

$$\tau_a = (\sigma_a + \alpha_a)H_a; \quad \sigma_a = \sigma_a(r_m)N_a; \tag{8.13}$$

H_a is the height of the homogeneous atmosphere for the aerosol;
N_a is the number of particles per unit volume.

In [14] the scattering cross section $\sigma_a(r_m)$ is tabulated as
a function of the mode radius r_m of the particle-size distribution
for the submicron fraction of the aerosol ($0.1 \leq r_m \leq 1.0$ µm). It
is seen from these data that smaller and larger particles do not
affect τ_a.

In [14, 15] the coefficient of aerosol absorption α_a is
represented as

$$\alpha_a = \sigma_a \frac{1-\omega_a}{\omega_a} ,$$

where $\hspace{10cm}$ (8.14)

$$\omega_a = \frac{1}{1+\dfrac{N_{ab}}{N_a} \cdot \dfrac{N_\Pi}{N_{\Pi b}}(1-\omega_{ab})}$$

Here subscript b denotes pure background (cloudless) conditions
($\omega_{ab} = 0.95$, see [15]), subscript Π denotes the number of absorbing
particles, subscript a denotes the aerosol, and quantities with
no subscript pertain to the cloud. Table 8.6 gives some calculation
results.

During the derivation of these relations the absorbing aerosol
was assumed to have a uniform (hydrocarbon) composition. It was
also assumed that small absorbing particles do not affect the
scattering, even in a cloudless atmosphere.

Using the above method, it becomes possible to ascertain the
effect on the energy parameters A and Π of an increase in the total
number of aerosol particles, of a change in their size, and of the
excess of the number of absorbing particles over the background.

Table 8.6. Parameters ω_a, λ_0, ω_{λ_0}, and integral albedos of cloud for $\mu_0 = 0.5$ and $\lambda_0 = 0.52$ µm.

σ, km^{-1}	50	50	50	50
$N_\Pi/N_{\Pi b}$	1	5	20	20
ω_0, λ_0	0.950	0.800	0.500	0.500
ω_{λ_0}	1.000	0.998	0.996	0.993
A for $\tau = 5$	0.568	0.466	0.454	0.435
A for $\tau = 20$	0.730	0.716	0.665	0.622

CHAPTER 9

INTEGRAL SOLAR RADIATION

9.1. Infrared albedo and absorptivity of cloud layers

The need to consider separately the infrared range was demonstrated in [7]. Because of the linear structure of the vibrational-rotational infrared absorption bands of gases, even over small spectral intervals the transmission functions are non-exponential, the concept of an absorption coefficient loses its meaning, and the radiation-transfer equation (steady-state, see [7]) is no longer applicable.

A means of surmounting this difficulty for homogeneous media was proposed in [11] (see also Part II and [7]).

Let $p_\lambda(z, r, l)$ be the probability density for the traversal of a path l by photons contributing to radiation of intensity $I_\lambda(z, r)$. We introduce the quantity

$$J_\lambda(z, r, l) = \overline{I}_\lambda(z, r)\, p_\lambda(z, r, l), \qquad (9.1)$$

where

$$\overline{I}_\lambda(z, r) = I_{0,\lambda}^{-1} I_\lambda(z, r).$$

Evidently,

$$\overline{I}_\lambda(z, r) = \int_0^\infty J_\lambda(z, r, l)\, dl. \qquad (9.2)$$

If over the path l there took place an amount of absorption described by the transmission function $P_\lambda(\rho l)$, where ρ is the density of the absorbing substance, then as a result of scattering and absorption the radiation intensity will be

$$I_\lambda^n(z, r) = \int_0^\infty I_{0,\lambda} J_\lambda(z, r, l)\, P_\lambda(\rho l)\, dl. \qquad (9.3)$$

Here and in the following, for simplicity the independent variable of the transmission function (namely ρl, the content of absorbing material over path l) will be written without the pressure correction. Actually, for absorbing gases in a homogeneous layer at height z, the effective mass is introduced (see Chap. 5 and [7]):

136

$$m_\lambda(z) = \rho l \left(p(z)/p(0) \right)^{n_\lambda}. \qquad (9.4)$$

Since the calculation methods and the laws describing the behavior of photon paths in a scattering medium have been considered in detail in Part II of this monograph and also in [7], let us confine ourselves here to just a few comments. The most practical means of calculating $J(l)$ is with the Monte-Carlo method [4, 6]. All the calculated values of the radiation characteristics of clouds presented below were obtained in the Monte-Carlo-method laboratory of the Computer Center at the Siberian Branch of the Soviet Academy.*)

In formula (9.3) the path-length distributions $J_\lambda(z, r, l)$ can be computed assuming conservative scattering or taking absorption by droplets into account. In each of these cases the absorption of radiation by water vapor is considered separately (assumed to be independent or separate in time). Such a separation is valid in practice only for weak absorption (see Part II). Nevertheless, formula (9.3) will be used below for the integral and spectral transmission functions of water vapor and water droplets, functions $J(l)$ being calculated assuming conservative scattering. By integrating both sides of (9.3) over λ in the interval $\Delta\lambda$, we obtain

$$I_{\Delta\lambda}^{(n)}(z, r) = \int_{\Delta\lambda} I_\lambda^{(n)}(z, r)\, d\lambda = \int_0^\infty I_{0,\,\Delta\lambda} J_{\Delta\lambda}(z, r, l)\, P_{\Delta\lambda}(\rho l)\, dl, \qquad (9.5)$$

if J_λ = const = $J_{\Delta\lambda}$ for $\lambda \in \Delta\lambda$, then the same applies for $I_{0,\lambda}$.

Now, integrating over the directions (with a weighting factor of cos (z, r)), we can arrive at similar expressions for the hemispherical radiation fluxes. In particular, for $z = 0$ and $z = H$ we get

$$A_{\text{п},\,\Delta\lambda} = \int_0^\infty J_{\Delta\lambda}^\uparrow(l)\, P_{\Delta\lambda}(\rho l)\, dl, \qquad (9.6)$$

$$T_{\text{п},\,\Delta\lambda} = \int_0^\infty J_{\Delta\lambda}^\downarrow(l)\, P_{\Delta\lambda}(\rho l)\, dl, \qquad (9.7)$$

$$\text{П}_{\Delta\lambda} = 1 - (A_{\text{п},\,\Delta\lambda} + T_{\text{п},\,\Delta\lambda}). \qquad (9.8)$$

Here $J_{\Delta\lambda}^\uparrow(l)$ is the path-length distribution in the fluxes of reflected light, and $J_{\Delta\lambda}^\downarrow(l)$ is the distribution in the transmitted

---

*) The calculations were carried out under the direction of B.A. Kargin.

Table 9.1. Albedo $A_{\Delta\lambda_j}$ and absorption $\pi_{\Delta\lambda_j}$ of inhomogeneous cloud allowing for $(\alpha_w \neq 0)$ and neglecting $(\alpha_w = 0)$ absorption by water

j	$\Delta\lambda_j$ μm band designation	$A_{\Delta\lambda_j}$ $\pi_{\Delta\lambda_j}$	α_w cm^2/g	$H=0.4$ km, $\sigma=50$ km^{-1} $\tau_0=20$ $A_s = 0.566(0.236)$			$H=1$ km, $\sigma=40$ km^{-1} $\tau_0=40$ $A_s = 0.73(0.45)$		
				0,1	1	10	0.1	1	10
2	0.712... 0.762 a	A Π	0	0.563 0.006	0.557 0.017	0.539 0.054	0.724 0.009	0.713 0.027	0.680 0.080
4	0.810... 0.865 0.8	A Π	0	0.564 0.004	0.559 0.014	0.544 0.044	0.725 0.007	0.716 0.022	0.688 0.066
6	0.80 ... 1.0 $\rho\delta\tau$	A Π	0	0.558 0.016	0.541 0.050	0.496 0.140	0.714 0.025	0.683 0.074	0.605 0.196
8	1.085... 1.21 φ	A Π		0.549 0.034	0.516 0.100	0.435 0.259	0.698 0.051	0.639 0.145	0.508 0.342
10	1.28... 1.535 ψ	A Π	0		0.438 0.251	0.322 0.461			
		A Π	16.25	0.430 0.285	0.362 0.398	0.267 0.570	0.452 0.465	0.376 0.561	0.270 0.689
11	1.535... 1.66	A Π	10.45		0.489 0.154			0.567 0.294	
12	1.66... 2.08 Ω	A Π	0		0.183 0.226	0.146 0.387			
		A Π	61.06	0.103 0.556	0.082 0.613	0.066 0.694	0.104 0.829	0.089 0.856	0.071 0.886
13	2.08.... 2.25	A Π	26.15		0.156 0.280			0.216 0.564	
14	2.25... 3.0 x	A Π	0		0.145 0.392	0.118 0.503			
		A Π	1008	0.006 0.994	0.005 0.995	0.004 0.996	0.001 0.999	0.001 0.999	0.001 0.999
15	3.0... 3.6 3.2	A Π	0		0.194 0.176	0.140 0.410			
		A Π	2167	0.004 0.996	0.003 0.997	0.003 0.997	0 1	0 1	0 1

Note. A_s is the albedo of conservative scattering for $\gamma_1(\varphi)$

layer in water-vapor absorption bands for ρ_v = 0.1, 1 and 10 g/m^3, droplets (w_{av} = 0.2 g/m^3).

$\mu_0 = 0.5$						$\mu_0 = 0.26$					
$H = 0.4$ km, $\sigma = 50$ km^{-1} $\tau_0 = 20$			$H = 1$ km, $\sigma = 40$ km^{-1} $\tau_0 = 40$			$H = 0.4$ km, $\sigma = 50$ km^{-1} $\tau_0 = 20$			$H = 1$ km, $\sigma = 40$ km^{-1} $\tau_0 = 40$		
$A_S = 0.81$						$A_S = 0.864$					
0.1	1	10	0.1	1	10	0.1	1	10	0.1	1	10
0.701 0.003	0.695 0.016	0.675 0.049	0.811 0.008	0.800 0.023	0.767 0.070	0.774 0.005	0.767 0.015	0.747 0.046	0.862 0.007	0.852 0.021	0.819 0.064
0.702 0.004	0.696 0.013	0.680 0.040	0.812 0.006	0.803 0.019	0.776 0.058	0.775 0.004	0.769 0.012	0.752 0.038	0.864 0.005	0.855 0.017	0.828 0.053
0.695 0.015	0.677 0.046	0.625 0.130	0.801 0.022	0.771 0.065	0.692 0.175	0.768 0.014	0.749 0.043	0.695 0.121	0.853 0.019	0.923 0.059	0.744 0.160
0.686 0.027	0.649 0.092	0.555 0.241	0.785 0.044	0.726 0.128	0.591 0.309	0.758 0.029	0.720 0.086	0.621 0.227	0.837 0.040	0.779 0.116	0.641 0.287
0.560 0.256	0.486 0.370	0.366 0.531	0.552 0.396	0.465 0.496	0.337 0.637	0.631 0.233	0.552 0.339	0.418 0.507	0.613 0.350	0.520 0.452	0.380 0.602
	0.638 0.125			0.669 0.240			0.713 0.111			0.731 0.204	
0.237 0.591	0.208 0.646	0.167 0.718	0.195 0.773	0.168 0.806	0.131 0.848	0.331 0.555	0.293 0.612	0.235 0.689	0.263 0.719	0.228 0.757	0.181 0.807
	0.358 0.314			0.357 0.517			0.475 0.292			0.450 0.466	
0.008 0.992	0.007 0.993	0.006 0.994	0.003 0.997	0.002 0.998	0.002 0.998	0.014 0.986	0.011 0.989	0.009 0.991	0.006 0.994	0.005 0.995	0.001 0.996
0.002 0.998	0.002 0.998	0.002 0.998	0.001 0.999	0.001 0.999	0.001 0.999	0.004 0.996	0.004 0.996	0.003 0.997	0.003 0.997	0.003 0.997	0.002 0.998

(values in parentheses for $\gamma_2(\varphi)$).

fluxes (see Part II).

In the particular case of $P(m) = 1$ formulas (9.6) and (9.7) give the albedo and transmittance for conservative scattering:

$$A = \int_0^\infty J^\uparrow (l)\, dl, \quad T = \int_0^\infty J^\downarrow (l)\, dl. \tag{9.9}$$

It should be noted that the foregoing all refers to a homogeneous medium, but it can also be generalized to include an inhomogeneous medium. Then individual photon trajectories have to be traced out, and along each of these it is necessary to calculate the mass of absorbing substance under the sign of the transmission function in (9.6) and (9.7), taking the height distribution of the substance into account and introducing a pressure correction for gaseous absorbers.

For a homogeneous medium another approach is also possible, based on the following theorem (see [7]):

$$\text{if} \quad P(m) = \sum_{i=1}^{n} a_i e^{-\alpha_i m}, \quad \text{then} \quad I^\Pi = \sum_{i=1}^{n} a_i I^{(i)}. \tag{9.10}$$

Here $I^{(i)}$ represents the solution of the transfer equation for an absorption coefficient α_i and with the same σ and $\gamma(\varphi)$ for all i.

A generalization to the case of layerwise-nonuniform absorption is given in Ref. [15] to Part II (Chaps. 6 and 7). For a practical application see Ref. [9] to Part II.

Formulas of the type of (9.6)-(9.8) were used to calculate the spectral albedo and total absorptivity of an inhomogeneous cloud layer (variant 1 of Section 9.2). The results are presented in Table 9.1. For these calculations the spectral coefficients of scattering and absorption by water droplets (see Table 4.1) and the spectral transmission functions for water vapor (see Table 5.5) were used.

We consider here the case of a wide droplet-size distribution with scattering functions γ_{1w} for $\lambda \leq 1.66$ μm and γ_{2w} for $\lambda >$ 1.66 μm. Similar data are given in Table 2.14 of [7] for a homogeneous layer for the parameters of a "narrow" droplet-size distribution.*) Since for albedo calculations functions γ_{1n} and γ_{1w} give similar results (see [7]), while an inhomogeneous layer does not differ from a homogeneous layer (see Section 9.2), the data of Table 2.14 of [7] and Table 9.1 given here are quite comparable.

---

*) In Table 2.14 of [7], by the way, the units of α_w are erroneously given as cm^2/g, rather than km^{-1}.

Evidently, for equal values of $\tau_0 = 20$, $\rho_v = 0.1$, 1.0, 10 g/m^3, and $\mu_0 = 1$, outside the absorption bands for water droplets the values of $A_{\Delta\lambda}$ and $\Pi_{\Delta\lambda}$ from [7] and from Table 9.1 are almost identical. The ensemble of these calculations represents quite completely the infrared albedo and absorption by cloud layers over a wide range of parameters μ_0, ρ_v, and τ_0. The variations of $A_{\Delta\lambda}$ and $\Pi_{\Delta\lambda}$ as functions of these parameters are delineated.

The separate effects of water vapor and water droplets are also considered. For $\lambda < 1$ and $\rho_v \leq 1$ g/m^3 the absorption is slight and the albedo is close to the albedo of conservative scattering. For $\rho_v = 10$ g/m^3 the absorption reaches 20% and has an appreciable effect on the albedo. For $\lambda \geq 1.5$ μm it is water droplets which affect the albedo and absorption most, while for $\lambda \geq 2.25$ μm practically all the radiation is completely absorbed. On the average, for $\lambda \leq 1.5$ μm water vapor plays the main part in absorbing radiation, whereas for $\lambda > 1.5$ μm water droplets are the main absorbers.

9.2. Integral albedo and absorptivity of cloud layers

In Section 9.1 a method for calculating the infrared radiation parameters of clouds was described. This method, when used together with any of those proposed in Chap. 8 to compute these same quantities for visible radiation, enables us to find the integral albedo and absorptivity of cloud layers. The calculations were carried out using the formula

$$A_\Sigma = \frac{\sum\limits_{j=1}^{15} I_{0,\Delta\lambda_j} A_{\Delta\lambda_j}}{\sum\limits_{j=1}^{15} I_{0,\Delta\lambda_j}} \tag{9.11}$$

and a similar one for the transmittance. Here $j = 1$ is the spectral interval of visible radiation, and $j = 3, 5, 7, 9$ are the spaces between the water-vapor bands, where absorption is absent (see Table 9.1). Conservative scattering is assumed in all these intervals (absorption by the aerosol and by ozone is neglected).

Calculations made using "standard" formulas of the type of (9.11) were compared with two variants of the simplified calculations:

$$A_\Sigma = \frac{A_v I_0^{(1)} + I_0^{(2)} \int\limits_0^\infty J^\uparrow (l) P_\Sigma \left[\rho_v (p/p_0) l, \, wl\right] dl}{I_0}, \tag{9.12}$$

$$A_\Sigma = A_v \frac{I_0^{(1)} + I_0^{(2)} P_\Sigma \left[\rho_v (p/p_0) l^\uparrow, \, wl^\uparrow\right]}{I_0}. \tag{9.13}$$

Similar formulas were used for transmission. Here $I_0^{(1)}$ and $I_0^{(2)}$ are the fractions of the solar constant I_0 pertaining to the visible and infrared ranges, respectively; $P_\Sigma(m_v, m_w)$ is the integral transmission function for water vapor and water droplets (see Chap. 5); A_v is the albedo of visible radiation.

Expression (9.12) was simplified and made less precise than (9.11) by employing the same scattering function $\gamma_1(\varphi)$ in the visible and infrared ranges. Moreover, when constructing P_Σ, the spectral dependence of parameter n in the expression for the effective mass (see (9.4)) was neglected. Nevertheless, the values of A_Σ found with (9.11) and (9.12) do not differ by more than 5%.

The calculation using formula (9.13) is even more simplified and even rougher: the random paths l under the sign of the transmission function are replaced by the mean path length $l^\uparrow$ (see Part II). The error introduced by this assumption does not exceed 10%.

Tables 2.13 and 2.15 of [7] present the values of A_Σ and Π_Σ for a homogeneous cloud layer over a wide range of parameters τ, μ_0, ρ_v, and w. Results of new calculations using formula (9.11), in which homogeneous layers are compared with inhomogeneous layers, will be given below.*)

Four models are considered:

1) $\sigma(z)$ from Section 4.4, the mean experimental profile of the extinction coefficient for stratiform clouds. The water content $w(z) = c\sigma(z)$;

2) homogeneous layer: $\sigma(z) = \text{const} = \sigma_{av}$ and $w(z) = w_{av}$ from model 1;

3) $\sigma(z)$ the same as in the first model, but $w(z) = w_{av}$;

4) $\sigma(z) = \sigma_{av}$, but $w(z)$ the same as in the first model.

In all cases it is assumed that: $\tau = 20$, $\sigma_{av} = 50$ km^{-1}, $w_{av} = 0.2$ g/m^3, $\rho_v = 10$ g/m^3, $\mu_0 = 1$.

Table 9.2 gives values of A_Σ and Π_Σ for all four models, allowing for ($\alpha_w \neq 0$) and neglecting ($\alpha_w = 0$) absorption by water droplets.

Inspection of the table shows that the integral albedos and absorptions of homogeneous and inhomogeneous layers for equal mean values of w and σ do not differ by more than 5%. It is quite possible to neglect the spectral variation of the scattering function when calculating the integral radiation parameters. This is because the contribution of solar radiation for $\lambda > 1.7$ μm, where the spectral dependence $\gamma_\lambda(\varphi)$ becomes appreciable (see Section 4.1), comprises only 10% of the solar constant.

Allowing for absorption of water droplets, with the aid of the data of Table 9.2 and [7], varies A_Σ by about 5%. Finally, absorption by water vapor over a wide range of humidity 0.1 g/m$^3 \leq \rho_v \leq 10$ g/m^3 varies the albedo, in comparison with the value for conservative scattering, by no more than 20%, and on the

____

*) Inhomogeneous only with regard to water droplets; $\rho_v(z) = \text{const}$.

Table 9.2. Integral albedo and absorption for four models

γ and λ, μm		α_w	Model			
			1	2	3	4
Albedo						
$\gamma_\lambda = \gamma_1$ $0.4 \leqslant \lambda \leqslant 3.6$		$\alpha_w \neq 0$	0.499	0.503	0.500	0.502
		$\alpha_w = 0$	0.520	0.524	0.520	0.524
$\gamma_\lambda = \gamma_2$ $0.4 \leqslant \lambda \leqslant 3.6$		$\alpha_w \neq 0$	0.206	0.219	0.206	0.219
		$\alpha_w = 0$	0.215	0.229	0.215	0.229
$\gamma_\lambda = \begin{cases} \gamma_1 & \lambda \leqslant 1.7 \\ \gamma_2 & \lambda > 1.7 \end{cases}$		$\alpha_w \neq 0$	0.493	0.497	0.494	0.496
		$\alpha_w = 0$	0.505	0.510	0.505	0.510
Absorption						
$\gamma_\lambda = \gamma_1$ $0.4 \leqslant \lambda \leqslant 3.6$		$\alpha_w \neq 0$	0.128	0.130	0.127	0.131
		$\alpha_w = 0$	0.088	0.091	0.088	0.091
$\gamma_\lambda = \gamma_2$ $0.4 \leqslant \lambda \leqslant 3.6$		$\alpha_w \neq 0$	0.126	0.128	0.126	0.128
		$\alpha_w = 0$	0.088	0.089	0.088	0.089
$\gamma_\lambda = \begin{cases} \gamma_1 & \lambda \leqslant 1.7 \\ \gamma_2 & \lambda > 1.7 \end{cases}$		$\alpha_w \neq 0$	0.128	0.130	0.127	0.131
		$\alpha_w = 0$	0.088	0.091	0.088	0.091

average introduces a correction of 10% in the actual variation range of parameters ρ_v, w, τ_0, and μ_0. The last two parameters have the greatest effect on A_Σ, as is evident from Fig. 9.1 and 9.2, and also from the tables of [7]. The dependence on τ_0 and μ_0 in [5] is approximated by the formula

$$A_\Sigma = A_v = 1 - (1 - A_{no\ cld}) \exp\left(-\frac{\alpha}{\mu_0^{\beta}}\tau_0^{\gamma}\right) \qquad (9.14)$$

with an error of 5–10% in the range $5 \leq \tau_0 \leq 100$; $0.25 \leq \mu_0 \leq 1$. Here $A_{no\ cld}$ is the albedo of the cloudless atmosphere, $\alpha = 0.14$, $\beta = 0.35$, $\gamma = 0.67$. This formula is apparently convenient for use in models of the general circulation of the atmosphere, where only indirect use is made of the scanty averaged information about the microstructure of clouds, so that the errors of the order of 10% introduced by neglecting absorption need not be taken into account.

However, just because the absorption effect can be ignored when determining the albedo, it by no means follows that the energy effect of the absorption can be neglected. The data in Tables 2.13 and 2.15 of [7] indicate that in cloud layers from 5 to 18% of the solar energy is absorbed, the average amount being about 10%. The latter heats a cloud layer 0.5 km deep by 1°C/h or, recalculating for a homogeneous atmosphere 8 km thick for

12 hours of daylight, it causes a heating of about 1.5°C/day. This is a very significant amount, as compared with the heating of a cloudless atmosphere by solar radiation and as compared with

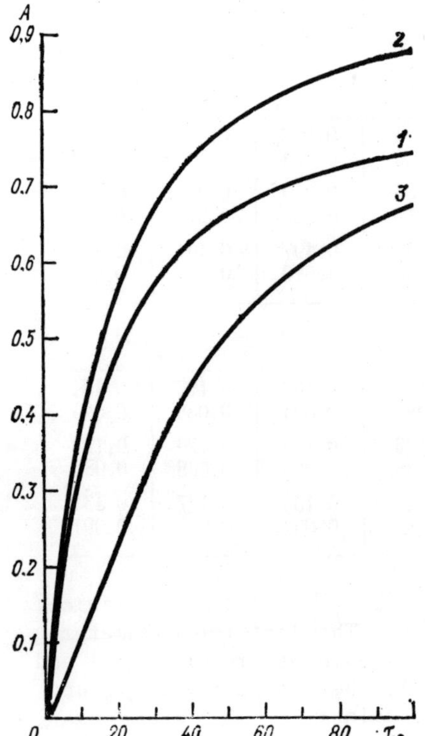

Fig. 9.1. Albedo as function of optical depth of layer τ_0, for σ_{av} = 50 km^{-1}, w_{av} = 0.2 g/m^3, ρ_v = 10 g/m^3, μ_0 = 1.
1) integral albedo A_Σ; 2 and 3) albedo of pure scattering A_v for "wide" functions γ_1 and γ_2, respectively.

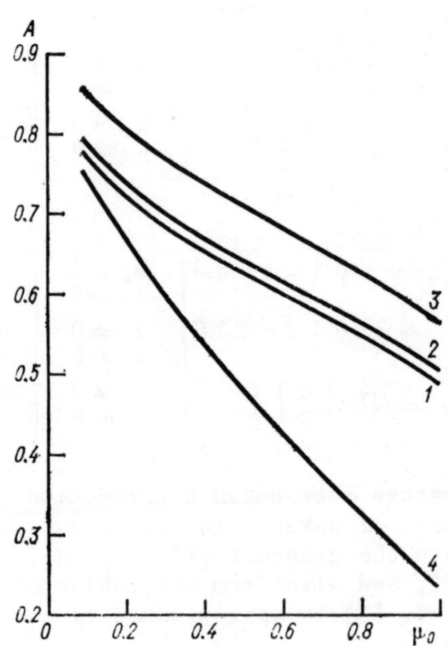

Fig. 9.2. Albedo as function of parameter μ_0, for σ_{av} = 50 km^{-1}, w_{av} = 0.2 g/m^3, ρ_v = 10 g/m^3. 1 and 2) A_Σ for $\alpha_w \neq 0$ and α_w = 0, respectively; 3 and 4) A_v for γ_1 and γ_2, respectively.

actual variations of the atmospheric temperature. Figures 9.3 and 9.4 show Π_Σ as a function of parameters τ_0 and μ_0; Π_Σ as a function of ρ_v and w, as well as of τ_0 and μ_0, can be found from Tables 2.13 and 2.15 of [7].

Interestingly enough, in [10] the maximum integral absorption in a cloud layer is also estimated to be 15 to 20% of the incident solar radiation.

It follows from Table 9.3 that, for a high tropical humidity

$\rho_{v} = 10$ g/m^3, 70% of the absorption is by water vapor and 30% by droplets, for $\tau = 20$. In the middle latitudes, that is, for $\rho_{v} \leq 5$ g/m^3, the effect of liquid water is obviously much greater.

Let us consider the distribution along the vertical of the

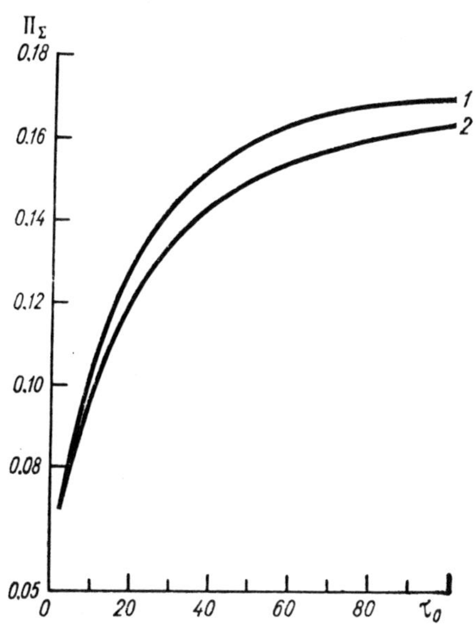

Fig. 9.3. Total absorptivity of cloud layer as function of τ_0, for $\sigma_{av} = 50$ km^{-1}, $w_{av} = 0.2$ g/m^3, $\rho_{v} = 10$ g/m^3, and $\mu_0 = 1$.
1) Π_{Σ} for $\alpha_w \neq 0$; 2) Π_{Σ} for $\alpha_w = 0$.

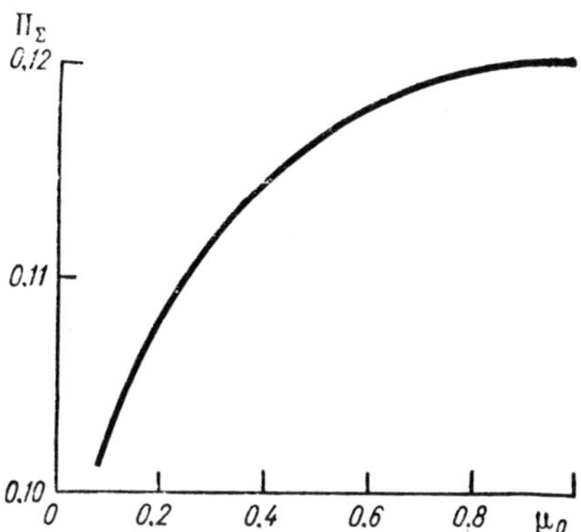

Fig. 9.4. Total absorptivity of cloud layer as function of μ_0, for $\sigma_{av} = 50$ km^{-1}, $w_{av} = 0.2$ g/m^3, $\rho_{v} = 10$ g/m^3, $\tau_0 = 20$, and $\alpha_w = 0$.

absorbed solar radiation inside a cloud layer. Figure 9.5 shows the distribution in homogeneous and inhomogeneous layers (see models 2 and 1, above), for the absorption bands of water droplets and water vapor. Inhomogeneity of the layer causes a substantial redistribution of the absorption over the height. In spectral intervals where the absorption is only by water vapor, the height distribution of the absorption is the same in both models.

Table 9.3 gives the probability density of the distribution with height of the integral absorption. Clearly, neglecting the stratification of σ and w introduces a certain error into an estimate of the absorption.

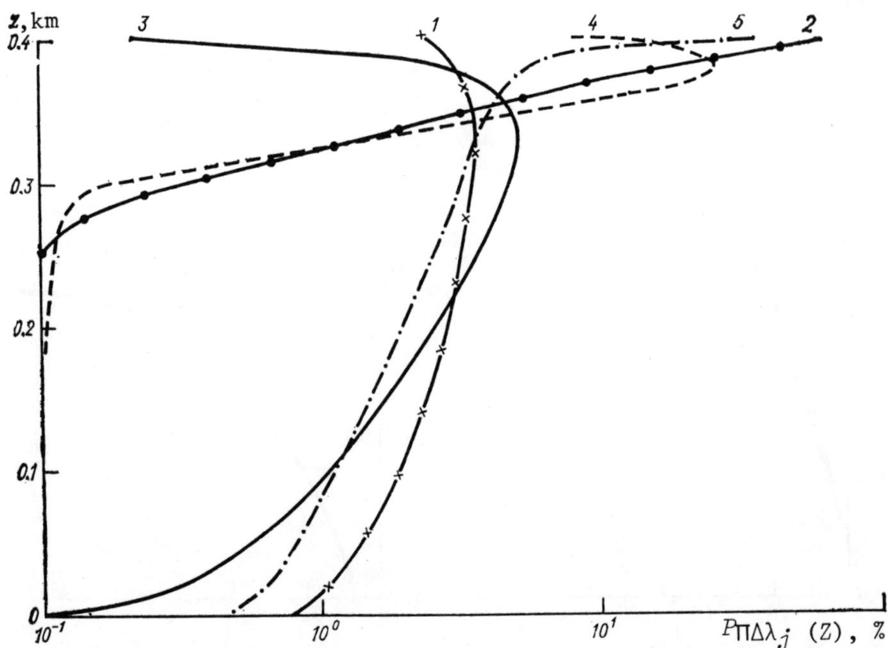

Fig. 9.5. Probability density for height distribution in homogeneous and inhomogeneous cloud layers, in different spectra intervals.
1 and 2) homogeneous layer; 3 and 4) inhomogeneous layer (bands
11 and 15, see Table 9.1); 5) both layers (band 2).

Table 9.3. Distribution of integral absorptivity (%) over thickness
of homogeneous and inhomogeneous cloud layers, for τ_0 = 20, H =
0.4 km, $\mu_\odot$ = 1, w_{av} = 0.2 g/m^3, ρ_v = 10 g/m^3.

Layer, m	Distribution		inhomo-geneous
	homogeneous		
	γ_1	γ_2	γ_1
380 ... 400	34.4	34.6	31.7
362 ... 380	12,7	11,6	16.2
321 ... 362	16.9	15.1	18.8
276 ... 321	11.2	10.3	12.5
232 ... 276	7.4	6.9	7,3
188 ... 232	5.7	5.9	4,9
128 ... 188	5.8	5,9	4,4
61 ... 128	3.6	3,7	2,8
30 ... 61	1.2	2,2	0.8
0 ... 30	1.2	1.2	0.1

9.3. Simplified method for calculating fluxes of infrared solar radiation in a cloudy atmosphere

Let us consider fluxes of infrared solar radiation $F^{\uparrow\downarrow}(z)$ in an atmosphere which is inhomogeneous over its height and which contains one or several cloud layers. In this case, as discussed in detail in [7], the flux calculations are plagued by fundamental difficulties associated with the use of nonexponential functions to describe the transmission of atmospheric gases. Since these functions do not obey the summation theorem, that is,

$$P\,(m_1 + m_2) \neq P\,(m_1)\,P\,(m_2), \qquad (9.15)$$

the flux calculation methods proposed in Chap. 8 for the ultraviolet and visible ranges are not applicable. Expressing the transmission function as a series of exponentials (see Chap. 5) does not eliminate the difficulties associated with the transition from an isolated homogeneous layer to an inhomogeneous atmosphere if the summation theorem does not hold true. For instance, in the absence of scattering, and neglecting for simplicity's sake the absorption in a cloudless atmosphere, for $z = z_{1b}$ we obtain (see (9.10))

$$F^{\downarrow}\,(z_{1b}) = I_0 \sum_{i=1}^{n} a_i e^{-\alpha_i m_0}.$$

However, for $z < z_{1b}$ the flux

$$F^{\downarrow}\,(z) \neq F^{\downarrow}\,(z_{1b}) \sum_{i=1}^{n} a_i e^{-\alpha_i m},$$

where m is the mass of absorbing substance in the $(z,\ z_{1b})$ layer, and m_0 is the mass in the $(z_{1b},\ z_{ub})$ layer.

In [7] an approximate method of calculating fluxes $F^{\uparrow\downarrow}(z)$ in a cloudy atmosphere was proposed. This method is as follows:

1) The atmosphere is divided into layers $(z_i,\ z_{i+1})$ with typical properties of radiation transfer; cloud layers, above-cloud layers, intercloud layers, and below-cloud layers. The atmosphere outside the clouds is stratified, with pressure and humidity that vary with height. There radiation is attenuated via molecular scattering by the aerosol, and via absorption by water vapor and the aerosol. The clouds are assumed to be homogeneous. Multiple scattering at droplets and absorption by water vapor and droplets take place in them.

2) The path of the photons contributing to the formation of fluxes $F(z)$ in the set of these layers is traced out.

3) The total reduced mass of absorbing substance on this path is calculated:

$$m(z) = \sum_{i=1}^{n-1} \overline{\sec \theta^{(i)}} m(z_i, z_{i+1}). \qquad (9.16)$$

Here $\overline{\sec \theta^{(i)}}$ characterizes the mean direction of the radiation reflected or transmitted by the ith layer. The layers are numbered in the sequence of propagation of solar radiation and $\theta^{(1)} = \zeta$. In the cloud layers, obviously $\sec \theta \approx l_{av}/H$ (see formula (9.13)), and outside the clouds we assume that $\overline{\sec \theta} = 1.7$. In [7] it was shown that a variation of $\overline{\sec \theta}$ from 1.5 to 3.0 has only an insignificant effect on the flux calculations.

4) The radiation flux

$$F^{\uparrow\downarrow}(z) = c I_0 \cos \zeta \, P \left[\sum_{i=1}^{n-1} \overline{\sec \theta^{(i)}} m(z_i, z_{i+1}) \right] \qquad (9.17)$$

is calculated, where $c = 1$ if no clouds were encountered along the radiation path, $c = A$ in light reflected from a cloud, and $c = T$ in transmitted light.

In this procedure the additivity of the reduced masses calculated using formula (9.16) is utilized, and the fact that the addition theorem does not apply for the transmission function is taken into account.

For single-layer cloudiness and zero albedo of the underlying surface A_S, the corresponding formulas are:

$$\frac{1}{I_0 \mu_0} F^{\downarrow}(z) = \begin{cases} P\left[\dfrac{1}{\mu_0} m(z, \infty)\right] & \text{for } z \geqslant z_{ub}, \\[2ex] TP\left\{\left[\dfrac{1}{\mu_0} m(z_{ub}, \infty) + l^{\downarrow}\rho_v \text{ cld} \left(\dfrac{p_{cld}}{p_0}\right)^n + \right. \right. \\[2ex] \left. \left. + \overline{\sec \theta}(z, z_{нr})\right], \; l^{\downarrow}w\right\} & \text{for } z \leqslant z_{ub}; \end{cases} \qquad (9.18)$$

$$\frac{1}{I_0 \mu_0} F^{\uparrow}(z) = \begin{cases} AP\left\{\left[\dfrac{1}{\mu_0} m(z_{ub}, \infty) + l^{\uparrow}\rho_v \text{ cld} \left(\dfrac{p_{cld}}{p_0}\right)^n + \right. \right. \\[2ex] \left. \left. + \overline{\sec \theta} m(z_{ub}, z)\right], \; l^{\uparrow}w\right\} & \text{for } z \geqslant z_{ub}, \\[2ex] 0. \end{cases} \qquad (9.19)$$

The principles of constructing the analogous formulas for multilayer cloudiness with $A_S \neq 0$ was described in [7] (see

Section 3.3.2 of that book). An expression for the flux of out-going radiation in the case of two-layer clouds was also given there.

9.4. Fluxes and influxes of infrared solar radiation in a cloudy atmosphere

Table 9.4 gives the relative infrared fluxes $F^\uparrow(\infty)$ and $F^\downarrow(0)$, that is, the fluxes of departing radiation at the upper boundary of the atmosphere and incident radiation at the ground, as functions of the height of the cloud layer. Thin $(\tau_0 = 10)$ and thick $(\tau_0 = 50)$ cloud layers are considered for $\sigma = 50$ km^{-1}, $\mu_0 = 0.5$, $A_s = 0$ and 0.8, $w = 0.2$, and $\rho_v = 5$ g/m^3. Clouds were located at levels with $z_{1b} = 0.5$, 1.5, 3, and 5 km. The table gives the distribution of water-vapor masses in the layers above the clouds m_1 and below the clouds m_2. The fluxes are expressed as fractions of $I_0\mu_0$. The calculations were carried out using formulas (9.18) and (9.19) and the analogous expressions for $A_s \neq 0$.

Table 9.4. Relative fluxes of incident $F^\downarrow(0)$ and departing $F^\uparrow(\infty)$ radiation for $A_s = 0$ and 0.8.

z_{1b}, km	τ_0										
	10		50		10		50		10	50	10, 50
	0	0.8	0	0.8	0	0.8	0	0.8			
	$F^\downarrow(0)$				$F^\uparrow(0)$				m_1		m_2
0.5	0.290	0.416	0.130	0.208	0.381	0.472	0.522	0.542	1.67	0.95	0.46
1.5	0.293	0.416	0.130	0.207	0.399	0.490	0.550	0.570	0.95	0.40	1.18
3	0.294	0.414	0.130	0.205	0.433	0.524	0.566	0.586	0.38	0.24	1.75
5	0.294	0.413	0.131	0.205	0.459	0.550	0.590	0.610	0.11	0.07	2.02

Inspection of the tables shows that $F^\downarrow(0)$ is independent of the height of the cloud layer, which is understandable in view of formula (9.18). In the assumed model, under the sign of the transmission function only the ratio between $\sec \zeta m_1$ and $\sec \theta m_2$ varies as a function of the cloud level. This variation is inconsiderable, however, because of the smallness of dP/dm_v for large impinging masses encountered as photons pass through the thickness of the atmosphere.

The flux $F^\uparrow(\infty)$ depends appreciably on the height of the cloud layer, since under the sign of the transmission function in formula (9.19) there is much less mass of absorbing substance than in (9.18), while dP/dm is large for small masses. The effect of the

Table 9.5. Heating R by short-wave radiation of individual atmospheric layers (z_1-z_2) in °C/h, as function of location (Δh) of lower-level clouds.

$\zeta = 2-°$

Layer	z_1-z_2 km	Δz km	Winter profile, $A_s = 0.6$			Summer profile, $A_s = 0.2$		
			R_{cld}	R_{1cld}/R_{2cld}	$R_{no\ cld}$	R_{cld}	R_{cld}/R_{1cld}	$R_{no\ cld}$
1	0 … 3	0.5 … 1	0.215	1.064	0.051	0,184	1.13	0.070
		1.5 … 2	0.202			0,162		
2	3 … 5	0.5 … 1	0.049	0.778	0.046	0.090	0.90	0.077
		1.5 … 2	0.063			0,100		
3	5 … 9	0.5 … 1	0.033	0.846	0.035	0.057	0.965	0.056
		1.5 … 2	0,039			0.059		
4	9 … 26	0.5 … 1	0.034	1.03	0,033	0.035	1.0	0.037
		1.5 … 2	0.033			0.035		

$\zeta = 60°$

Layer	z_1-z_2 km	Δz km	Winter profile, $A_s = 0.6$			Summer profile, $A_s = 0.2$		
			R_{cld}	R_{1cld}/R_{2cld}	$R_{no\ cld}$	R_{cld}	R_{cld}/R_{1cld}	$R_{no\ cld}$
1	0 … 3	0.5 … 1	0.103	1.05	0.026	0.096	1.14	0.036
		1.5 … 2	0.098			0.084		
2	3 … 5	0.5 … 1	0.032	0.81	0.029	0.052	0.91	0.048
		1.5 … 2	0.038			0.057		
3	5 … 9	0.5 … 1	0.024	1.0	0.024	0.037	1.0	0.037
		1.5 … 2	0.024			0.037		
4	9 … 26	0.5 … 1	0.021	0.96	0.023	0.024	1.0	0.026
		1.5 … 2	0.022			0.024		

cloud height is comparable to that of the optical depth of the cloud, for a high albedo of the underlying surface.

Of course, the influxes of infrared solar radiation to the total thickness of a cloudy atmosphere, or to individual layers of it, depend on the height of the cloud layer.

Quantitative experiments were performed in order to study the profile of the short-wave heat influx in a cloudy atmosphere as a function of the height of the cloud layer, the albedo of the underlying surface, and the zenith angle of the Sun. A four-layer model of the atmosphere was considered, and the effect of a movement of the lower-level cloudiness (depth 0.5 km) on the influxes in all the layers was ascertained.

The fluxes in a cloudy atmosphere were computed with the aid of a method, described in [2, 3], allowing for absorption by water vapor, carbon dioxide, and ozone, as well as scattering by air and aerosol molecules [8], scattering and absorption by clouds [7], and lengthening of photon paths in cloud and subcloud layers for $A_s > 0$ [1].

Data on the heating of individual atmospheric layers are given in Table 9.5. Calculations were carried out for two mean humidity profiles, for winter and summer in the middle latitudes. The numerator is the influx to the corresponding layer for a cloud with a lower boundary at 0.5 km, and the denominator is the influx for a cloud with a lower boundary at 1.5 km. The ratios of these influxes $(R_{1\,cld}/R_{2\,cld})$ for each layer are also given, as well as the influxes to the layers under clear-sky conditions.

The table shows that a shift in the position of the lower boundary of the cloud layer from 0.5 to 1.5 km varies the heating of the 1st layer by 6% in winter and 13–14% in summer. The influxes in the 2nd layer also depend on the location of the lower-level clouds. Under winter conditions the influx variations are within limits of 15 to 20%, and in summer they are within 10%. The influxes in the 3rd and 4th layers are practically independent of the location of the lower-level clouds. Inspection of the table also shows that an atmospheric layer with clouds in it is subject to the greatest increase in heating, in comparison with the same layer under cloudless conditions.

CHAPTER 10

EXPERIMENTAL AIRCRAFT STUDIES OF SOLAR FLUXES IN THE PRESENCE OF
STRATIFORM CLOUDS

The first aircraft measurements of radiation fluxes in a
cloudy atmosphere date back to 1919 [26], but detailed measurements
were resumed only about 30 years ago [23, 24, 27, 28]. The first
Soviet measurements of the radiation characteristics of clouds
were carried out at the Central Aerological Observatory [22], and
a number of Soviet works on this subject were subsequently published
[1, 13-16]. However, in view of the diversity of the data, the
difference in measurement methods, and the insufficient volume of
experimental data, it was not yet possible to establish the patterns of the radiation regime of a cloudy atmosphere. This problem
became solvable only with the availability of sufficient experimental material at the Ukrainian Hydrometeorological Scientific-
Research Institute, the Voeikov Main Geophysical Observatory, and
the Central Aerological Observatory. The main results of the studies
carried out at these institutions will be presented below.

10.1. Regime of integral solar radiation for St-Sc clouds

The measurements were made with standard actinometric instruments (pyranometers, actinometers, Yanishevskii balance meters),
mounted aboard an Il-14 (or other) aircraft. The measurement
techniques and data-processing methods were described in [6, 13,
14].
The data were used to find the albedo A at the upper (ub)
and lower (lb) boundaries of the cloud layer, the transmission
coefficient T, and the relative absorptivity Π of the cloud layers:

$$A_{ub} = \frac{F^{\uparrow}_{s.u}}{F^{\downarrow}_{s.u}}, \qquad A_{lb} = \frac{F^{\uparrow}_{s.l}}{F^{\downarrow}_{s.l}}, \qquad T = \frac{F^{\uparrow}_{s.l}}{F^{\downarrow}_{s.u}}, \qquad (10.1)$$

$$\Pi = \frac{R_s(z_u,\, z_l)}{F^{\downarrow}_{s.u}},$$

where $F^{\downarrow}_s$ and $F^{\uparrow}_s$ are the downward and upward fluxes of solar (short-
wave) radiation at the upper (u) and lower (l) cloud boundaries,
and $R_s(z_u, z_l)$ is the influx of short-wave radiation to the cloud
layer, defined as

$$R_s(z_u,\ z_l) = \left[F^{\downarrow}_{s.u} - F^{\uparrow}_{s.u}\right] - \left[F^{\downarrow}_{s.l} - F^{\uparrow}_{s.l}\right]. \qquad (10.2)$$

Since clouds are semitransparent to short-wave radiation, coefficients A, T, and Π are influenced appreciably by the multiple reflection of radiation between the lower boundary of the cloud layer and the underlying surface. This was pointed out in a number of works [9, 12, 29, 30], but not allowed for, so that the values of A and T given by these investigators pertain not to the cloud proper, but rather to the system consisting of the cloud plus the Earth's surface. In accordance with our discussion in [5, 6], a different method for processing the measurement results was proposed, and then applied in [7, 9, 10]. It essentially involves reducing the values of A_{ub}, T, and Π measured for specific value of A_s (the albedo of the Earth's surface) to $A_s = 0$. These values were called the "true" values, being typical of the cloud itself, rather than of the system as a whole. The reduction was carried out using the formulas:

$$A^* = \frac{A_{ub} - T^2 A_s}{1 - T^2 A_s^2}, \quad T^* = T(1 - A_s A^*),$$

$$\Pi^* = 1 - A^* - T^*. \qquad (10.3)$$

In order to describe the absorptivity of a cloud, the ratio $\Pi_{eff} = \dfrac{R_s(z_u,\ z_1)}{F_{ent}}$, the effective absorptivity, is also useful. Here F_{ent} is the short-wave radiation entering the cloud via the upper and lower cloud boundaries. It is defined, according to [7], as

$$F_{ent} = F^{\downarrow}_u \left(1 - A^*_{ub}\right)\left(1 + \frac{T^* A_s}{1 - A^* A_s}\right). \qquad (10.4)$$

Figures 10.1 and 10.2 show the "true" albedo and transmission as functions of the main cloud parameters: depth H and mean water content $\bar{w}$. These graphs, which are based on experimental data, reveal a pronounced nonlinear dependence of A^* and T^* on these parameters. The dependences of A^* and T^* on the cloud depth H, the water content w, and the water reserves $m_w = Hw$ were approximated by the functions:

$$T^* = \exp\left[-(5.5 - 3.7H)H\right],$$
$$A^* = 1 - \exp\left[-(4.7 - 3.2H)H\right], \qquad (10.5)$$

$$T^* = \exp\left(-3.1\sqrt[3]{\overline{w}}\right), \quad A^* = 1 - \exp\left(-2.5\sqrt[3]{\overline{w}}\right), \quad (10.6)$$

$$T^* = \exp\left(-0.56\sqrt[4]{m_w}\right), \quad A^* = 1 - \exp\left(-0.46\sqrt[4]{m_w}\right). \quad (10.7)$$

Formulas (10.6) and (10.7) show a good fit with actual data over the whole range of real values of $\overline{w}$ and m_w, but formula (10.5) is only valid for $H \leq 1$ km.

The height of the Sun $h_\odot$ is one of the factors affecting

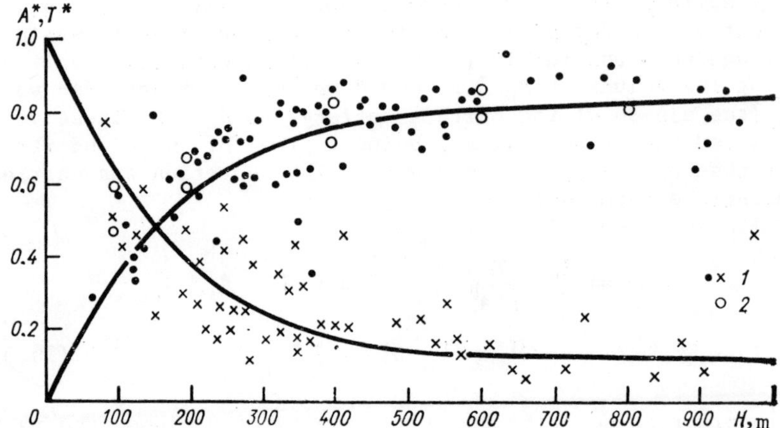

Fig. 10.1. "True" albedo A^* and transmittance T^* of clouds as functions of cloud depth H. Curves plotted using formulas (10.3). 1) experimental data, 2) calculated, for $\zeta = 60$ and 75° (see Section 10.5).

the albedo and transmittance of a cloud layer. An analysis of the data indicated that a 10% decrease in $h_\odot$ leads to a 3% increase in the cloud albedo, on the average, and that the transmittance is reduced by approximately the same amount (Table 10.1). The greatest changes in A^* and T^* with a variation in $h_\odot$ are observed at low cloud depths and water contents. The albedo increases with a decrease in the Sun's height to $h_\odot = 10$-12°. A further decrease in $h_\odot$ is attended by a slight drop in A^*, evidently because of shading of parts of the upper boundary by its irregularities.

Table 10.1 also portrays the absorbing powers Π^* and Π^*_{eff} as functions of cloud depth and Sun height. Parameter Π^* is seen to be practically independent of H, whereas Π^*_{eff} depends appreciably on this parameter. The dependence of the effective absorptivity Π^*_{eff} on the water content and water reserves is also quite evident,

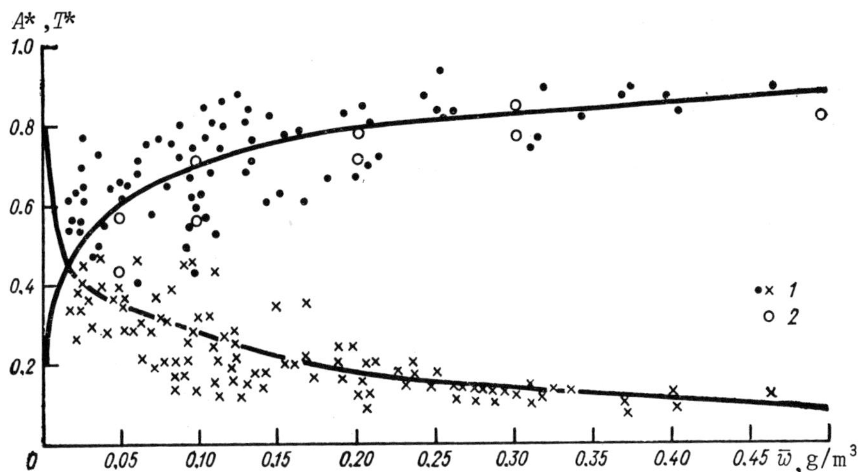

Fig. 10.2. "True" albedo A^* and transmittance T^* as functions of mean water content of clouds (notations same as in Fig. 10.1).

the relationship with the water reserves being approximated by the formula:

$$\Pi_{eff}^* = 1 - \exp\left(-0.11 \sqrt[4]{m_w}\right). \qquad (10.8)$$

If we know the absorptivity Π^* and the flux $F_u^{\downarrow}$, we can find the heat influx to the cloud layer due to the absorption of solar radiation. For a mean height of the upper cloud boundary z_{ub} = 1 km and for known mean values of F_u^* and Π^*, the amount of radiation absorbed by the cloud $R_s^*(z_u, z_1)$ and the rate of radiative heating of the cloud $(dT/dt)_s$ were found to have the values given in Table 10.2.

Inspection of the table shows that $R_s^*(z_u, z_1)$ and $(dT/dt)_s$ both increase with an increase in $h_\odot$. As the cloud depth grows, $R_s^*(z_u, z_1)$ first increases and then, for $H \geq 350$ m, there is a certain tendency for it to decrease.

The rate of radiative heating of clouds more than 250 m deep is curtailed considerably as H increases, since practically the same amount of radiation $R_s^*(z_u, z_1)$ becomes distributed throughout an increasingly large cloud volume.

To solve many problems, it is necessary to know the radiative heating over certain time intervals: 12 h, 24 h, a month, etc.

Table 10.3 gives diurnal (24-hour) values of the radiative heating of cloud layers of different depths for the middle latitudes during the cold season, obtained by integrating the dependence of

Table 10.1. Parameters A^*, T^*, Π^*, and Π^*_{eff} (%) as functions of Sun height and cloud depth.

Depth H m	Radiation character-istic	$h^\circ_\odot$				
		10	20	30	40	50
175	T^*	37.0	43.5	48.8	53.2	56.0
	A^*	59.5	52.4	47.0	43.6	41.0
	Π^*	3.5	4.1	4.2	3.2	3.0
	Π^*_{eff}	8.6	8.6	8.0	5.7	5.1
260	T^*	23.2	27.8	31.0	34.5	37.2
	A^*	72.0	67.0	62.8	59.2	57.0
	Π^*	4.8	5.2	6.2	6.3	5.8
	Π^*_{eff}	17.2	15.7	16.6	15.4	13.5
350	T^*	18.5	19.8	22.3	23.8	25.6
	A^*	75.5	73.2	70.8	68.8	67.4
	Π^*	6.0	7.0	6.9	7.4	7.0
	Π^*_{eff}	24.5	26.1	23.6	23.7	21.5
450	T^*	14.5	16.0	17.5	19.0	20.5
	A^*	80.0	77.5	75.8	74.2	72.8
	Π^*	5.5	6.5	6.7	6.8	6.7
	Π^*_{eff}	27.5	28.8	27.7	26.4	24.6
610	T^*	11.5	12.5	14.0	15.7	17.0
	A^*	83.6	81.2	79.7	78.2	77.0
	Π^*	4.9	6.3	6.3	6.1	6.0
	Π^*_{eff}	29.8	33.4	31.0	27.9	26.0
900	T^*	10.0	11.0	11.7	12.3	13.0
	A^*	85.6	83.2	82.0	81.0	80.0
	Π^*	4.4	5.8	6.3	6.7	7.0
	Π^*_{eff}	30.5	34.5	35.0	35.2	35.0

Table 10.2. Radiation absorbed by clouds $R^*(z_u, z_1)$ in watts/m^2 and rate of radiative heating $(dT/dt)_s$ in °C/h, as functions of $h_\odot$ and H for $A_s = 0$.

Interval of H, km	Parameter	$h_\odot$				
		10	20	30	40	50
0.09–0.20	$R_s^*(z_u,z_1)$	4.89	13.3	20.9	21.6	23.7
	$(dT/dt)_s$	0.08	0.22	0.35	0.36	0.39
0.2–0.3	$R_s^*(z_u,z_1)$	6.98	16.8	31.4	41.9	46.1
	$(dT/dt)_s$	0.08	0.19	0.35	0.47	0.52
0.3–0.4	$R_s^*(z_u,z_1)$	8.38	23.03	34.9	48.9	55.1
	$(dT/dt)_s$	0.07	0.19	0.29	0.40	0.46
0.4–0.5	$R_s^*(z_u,z_1)$	7.68	21.6	33.5	47.5	53.05
	$(dT/dt)_s$	0.05	0.14	0.22	0.26	0.34
0.5–0.7	$R_s^*(z_u,z_1)$	6.98	20.9	32.1	40.5	47.5
	$(dT/dt)_s$	0.03	0.10	0.15	0.19	0.23
0.7–1.2	$R_s^*(z_u,z_1)$	6.28	18.8	33.5	44.7	55.1
	$(dT/dt)_s$	0.02	0.06	0.10	0.14	0.18

$R_s^*(z_u, z_1)$ on $h_\odot$ given by Table 10.2. During the warm season the values of $R_s^*(z_u, z_1)$ will be higher than those in Table 10.2, but at the same time St–Sc cloud layers are rarely observed during this season.

Table 10.3. Radiative heating (°C/h) of St-Sc clouds of various
depths during cold season.

Lati-tude,°	Month	Depth, m					
		175	260	350	450	610	900
55	October	1.8	1.7	1.6	1.2	0.8	0.5
	November	0.8	0.7	0.7	0.5	0.3	0.2
	December	0.5	0.4	0.4	0.3	0.2	0.1
	January	0.6	0.6	0.5	0.4	0.2	0.1
	February	1.3	1.2	1.1	0.8	0.6	0.4
	March	2.6	2.6	2.3	1.7	1.2	0.8
	April	3.6	4.1	3.5	2.6	1.7	1.2
50	October	2.3	2.2	1.9	1.4	1.0	0.6
	November	1.3	1.1	1.1	0.8	0.5	0.3
	December	0.8	0.8	0.7	0.5	0.3	0.2
	January	1.0	0.9	0.8	0.6	0.4	0.3
	February	1.8	1.7	1.6	1.2	0.8	0.5
	March	2.8	3.0	2.6	1.9	1.3	0.9
	April	3.7	4.3	3.8	2.8	1.8	1.3
45	October	2.6	2.8	2.4	1.8	1.2	0.8
	November	1.7	1.6	1.5	1.1	0.7	0.5
	December	1.2	1.1	1.0	0.8	0.5	0.3
	January	1.4	1.3	1.2	0.9	0.6	0.4
	February	2.3	2.2	1.9	1.4	1.0	0.7
	March	3.0	3.4	2.9	2.2	1.5	1.0
	April	3.8	4.5	3.9	2.9	2.0	1.0

10.2. Experimental model of "average" St-Sc cloud

Radiation measurements indicate that even in the simplest
case of a single-layer cloud the radiation field is quite variable.
In addition, random errors in the measurements distort the true
picture. Therefore, in order to ascertain the most typical properties of the vertical structure of the radiation fluxes, we have
to average the data of individual measurements. Such an averaging
yields an experimental model of an "average" stratiform cloud.

To devise the model, 25 vertical actinometric soundings made
with aircraft were used. These soundings were made when single-
layer St-Sc clouds were present and all other cloud types were
absent.

The distribution of meteorological parameters and radiation
characteristics in the ranges of solar (sw) and thermal (lw)
radiation is portrayed in Table 10.4 and Fig. 10.3. The average
stratiform cloud has the following parameter values: height of

Table 10.4. Vertical distribution of meteorological elements and radiation parameters in cloud-forming layer containing "average" stratiform cloud ($A_s = 0$, $h \odot = 40°$).

z, m	$t°C$	g/kg q	g/m³ w	E_{eff}	B_{sw}	B	$R_{1w}(z)$	$R_{sw}(z)$	$R(z)$	dT/dt_{1w}	dT/dt_{sw}	dT/dt
				watts/m²			watts/m³·10⁴			°C/h		
100	−0.5	3.4		14	131	117	0	140	140	0.00	0.03	0.03
420	−3.3	3.0	0.00	15	135	120	279	558	837	0.08	0.16	0.24
470	−3.7	3.0	0.05	14	138	124	627	836	1463	0.20	0.24	0.44
520	−4.1	2.9	0.09	10	142	130	70	976	1046	0.02	0.29	0.31
570	−4.4	2.8	0.15	9	147	138	0	977	977	0.00	0.28	0.28
620	−4.8	2.8	0.20		151	142	−488	1394	906	−0.14	0.44	0.30
670	−5.2	2.7	0.25	11	158	147	−1 394	1533	139	−0.40	0.44	0.04
720	−5.6	2.6	0.29	19	166	147	−3 694	1673	−2021	−1.12	0.48	−0.17
770	−5.8	2.5	0.28	31	175	144	−10 734	2091	−8643	−3.55	0.63	−2.92
820	−3.2	2.5	0.00	91	185	94						
860	+1.2	3.0										
970	+1.3	2.0										
1900	−2.1	1.6		106	206	100	−209	209	0	−0.06	0.07	0.01
2800	−7.2	1.4		129	226	97	−279	209	−70	−0.09	0.07	−0.02

lower boundary 430 m, depth H = 400 m, and water content w = 0.2 g/m . The main parameters of the average cloud, namely its

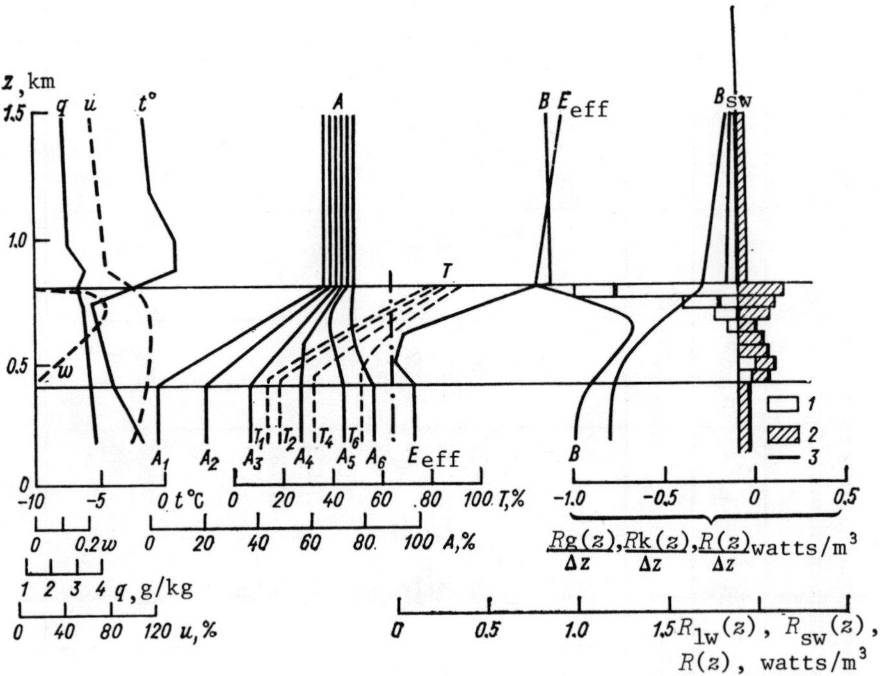

Fig. 10.3. Experimental model of "average" stratiform cloud. w is water content, u, q are relative and specific air humidity; t is air temperature; E_{eff} is effective radiation, and B_{sw} and B are short-wave and total radiation budgets for $h_{\odot}$ = 40°. The radiation influxes are: 1) long-wave, 2) short-wave, 3) total. A_i and T_i are albedos and transmittances for various albedos of Earth's surface: A_s = 0, 20, 40, 60, 75, and 85%, with i = 1, 2,...,6.

depth and water reserves, are close to the average characteristics obtained for internal St-Sc clouds on the basis of considerable statistical material. The profiles of the water content, temperature, and humidity are typical for these cloud types.

The radiation parameters in Table 10.4 are defined as follows:

$$E_{eff} = F_{lw}^{\uparrow} - F_{lw}^{\downarrow}, \qquad B_{sw} = F_{sw}^{\uparrow} - F_{sw}^{\downarrow}, \qquad B = B_{sw} - E_{eff},$$

$$R_{\text{lw}}(z) = \frac{E_{\text{eff}}(z) - E_{\text{eff}}(z + \Delta z)}{\Delta z},$$

$$R_{\text{sw}} = \frac{B_{\text{sw}}(z + \Delta z) - B_{\text{sw}}(\Delta z)}{\Delta z},$$

$$R = R_{\text{lw}} + R_{\text{sw}}.$$

The radiation (sw) data in Table 10.4 were determined in the following manner. To construct the vertical profile of the short-wave radiation fluxes for the average cloud, the "true" values of the albedo $A^*(z)$, transmission coefficient $T^*(z)$, and absorptivities $\Pi^*(z)$ and $\Pi^*_{\text{eff}}(z)$ at various levels z inside the cloud were used. These quantities were calculated with the aid of the formulas:

$$A^*_{z,\,h} = \frac{A_{z,\,h} - T^2_{z,\,h}A_s}{1 - T^2_{z,\,h}A^2_s}, \tag{10.9}$$

$$A^*_{H,\,z} = \frac{A_v - T^2_{H,\,z}A_{z,\,h}}{1 - T^2_{H,\,z}A^2_{z,\,h}}, \tag{10.10}$$

$$T^*_{H,\,z} = \frac{T_{H,\,z}\left(1 - A_{z,\,h}A^*_{H,\,z}\right)}{1 - A^*_{z,\,h}A^*_{H,\,z}}, \tag{10.11}$$

$$\Pi_{H,\,z} = (1 - A_v) - T_{H,\,z}(1 - A_{z,\,h}), \tag{10.12}$$

$$\Pi^*_{\text{eff},\,H,\,z} = \frac{R_{\text{sw}}(H,\,z)}{E_{\text{ent}}(H,\,z)} = \frac{\Pi_{H,\,z}}{\left(1 - A^*_v\right)\left(1 + \dfrac{T^*_{H,\,z}A_{z,\,h}}{1 - A^*_s A_{z,\,h}}\right)}. \tag{10.13}$$

Quantities without asterisks pertain to direct measurements, and subscripts $(H,\,z)$ and $(z,\,h)$ indicate, respectively, cloud layers from the upper boundary (H) to level z in the cloud and from level z to the lower boundary (h).

The profile of the transmission coefficients $T^*_{H,\,z}$ was obtained by arithmetic averaging of the transmittance values at the corresponding levels. The profile of the albedo inside the cloud was obtained from the dependence of $A^*_{z,\,h}$ on the water reserves of the lower cloud layer $m_w(z,\,h)$. This dependence was used to find the albedos at various levels, in accordance with the water reserves of the lower layer of the average cloud. In the layer above the cloud the profile of the albedo was constructed on the basis of A^*_v and the mean gradient of the albedo in this layer (γ_A).

The calculated values of $T^*_{H,\,z}$, $A^*_{z,\,h}$, and $\Pi^*_{H,\,z}$, $\Pi^*_{\text{eff},\,H,\,z}$ are given in Table 10.5, from which it is seen that the average cloud attenuates the flux of solar radiation incident upon its upper surface by 81%, of which 41% is attributable to the top 100 meters. Over a nonrefelecting surface the average cloud absorbs 7% of the short-wave radiation incident upon its upper surface and 27% of that entering it; the upper 100 meters of the cloud account for 38% of the total radiation absorbed by the average cloud.

In order to determine the absolute amounts of radiation absorbed, we must first know the total radiation incident upon the upper cloud boundary.

Table 10.5. Transmittance, absorptivity, and albedo inside "average" cloud for $A_s = 0$.

$H - z$	$T^*_{H,z}$		$A^*_{z,h}$		$\Pi^*_{H,z}$	$\Pi^*_{eff,H,z}$
	mean	no. of cases	mean	no. of cases		
0	1.000		0.735	24		
100	0.587	50	0.595	24	0.027	0.042
200	0.397	46	0.454	38	0.048	0.121
300	0.265	37	0.232	45	0.062	0.203
400	0.193	24	0.000		0.072	0.272

Table 10.6 shows the mean values of $F^{\downarrow}_{sw}$ for various $h_{\odot}$ at $z = 1$ km, and also at $z = 2$ and 3 km. Together with the data in Table 10.5, these data enabled us to find B_{sw} and R_{sw} in Table 10.4, as well as R_{sw} (see Table 10.6) for a specified albedo of the above-cloud atmosphere at the level $z = 3$ km.

Table 10.6. Dependence of $F^{\downarrow}_{sw}$ and R_{sw} (1 and 3 km) in watts/m^2 on Sun's height.

	$h^{\circ}_{\odot}$							
	10	15	20	25	30	35	40	50
$F^{\downarrow}_{sw}(z)$ for $z=1$ km	153	251	335	433	524	614	705	838
$z=2$ km	161	251	349	447	537	628	719	866
$z=3$ km	174	272	370	461	558	656	747	893
$R_{sw}(1;3km)$	14.0	20.9	20.9	27.9	34.9	41.9	41.9	55.8

Table 10.7. Radiation characteristics of individual layers inside "average" stratiform cloud for various albedos of Earth's surface ($F^{\downarrow}_{sw,v}$ = 698 watts/m^2).

Δz, m	Parameter	Albedo of Earth's surface					
		0.00	0.20	0.40	0.60	0.75	0.85
UB ... 100	$A_{100,h}$	0.595	0.616	0.652	0.703	0.756	0.804
	$T_{H,100}$	0.587	0.598	0.618	0.650	0.689	0.728
	$R_{sw}(\Delta z)$watts/m^2	18.8	19.5	20.9	22.3	23.7	25.8
	$R_{sw}(\Delta z)/R^{*}_{sw}(\Delta z)$	1.00	1.035	1.11	1.18	1.26	1.37
	$R_{sw}(\Delta z)/R_{sw}(cld)$	0.375	0.373	0.366	0.368	0.358	0.352
100 ... 200	$A_{200,h}$	0.454	0.505	0.571	0.660	0.745	0.810
	$T_{H,300}$	0.397	0.421	0.451	0.501	0.560	0.609
	$R_{sw}(\Delta z)$watts/m^2	14.7	14.7	15.4	16.0	17.4	20.2
	$R_{sw}(\Delta z)/R^{*}_{sw}(\Delta z)$	1.00	1.00	1.05	1.09	1.19	1.38
	$R_{sw}(\Delta z)/R_{sw}(cld)$	0.292	0.280	0.268	0.264	0.263	0.276
200 ... 300	$A_{300,h}$	0.232	0.346	0.471	0.611	0.740	0.830
	$T_{H,300}$	0.265	0.293	0.333	0.396	0.465	0.550
	$R_{sw}(\Delta z)$watts/m^2	9.77	10.5	11.9	13.3	15.4	16.0
	$R_{sw}(\Delta z)/R^{*}_{sw}(\Delta z)$	1.00	1.07	1.21	1.36	1.57	1.64
	$R_{sw}(\Delta z)/R_{sw}(cld)$	0.194	0.200	0.207	0.218	0.230	0.219
300 ... 400	$R_{sw}(\Delta z)$	6.98	7.68	9.07	9.07	9.77	11.17
	$R_{sw}(\Delta z)/R^{*}_{sw}(\Delta z)$	1.00	1.10	1.30	1.30	1.40	1.60
	$R_{sw}(\Delta z)/R_{sw}(cld)$	0.139	0.147	0.159	0.150	0.149	0.153
Entire cloud	A_{cld}	0.735	0.743	0.755	0.775	0.798	0.818
	T_{cld}	0.193	0.227	0.271	0.345	0.430	0.515
	$R_{sw}(cld)$watts/m^2	50.3	52.4	57.2	60.7	66.3	73.3
	$R_{sw}(cld)/R^{*}_{sw}(cld)$	1.00	1.04	1.12	1.21	1.32	1.46
	Π	0.072	0.075	0.082	0.087	0.094	0.105
	Π_{eff}	0.272	0.274	0.272	0.272	0.271	0.273

These data indicate that solar radiation is absorbed throughout the entire depth of the average cloud, with even the bottom 100-meter layer accounting for 14% of the total radiation absorbed by the cloud. Most of the absorption is concentrated in the upper portion of the cloud. Here the radiation heating is about 0.6°C/h; in the interior of the cloud it gradually diminishes, falling to 0.16°C/h at the lower boundary.

Absorption of the solar radiation has an appreciable effect on the total radiation budget of the average cloud. The radiation heating in the lower part of the cloud is enhanced considerably and reaches 0.24°C/h, while the radiation cooling in the upper layers decreases. The contribution of short-wave radiation is inconsiderable only in the regime of the upper 50-meter layer, where the radiation cooling remains intense (-2.9°C/h). On the whole, the average cloud is subject to radiation cooling (R_{cld} = 28 watts/m^2) even for considerable heights of the Sun (about 40°).

These, then, are the characteristics of the short-wave radiation regime in an average stratiform cloud, under ideal conditions. Under actual conditions ($A_s > 0$), as A_s increases, there will be a rise in both the albedo and the transmittance of all the layers of the average cloud (see Fig. 10.3, Table 10.7).

For $A_s > 0$ the absorption increases somewhat. The ratio $R_{sw}(\Delta z)/R_{sw}^*(\Delta z)$ can serve as a measure of the variation in absorption, where R_{sw} and R_{sw}^* are the amounts of radiation absorbed by a layer Δz inside the cloud for, respectively, $A_s \neq 0$ and $A_s = 0$. Table 10.7 gives this ratio, along with other parameters.

10.3. Spectral and integral radiation characteristics of stratiform clouds

Within the framework of the cloud program of the KENEKS experiment [2, 19], measurements were made of the spectral and integral albedos of clouds, as well as of their optical depths, water contents, and particle sizes. The experiment was carried out aboard an Il-18 flying laboratory (belonging to the Main Geophysical Observatory) in 1971-1972 in the presence of solid layers of stratiform clouds. The flights reached 200 m above the upper cloud boundary for Sun heights from 16 to 56° above the water surface of the Black Sea, the Sea of Azov, and the Kara Sea, and Lake Ladoga.

The radiation equipment comprised pyranometers (B. P. Kozyrev's design) measuring the integral radiation fluxes in the wavelength range from 0.3 to 3.0 μm, and two K-2 diffraction spectrometers, measuring the descending and ascending spectral radiation fluxes in the range from 0.35 to 0.95 μm. A detailed description of the equipment and the measurement technique is given in [19]. The relative error in determining the albedo in the visible range was 6%, while in the ultraviolet and near infrared it was 10%.

Figure 10.4 shows curves of the spectral albedo of clouds over the sea for a Sun height of about 52°. Table 10.8 describes the experimental conditions: flight altitude z, geometrical thickness H and type of clouds, and it gives values of the microstructural cloud parameters, averaged over the flight time: concentration N, mean cloud-droplet size r, water content w, optical depth τ_0, and integral albedo A_Σ in wavelength interval from 0.3 to 3.0 μm.

Inspection of Fig. 10.4 and Table 10.8 shows that the cloud

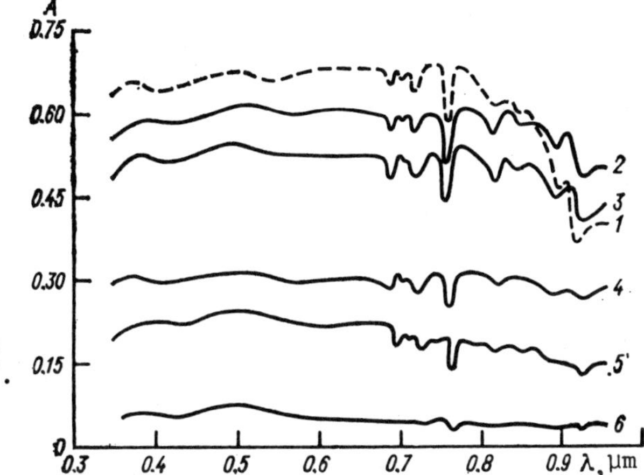

Fig. 10.4. Spectral albedo of clouds for Sun height $h_\odot \approx 52°$. Numbering of curves in accordance with Table 10.8.

Table 10.8. Albedos of solid clouds and their microphysical and macrophysical parameters, according to measurements made in April 1971 over Black Sea (Sun height $h_\odot = 52°$, for Cs $h_\odot = 48°$).

No. n/n	Cloud type	z, km	H, km	N, cm^{-3}	r, μm	w, g/m^{-3}	τ_0	A_Σ
1	Ac	4.8	0.2	90	8	0.2	< 10	0.63
2	St	1.0	0.45	220	8	0.4	20	0.56
3	St	1.0	0.4	100	10	0.4	15	0.49
4	St	0.4	0.15	70	8	0.13	5	0.27
5	Cs	8.4	1.5	—	—	—	2	0.20
6	Sea	0.2						0.04

albedo and the visible range is higher than the integral albedo and the albedo in the infrared. The latter is selective in the region of molecular-absorption bands and depends on the microstructural parameters of the cloud.

The high albedos of altocumuli (Ac) of small optical depth
are due to the effect of clouds lying under them for $A_\Sigma \approx 0.45$.
In all other cases the measurements were carried out in the absence of underlying clouds. The spectral albedo of the sea,
measured from a height of 0.2 km, is shown in Fig. 10.4
(curve 6).

The albedo of an ice-crystal Cs cloud is low, because of
its small optical depth. The appreciable drop in albedo of this
cloud toward higher wavelengths can be attributed to the effect
of the lower-lying layer of the atmosphere and the sea surface.

Measurements of albedos of stratiform clouds of approximately equal optical depth, carried out over the sea for various
Sun heights, provided an approxi-
mate picture of the variation of
the cloud albedo as a function of
the Sun's height $\Delta A/\Delta h_\odot$, which
turned out to be about $0.004~(°)^{-1}$
for low Sun heights (8–20°). The

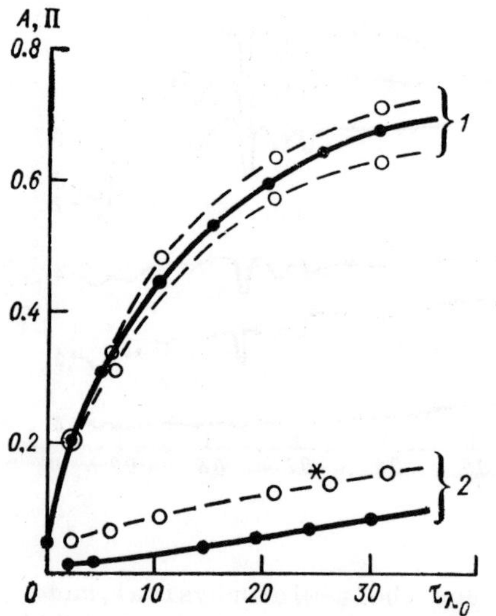

Fig. 10.5. Albedo (1) and absorptivity (2) of stratiform cloud
as functions of optical depth
τ_{λ_0}, for $h_\odot = 52°$ and $\lambda_0 =$
0.7 μm. Asterisk denotes absorp-
tivity for heightened aerosol content. Upper dashed curve 1 shows
calculated A_V, lower curve shows
calculated A_Σ; dashed curve 2
shows absorption (see Section
10.5).

variability of the albedos of stratus clouds in the range of Sun
heights from 45 to 55° was practically zero.

These results are in accordance with the estimates made in
Section 10.1, as well as with theoretical estimates (see Section
10.5).

The presence of atmospheric haze above the clouds reduces
the albedo of the cloud-atmosphere system, and this is especially
noticeable in the range from 0.75 to 2.5 μm [12, 19] (see also
Section 10.2).

In an experiment over the Sea of Azov for a Sun height of
around 38° [3], the upper part of the cloud was found to play an
active part in the absorption of radiation: the absorptivities of
the upper half of the cloud and the entire cloud layer were close
in value: the cloud layer as a whole absorbed 91 watts/m^2 of solar
energy, of which its upper half accounted for 84 watts/m^2.

At wavelengths from 0.3 to 3.0 μm the absorptivities of the
clouds studied varied from 1 to 20%, increasing when embedded
aerosols were present (see Section 10.4). The absorptivities of

Table 10.9. Integral albedo A_Σ and its ratio to spectral albedo A_Σ/A_{sp} (columns 3–10).

Date, cloud type, optical depth, Sun height, measurement site	A_Σ	λ, µm							
		0.369	0.391	0.457	0.676	0.4…0.65	0.6…0.92	0.76	0.69…0.74
1	2	3	4	5	6	7	8	9	10
4/IV 1971 Cs H=1500 m; $h_\odot$=48.3°; τ_0=2 Black Sea	0.20	0.83	0.90	0.86	0.89	0.87	1.08	1.33	1.05
6/IV 1971 St H=50 m; $h_\odot$=51.3°; τ_0=5 Lake Ladoga	0.27	0.87	0.96	0.89	0.91	0.90	0.93	0.99	0.93
9/IV 1971 As, St H=200/400 m; $h_\odot$=52.2°; τ_0=10/15 Kara Sea	0.63	0.94	0.97	0.95	0.94	0.95	1.02	2.05	0.95
10/IV 1971 St H=450 m; $h_\odot$=54.6°; τ_0=20 Sea of Azov	0.59	0.95	0.98	0.94	0.94	0.94	0.96	1.12	0.98
10/IV 1971 St H=400 m; $h_\odot$=52.7°; τ_0=20 Sea of Azov	0.56	0.98	0.97	0.93	0.93	0.93	0.96	1.12	0.98

Date, cloud type, optical depth, Sun height, measurement site	A_Σ	λ, µm							
		0.369	0.391	0.457	0.676	0.4...0,65	0.6...0.92	0.76	0.69...0.74
1	2	3	4	5	6	7	8	9	10
11/IV 1971 St H=350 m; $h_\odot$=52.1°; τ_0=15 Black Sea	0.49	0.94	0.94	0.92	0.94	0.94	0.98	1.13	0.97
24/IX 1972 As, St H=1900/1200 m; $h_\odot$=26°; τ_0=30/50 Lake Ladoga	0.71	0.92	0.94	0.98	0.98	0.98	1.02	1.59	1.08
1/X 1972 Sc H=300 m; $h_\odot$=16°; τ_0=15 Kara Sea	0.66	0.93	0.93	0.97	0.93	0.98	0.92	1.05	0.92
5/X 1972 Sc H=450 m; $h_\odot$=38°; τ_0=20 Sea of Azov	0.65	0.89	0.91	0.97	0.98	0.99	1.00	1.21	1.00
Mean A_Σ/A_{sp}		0.91	0.95	0.94	0.95	0.95	0.99	1.18	0.98
σ^2		0.0023	0.0085	0.0021	0.0023	0.0020	0.0025	0.125	0.0028

ice-crystal clouds did not exceed 0.01-0.02. The absorption in clouds of different kinds ranged from 15 to 150 watts/m^2, and the radiation heating was from 0.1 to 1.0°C/h.

Figure 10.5 shows the cloud albedo and absorptivity as functions of the optical depth τ_λ for λ_0 = 0.7 µm, according to measurements over the Sea of Azov for $h_\odot$ = 52°.

For a number of sets of measurements, Table 10.9 gives the integral albedo and the ratio of it to the spectral albedo A_{sp}. The values of A_Σ were averaged over at least 2 min of flying time, and the A_{sp} values were each based on three pairs of 10-second flux spectrograms, also averaged over at least 2 min. The table also gives the mean ratios A_Σ/A_{sp} and their variances σ^2.

From these data a conclusion which is important for radiation kinetics can be drawn. This conclusion, obtained earlier on the basis of numerical modeling (see Section 9.2 and [21]), states that the integral albedo of the clouds can be equated to the albedo in the visible range (outside of the molecular-absorption bands) with an error of the order of 10%.

Table 10.9 shows that ratios A_Σ/A_{sp} are stable and practically independent of τ.

10.4. Radiation properties of urban clouds

The presence of an aerosol in clouds may alter their radiation characteristics and increase the absorptivity in the visible range [19, 20, 25, 29]. In [20, 25, 29] direct measurements of the aerosol were not carried out, its influence being estimated only indirectly or by inverse-problem methods (see Section 4.6).

Within the framework of the GAARS program [11], in December 1978 aircraft measurements were made over major industrial centers (the cities of Zaporozhe and Donetsk) in order to investigate the effect of a city on the radiation regime of the atmosphere in the presence of subinversion clouds. These studies were carried out with two flying laboratories of the Main Geophysical Observatory (one Il-18 and one Il-14). The Il-18 was intended mainly for radiation measurements, and the Il-14 mainly to study the microphysical and macrophysical parameters of the clouds, the aerosol, and the condensation nuclei. The two aircraft operated simultaneously.

The aircraft performed vertical soundings at heights from 500 to 7200 m over the city and beyond it, on both the windward and leeward sides. Radiation measurements were made 200-300 m above the upper boundary, inside the cloud, and along its lower boundary. The effect of the albedo of the underlying surface was allowed for using a method proposed in [10] (see formula (10.3)). During the measurements single-layer stratiform clouds were present (6 and 11 December) or the skies were clear (8 and 9 December) over a snow-covered surface, or else clouds were present (16, 18, and 20 December) or the skies were clear (19 December) over a weakly reflecting surface. The temperature inside the cloud varied from

1 to 13°C, the relative humidity was from 60 to 100%, and the water content was from 0.9 to 0.13 g/m^3.

A chemical analysis of the samples of cloud water and snow, carried out on water-soluble admixtures [pollutants], revealed a quite high mineralization of the cloud water. There was a definite drop in the admixture concentration of the cloud water with a raising of the sampling altitude. The sample taken at a height of about 900 m was the most polluted. The SO_4^{2-} ions predominated in it, these accounting for more than 90% of the total anions. The main cation was NH_4^+, about 50% of the total cation count.

The smell of anthropogenic sulfur compounds was perceptible 500 m above the city. An analysis of soluble cloud-water components revealed the presence of the sulfate ion SO_4^{2-}, ammonium sulfate $(NH_4)_2SO_4$, and $CaSO_4$, and also the possible presence of dilute solutions of sulfuric acid, nitrogen peroxide, and other ions. The mineralization level and the content of organic substances in clouds over rural areas proved to be 1/2.3 of the values for clouds over a city.

The background concentration of Aitken nuclei was $1\cdot10^2$ to $3\cdot10^2$ cm^{-3}. At a height of 300 m in clear weather over the city the concentration was $3\cdot10^4$ to $21\cdot10^4$ cm^{-3}, in the subcloud layer over the city it was $1\cdot10^4$ to $3\cdot10^4$ cm^{-3}, and at a height of more than 2 km it was $1\cdot10^3$ to $2\cdot10^3$ cm^{-3}. The spectrum of aerosol particles over a city is close to lognormal.

Clouds over a city are characterized by a higher content of condensation nuclei, compared to that on the windward side of the city, and thus they also have a higher droplet concentration. This concentration rise is promoted by the presence of submicron-sized particles of H_2SO_4 solution, for high relative air humidities. Chemical analyses of precipitation also attest to a significant pollution of the atmosphere. For example, two samples of freshly fallen snow collected in the Dnepropetrovsk region revealed a lower mineralization compared to the water of subinversion clouds, but at the same time SO_4^{2-}, NH_4, and Ca^{2+} ions, primarily of anthropogenic origin, were preponderant in these samples. The data on the chemical composition of water in subinversion clouds is quite typical of the southwestern part of the European USSR and is in good agreement with the results of earlier studies [18].

The cloud-water samples were dark in color and, in addition to the high concentration of soluble admixtures, they contained a large amount of insoluble black soot particles as well as other organic and inorganic pollutants. The amount of organic material, which included carbon compounds, was quite significant, being 30-50% of the mass of the entire insoluble precipitation. Raman spectroscopy of the urban aerosol and combustion products from boiler-room furnaces and automobile engines indicates that the graphitic components of the aerosol are identical in the visible part of the spectrum. Precipitation spectrograms obtained after calcination clearly show absorption lines of silicon, iron, copper and aluminum.

Figure 10.6 shows the integral albedo of a system consisting of a cloud plus a subcloud layer as a function of the geometrical depth H of the cloud layer and the optical depth τ_0, for Sun heights $h_\odot$ from 19 to 20°. Curve 1 pertains to conditions over a city and curve 2 to a place on the upwind side of a city. The conversion from H to τ_0 was carried out using data of [17] (see Section 4.4). The figure also shows the albedo of a cloud–atmosphere system over a city (curve 3) and outside of it (curve 4) as a function of the flight altitude z. These curves reveal definite albedo differences (greater than the error in the albedo determination) at least up to a level of 4.5 km. It should be noted

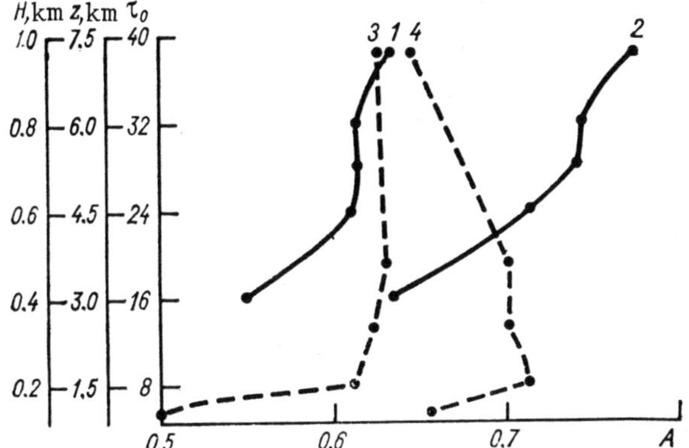

Fig. 10.6. Mean dependence of integral albedo of system consisting of cloud and subcloud layer on geometrical depth H of cloud layer and optical depth τ_0 for Sun heights of 19–20° (curve 1 over city; curve 2 outside of city.) Curve 3 (over city) and curve 4 (outside of city) show albedo of cloud–atmosphere system as function of height z (6 December 1978).

that over the city there were certain narrow regions of elevated albedo of the cloud–aerosol layer, associated perhaps with a predominant influence of sources of hygroscopic particles and with an increase in the concentration of the finely dispersed fraction of cloud droplets. The albedo of the cloud–atmosphere system decreases with observation height by 0.01–0.02, on the average, for each kilometer of layer thickness.

Figure 10.7 shows the ratio of the albedo over the city A_c to the albedo outside of the city A_0 as a function of H and τ_0 (curve 1), as well as the contribution of the atmospheric layer above the cloud (curves 2–5) to A_c/A_0. The measurements were carried out over a snow-covered urban surface and over a place outside of town with mean albedos (from a height of 500 m) of 0.57 and 0.63 (curve 4), as well as after the sbow had melted over the city and over a ploughed field with albedos of 0.13 and 0.09 (curves 2, 3, 5).

Table 10.10 gives the radiation characteristics of clouds above a city (subscript c) and outside of it (subscript o): albedo A, absorptivity Π, heat influx due to absorption of solar radiation ΔB_{sw}, rate of radiation-caused temperature variation $(dT/dt)_{sw}$, heat influx due to absorption of long-wave radiation ΔB_{lw}, total radiation-caused heat ΔB_t, and total rate $(dT/dt)_t$. Ratio A_c/A_o = 0.82–0.89 for τ = 16 to 38 and ζ = 70°. Calculations show that for the city N_p/N_{pb} = 20 and w = 0.5, whereas outside the city N_p/N_{pb} = 5 and $\tilde{w}_a$ = 0.8, so that A_c/A_o = 0.85–0.9 for τ = 20 and $\cdot\zeta$ = 70°. The radiation measurements were made with Sun heights of 19 to 20°, that is, in the angle region at the limit of the deviation from the cosine law for the pickup areas of the radiation instruments. Consequently, a statistical processing of the data gave albedo values with a relative error of not less than 6%, and it gave the radiation budget with an error up to 30%, and the heat influx up to 70 or 80%; the data on the rate of the radiation-caused temperature variation are tentative. Nevertheless, these

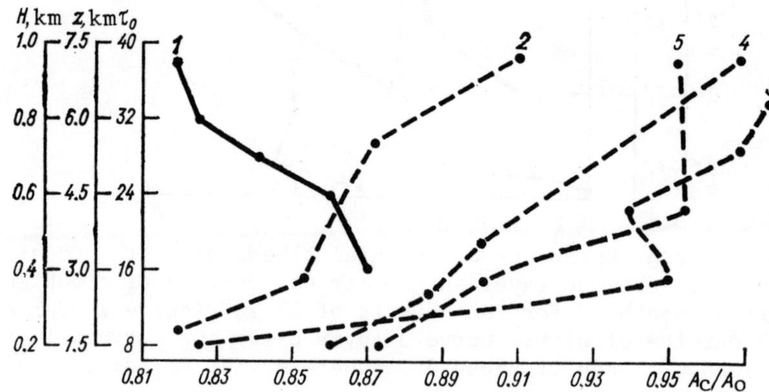

Fig. 10.7. Ratio of albedo over city A_c to albedo outside of it A_o as function of H and τ_0 (1) and as function of z, on 18 (2), 20 (3), 6(4), and 16 (5) December 1978.

data show clearly that anthropogenic urban factors make the absorptivity and radiation heating of clouds approximately twice as high as outside of the city. Assuming identical cloud depths in the measurement region, this difference can be attributed to the presence of the anthropogenic aerosol, the imaginary part $\varkappa$ of the complex refractive index of which is several orders of magnitude greater than the corresponding values for pure water. Aircraft measurements of the absorption by a cloudless atmosphere at a wavelength of 0.5 μm indicate that $\varkappa$ may reach 0.03 over Zaporozhe [4]. The optical constants of a cloud-aerosol medium have not yet been sufficiently studied.

Figure 10.8 shows the measured and calculated spectral and integral absorptivities of stratiform clouds as functions of the

Table 10.10. Radiation characteristics of clouds over city and outside it for various geometrical and optical depths of clouds with Sun height of 19 to 20°. December 1978.

Characteristic	16 XII St	6 XII Sc	11 XII Sc	20 XII Sc	18 XII Sc
H_{ub} m	600	1100	1500	1400	1500
H m	400	600	700	800	950
τ	16	24	28	32	38
A_c	0.56	0.61	0.61	0.61	0.63
A_o	0.63	0.71	0.72	0.74	0.77
Π_c	0.11	0.09	0.15	0.23	0.27
Π_o	0.07	0.05	0.08	0.11	0.11
$\Delta B_{sw,c}$ watts/m²	41.88	76.78	62.82	76.78	125.64
$\Delta B_{sw,o}$ watts/m²	27.92	20.94	20.94	62.82	76.78
$(dT/dt)_{sw,c}$ °C/h	0.12	0.22	0.16	0.23	0.31
$(dT/dt)_{sw,o}$ °C/h	0.08	0.06	0.05	0.18	0.19
ΔB_{lw} watts/m²	−20.94	−69.8	−48.86	−90.74	−104.7
$\Delta B_{t,c}$ watts/m²	20.94	6.98	13.96	−13.96	20.24
$\Delta B_{t,o}$ watts/m²	6.98	−48.86	−27.92	−27.92	−13.96
$(dT/dt)_{t,c}$ °C/h	0.06	0.02	0.04	−0.04	0.05
$(dT/dt)_{t,o}$ °C/h	0.02	−0.14	−0.03	−0.08	−0.04

relative optical depth. The latter is normalized to the mean optical depth of St–Sc clouds, equal to 20 according to [10] (see Sections 4.1 and 10.2).

The spectral data in Fig. 10.8 are prelimary. Almost all the measurement data give measured absorptivities that are higher than

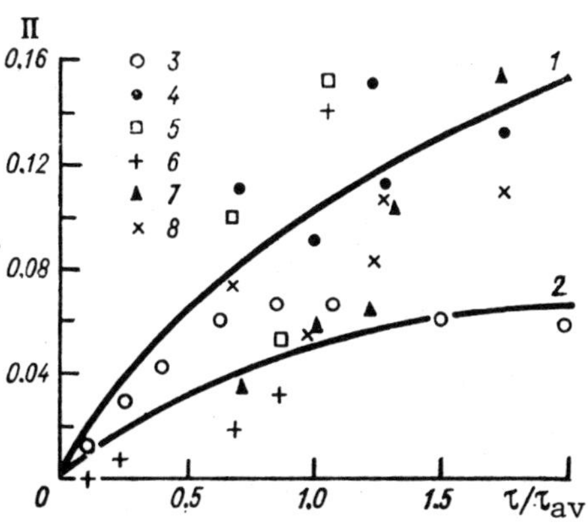

Fig. 10.8. Absorptivity Π as function of relative optical depth τ_o/τ_{av}.
1) calculation allowing for aerosol absorption, 2) molecular integral absorption in cloud, 3) integral data according to [10], 4) pyranometric data (last two points on right have values twice as high as those in figure, 5) integral data, 6) at wavelength of 0.5 μm (over Black Sea), 7) wavelength of 0.5 μm (over Zaporozhe), 8) actinometric data outside city.

the calculated values corresponding just to absorption by water vapor. The measured absorptivities of the aerosol in clouds at $\lambda = 0.5$ and 0.7 μm are commensurate with the molecular absorption. The absorption over an industrial center is generally higher than over the Black Sea. On the whole, the results obtained over the cities of Zaporozhe and Donetsk attest to the presence of an optically active aerosol in the clouds, which causes a sizable drop in the albedo of the cloud-aerosol layer, in comparison with a cloud upwind from the city, and an increase in absorptivity. The radiation cooling of the entire cloud depth outside of the city may give way to a slight heating of the same cloud over the city.

10.5. Comparison of calculations with data of aircraft measurements

The measurement data of Sections 10.1–10.3 were compared with the theoretical results of Chaps. 8 and 9, with the data of [21], and also with certain other calculations (see, for example, Fig. 10.8). The mean radiation model of St-Sc clouds described in Section 10.2 is the most appropriate for comparison with theory. These clouds are assumed to have the following parameter values: $H = 0.4$ km, $z_{ub} = 0.8$ km, $t = -4.0°C$, $\rho v = 3$ g/m^3, $\bar{w} = 0.2$ g/m^3 (see Table 10.4), $\sigma = 50$ km^{-1}, $\tau_0 = 20$ (see Chap. 4). The measurements were made with $40° \leq \zeta \leq 75°$ and a mean zenith angle $\bar{\zeta} = 50°$.

The albedo of a so-called average cloud $A^* = 0.74$, and the absorptivity of its entire thickness $\Pi^* = 0.072$ (see Table 10.5). Calculations for these same cloud parameters, for $\zeta = 50°$ and allowing for attenuation of solar radiation in the atmosphere above the clouds, give $A^* = 0.72$.

The calculation was carried out using the approximate method of Section 9.3 (see formulas 9.18–9.19). The working formula has the form

$$A^* = \frac{F_{ub}^{\uparrow}\left(\zeta,\ \tau^{(1)},\ \tau_0,\ m_v^{(1)},\ m_w\right)}{F_{ub}^{\downarrow}\left(\zeta,\ \tau^{(1)},\ m_v^{(1)}\right)} \approx$$

$$\approx \frac{1}{I_0} A_v \left\{ I_0^{vis} + I_0^{ir}\ \frac{P\left\{\left[\sec \zeta m_v^{(1)} + \rho_v l^{\uparrow} \frac{p}{p_0}\right]\right\},\ \overline{w}\, l^{\uparrow}}{P\left\{\sec \zeta m_v^{(1)}\right\}} \right\}. \quad (10.14)$$

Here $I_0^{vis} + I_0^{ir} = I_0$ is the solar constant, A_v is the "visible" albedo, and $l^{\uparrow}$ is the mean photon path length in the cloud, for reflected radiation. For $\tau_0 = 20$, $\sigma = 50$ km^{-1}, and $\zeta = 50°$ the data in Table 8.4 give $A_v = 0.80$; according to Table 6.1 $l^{\uparrow} = 0.80$ km. Parameter $m_v^{(1)}$ in (10.14) is the water-vapor content in the atmosphere above the clouds. According to Table 10.4, $m_v^{(1)} = 1.35$ cm. The effect of reflection from the Earth's surface is not taken into account in formula (10.14), since the calculation result is compared with the "true" albedo (see Section 10.1).

The calculated integral gaseous absorption in an average cloud was found to be 0.06.

In Figs. 10.1 and 10.2 calculations made with formula (10.14) for $m_v^{(1)}$ = 0 and ρ_v = 5 g/m^3 are compared with mean data of albedo measurements as a function of the depth and water content of the cloud layer. The theoretical values of A_V in formula (10.14) as a function of τ were taken from Table 8.4; then the relation $\tau = \sigma H w$ was invoked for σ = 2500 cm^2/g (see Chap. 4) and w = 0.2 g/m^3 for variable H (see Fig. 10.1) or H = 0.4 km for variable w (see Fig. 10.2). Similar comparisons are shown in Fig. 10.5. Here the measured integral albedo is compared with A_V and A_Σ as a function of τ_{λ_0}, the optical depth at λ_0 = 0.7 μm. The calculated gaseous absorption is also compared with the measured absorption.

The albedo of a cloud layer varies appreciably as a function of the height of the Sun. In Section 10.1 it was shown that $dA/d\zeta$ = 0.003 deg^{-1} (the measurements were carried out for $\zeta \geq 40°$). According to the data of Section 10.3, $dA/d\zeta$ = 0.004 deg^{-1} for $\zeta \geq$ 60° and $dA/d\zeta \approx 0$ for large Sun heights. Theoretical data (see Chaps. 8 and 9) give 0.001 $\leq dA/d\zeta \leq$ 0.003 for 50 $\geq \tau \geq$ 5. For a low Sun (ζ = 60-75°) we have 0.004 $\leq dA/d\zeta \leq$ 0.009 in this same range of τ_0.

The calculated and measured radiation heating of an average cloud layer are compared in [21]. There is a good fit between calculation and measurement everywhere except in the vicinity of the upper boundary. The comparatively low measured absorption there may be associated with the horizontal inhomogeneity of the upper boundary of the cloud layer. Moreover, absorption in the atmosphere above the clouds evidently has an effect, not taken into account in these calculations.

These comparisons, together with the diverse results of similar comparisons described in [21], reveal a satisfactory correspondence between the mean measured cloud albedos and the calculated values. Naturally, there may be sizable discrepancies between individual measured and calculated values [21]. A low absorptivity of the clouds, that is, a low difference 1 - ($A + T$), is obtained from the measured and calculated values of A and T with considerable error. Nevertheless, as Fig. 10.5 shows, and also certain data in [21], the calculated and measured absorptions agree within reasonable limits of error. The above-described calculations did not allow for aerosol attenuation of the radiation in the atmosphere above the clouds; however, this attenuation may well be inconsiderable, since it enters almost equally into the numerator and denominator of formula (10.14). Aerosol absorption in clouds is not taken into account either. The latter reduces A_V in accordance with Table 8.4 and tends to make the calculated A_Σ too low, in comparison with the measured value, an effect which is only slightly perceptible in Figs. 10.1, 10.2, and 10.5.

In view of our findings, it is strange indeed that in non-Soviet publications (see, for instance, [29, 30]) it is often stated that the calculated values of A are too high compared with the measured values, and the calculated Π are too low. The discrepancy between measurements and calculations cited in these and other works is apparently due to various shortcomings related to the statement of the theoretical problem: insufficient initial information; ignoring the multiple lengthening of photon paths in clouds, under clouds, and between clouds; and, finally, the fact that real clouds differ from the ideal plane layers considered in the calculations. As shown in Section 4.8, taking fluctuations of the optical depth into account reduces the albedo.

CHAPTER 11

SOLAR FLUXES IN THE PRESENCE OF CUMULUS CLOUDS

The transfer of radiation in the presence of cumulus clouds depends both on the optical properties of individual clouds and on the structure of the cloud field. A theoretical computation of the random size distribution of cumuli and their distribution in space leads to a radiation-transfer equation the parameters of which are random functions. Fundamental difficulties arise in the solution of such an equation.

So far only model problems have been considered, and these can be divided into the following two types:

1. Models based on a calculation of the optical and radiation properties of individual cumulus clouds. The radiation regime of the cloud field is calculated approximately by adding up the radiation fields of individual clouds, taking into account their size distribution and their distribution in space (see Sections 11.1 and 11.6).

2. Models based on solution of a statistical transfer equation. In this case, too, simplifications are unavoidable, consisting mainly in approximately allowing for both multiple scattering in individual clouds and the spatial distribution of clouds (see Sections 11.1-11.3 and also [4, 6]). Without invoking transfer theory, it is possible to construct an experimental-theoretical statistical model of the radiation regime of the atmosphere in the presence of cumulus clouds, based on the chacateristics of these clouds described in Chap. 3 and on the methods of the theory of random fluctuations (see Section 11.2). Finally, the empirical structure of the radiation field in the presence of cumuli can be studied directly with the aid of measurement data (Sections 11.3 and 11.4).

11.1. Radiation regime of isolated cumulus cloud

As a first approximation, a cloud can be considered to be a volume filled with a homogeneous scattering medium and bounded by a convex surface W (for instance, a cylinder, sphere, truncated paraboloid, or cube). The radiation transfer inside such a volume is calculated with the aid of the Monte-Carlo method, the application techniques of which are discussed in [2]. All calculations were carried out for a scattering coefficient $\sigma = 30$ km^{-1}, unless otherwise indicated. The scattering function $\gamma_1(\mu)$ is "narrow" (see Chap. 4). The optical depth of the cloud $\tau = \sigma H$, where H is the cloud height. Absorption of radiant energy is not taken into account.

177

The calculations give the following characteristics of the cloud's radiation regime:

a) the relative emissivity of the cloud surface

$$E(P) = \frac{1}{2\pi} \int_0^{2\pi} \int_0^1 J(P, \mu, \psi)\, \mu\, d\mu\, d\psi, \qquad (11.1)$$

where $J(P, \mu, \psi)$ is the brightness of the surface, relative to the incident flux of direct solar radiation; $\mu = \cos\theta$, angle θ being reckoned from the normal to the surface at point P; angle ψ describes the rotation about this normal;

b) the scattering function of the cloud, considered as some point scattering element

$$\Gamma(\mu, \psi) = \int_{W'} J(P, \mu, \psi)\, dP, \qquad (11.2)$$

where W' is the set of points P of surface W for which rays drawn in direction (μ, ψ) from points P are directed outward;

c) the asymmetry coefficient, given by the formula

$$\gamma = \int_0^{2\pi} \int_0^1 \Gamma(\mu, \psi)\, \mu\, d\mu\, d\psi \bigg/ \int_0^{2\pi} \int_{-1}^0 \Gamma(\mu, \psi)\, \mu\, d\mu\, d\psi; \qquad (11.3)$$

d) the reflectivity (albedo) A of the cloud and its transmittance T:

$$A = \frac{1}{2\pi} \int_0^{2\pi} \int_0^1 \Gamma(\mu, \psi)\, \mu\, d\mu\, d\psi, \qquad (11.4)$$

$$T = \frac{1}{2\pi} \int_0^{2\pi} \int_{-1}^0 \Gamma(\mu, \psi)\, \mu\, d\mu\, d\psi. \qquad (11.5)$$

The effect of the geometrical shape of the clouds was studied for these principal radiation characteristics. Figure 11.1 shows the albedos of clouds of various shapes as functions of the zenith distance of the Sun $\theta_\odot$, assuming a constant optical cloud volume $V_{opt} = c\tau_0^3$, where coefficient c depends on the cloud shape. The effect of the shape is greatest for low $\theta_\odot$, and for intermediate or high values it does not exceed 20%. The effect of the cloud shape also diminishes with a reduction of τ_0 (Fig. 11.2). The optical regimes for cylindrical, cubic [28, 30], and paraboloidal (for $\theta_\odot > 20°$) clouds resemble one another most. In the subsequent calculations the cloud is assumed to be paraboloidal, unless otherwise indicated.

For high τ_0 the value of A ranges from 0.5 to 0.6 and $\lim\limits_{\tau_0 \to \infty} A < 1.$

This is due both to edge effects and to the presence of nonhorizontal bounding surfaces. Since for a spherical shape edge effects are unimportant, the albedo is considerably higher.

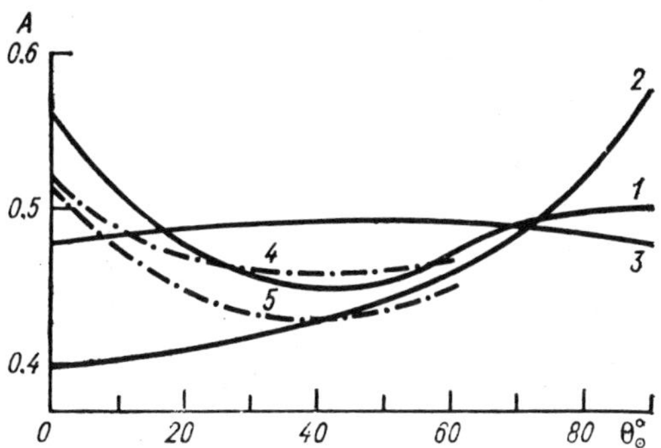

Fig. 11.1. Cloud albedo as function of Sun's position. $V_{opt} = 2 \cdot 10^4$.
1) cylinder, 2) paraboloid, 3) sphere, 4) cube [28], 5) cube [30].

Above we considered clouds for which the ratio of the height H to the base diameter D was equal to unity. Fig. 11.3 shows the albedo as a function of cloud diameter for heights from 0.1 to 2 km.

Depending on the ratio between the vertical and horizontal cloud dimensions, the reflectivity can vary over a very wide range, especially for low σ. Albedos of cumuli are shown as functions of ratio D/H in Fig. 11.4, together with albedos of layers with the corresponding optical depth. The albedo of a finite cloud approaches the albedo of a layer more rapidly for low optical depths and low zenith distances of the Sun than for high values of these parameters.

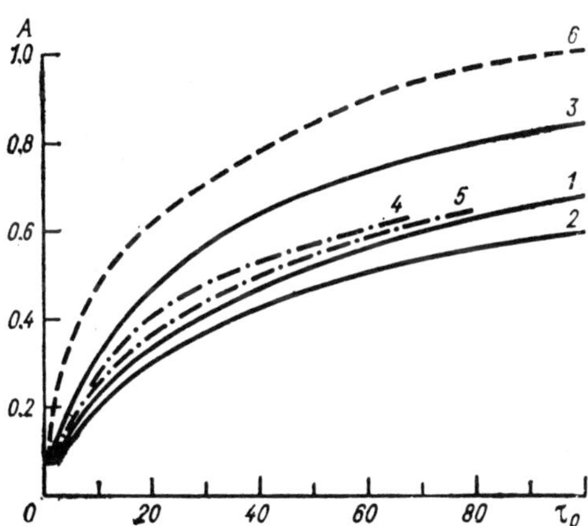

Fig. 11.2. Albedo as function of optical depth for $\theta_\odot = 30°$. Curves 1–5 pertain to cumulus clouds (for arbitrary notation see Fig. 11.1), and curve 6 to a stratiform cloud.

If a cumulus cloud with a parallel flux incident upon it is surrounded by identical clouds capable of redistributing the radiation scattered by the first cloud, then the optical regime of this cloud will vary appreciably. For instance, according to [28], the presence of four additional clouds each located at a distance equal to the diameter

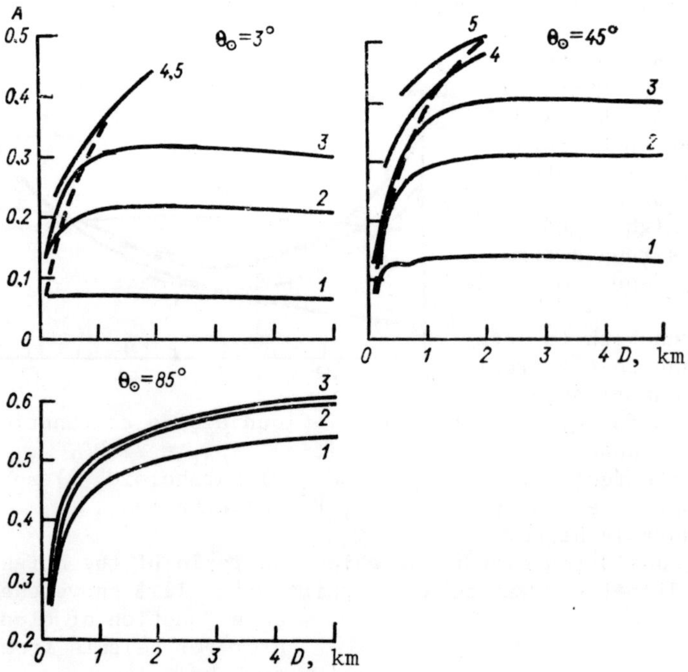

Fig. 11.3. Cloud albedo as function of base diameter, for $\sigma =$ 30 km^{-1}.
1) $H = 0.1$ km; 2) $H = 0.3$ km; 3) $H = 0.5$ km; 4) $H = 1$ km; 5) $H =$ 2 km. Dashed curve $D/H = 1$.

of the central cloud increases the albedo by 20–25% for an optical depth of the clouds for $\tau_0 = 4.9$, and by 12–15% for $\tau_0 = 49$. The surrounding clouds stop having an effect on the central cloud when they are 5 diameters or more away from it.

Vertical profiles of the scattering coefficient of a cumulus cloud (see Chap. 4) are given in [14, 24]. For these profiles calculations were made and the characteristics of the optical regime were compared for homogeneous and inhomogeneous clouds. The corrections for inhomogeneity turned out to be small, so that as a first approximation a model of a homogeneous cloud can be used to calculate the integral radiation characteristics.

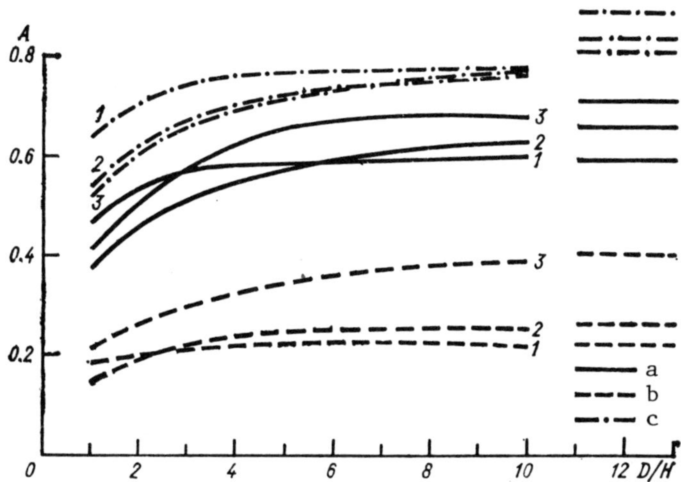

Fig. 11.4. Cloud albedo as function of ratio D/H.
a) $\tau_0 = 20$ (paraboloid), b) $\tau_0 = 5$ (cube) [28], c) $\tau_0 = 50$ (cube) [28].
1) $\theta_\odot = 0$, 2) $\theta_\odot = 30°$, 3) $\theta_\odot = 60°$. Line segments at right give albedos of layers under corresponding conditions.

11.2. Statistical structure of fluxes for broken cloudiness

Let us consider the variability of radiation fluxes, making use of information about the statistical structure of cumulus fields (see Chap. 3 and [13, 18]). Fluxes of solar radiation exhibit a pronounced diurnal variation. In order to exclude this effect, during the experimental studies [10, 11, 13] dimensionless relative fluxes were analyzed (see Section 11.4). For the theoretical analysis, since the correlation radii for the fluxes do not in most cases exceed 10 min, the absolute fluxes can also be used [18].

11.2.1. Direct radiation

As shown in [13, 17, 18, 19, 20], the mean flux of direct solar radiation in the presence of cumuli can as a first approximation be written as

$$S_{cld} = S_{no\,cld}\,[1 - n\,(\zeta)], \qquad (11.6)$$

where the amount of cloudiness in the Sun's direction $n(\zeta)$ can be computed using formula (3.1) or (3.3), and $S_{no\ cld}$ is the flux of direct radiation for a cloudless sky.

The variance of the direct flux is given by a formula allowing for the "semitransparency" of the peripheral parts of the clouds [13, 19]. According to the data of [13],

$$\sigma_S^2 = 0.64 \frac{\varkappa\,[n\,(\zeta)]}{\varkappa\,[n\,(\zeta) = 0.5]}\,S_{no\ cld}^2\,\sigma_{n\,(\zeta)}^2, \qquad (11.7)$$

where $\varkappa$ and $\sigma_{n(\zeta)}^2$ are found using formulas (3.7) and (3.15).

If $n(\zeta)$ varies from 0.2 to 0.9, the normalized correlation functions for the direct fluxes are approximated well by formula (3.16), and with a lower accuracy by an exponential (see Section 11.4).

11.2.2. Scattered radiation

The mean fluxes of scattered radiation depend appreciably on the amount of cloudiness and on the optical depth of the clouds, which varies from case to case. Consequently, more precise algorithms, taking parameters n and τ_0 into account, have to be used when computing the mean fluxes of scattered radiation (see preceding section of this chapter).

In order to gauge the variability, it is convenient to write the flux of scattered radiation as [13]

$$D_{cld}(x) = 2\pi \int_0^{\pi/2} [J_{cld}(\theta) - J_{no\ cld}(\theta)]\,n\,(\theta,\ x)\,d\cos\theta + D_{no\ cld}, \qquad (11.8)$$

where the intensity of scattered radiation for a cloudless sky $J_{no\ cld}(\theta)$ and for cloudiness $J_{cld}(\theta)$ are integrated over the azimuth, and x is the space coordinate. The difference $J_{cld}(\theta) - J_{no\ cld}(\theta)$ is obviously most significant at the zenith ($\theta = 0$) and decreases toward the horizon. As a first approximation

$$J_{cld}(\theta) - J_{no\ cld}(\theta) = [J_{cld}(0) - J_{no\ cld}(0)]\cos\theta. \qquad (11.9)$$

It should be noted that other dependences of the brightness difference on the zenith angle are considered in [13, 18].

From (11.8) and (11.9), averaged over x, we obtain the following expression for the flux of scattered radiation:

$$D_{cld} = \frac{2}{3}\,\pi\,[J_{cld}(0) - J_{no\ cld}(0)]\,n\,(0) + D_{no\ cld}. \qquad (11.10)$$

The correlation function of the scattered flux is obtained with the aid of the inverse Hankel transform of the filtered spectral density

$$r_D(x, z) = \frac{2\pi}{K_e(z)} \int_0^\infty H_2^2(\omega) S_{2, n(0)}(\omega) J_0(\omega x) \omega \, d\omega, \qquad (11.11)$$

where $S_{2, n(0)}(\omega)$ is found using formula (3.18); z is the height of the lower boundary of the cloud layer; $J_0(\omega x)$ is a zero-order Bessel function; and $K_e(z)$ is an efficiency factor characterizing the decrease in the variance of the scattered radiation $\sigma_D^2(z)$ with increasing z:

$$K_e(z) = \frac{\sigma_D^2(z)}{\sigma_D^2(z=0)} = \frac{\int_0^\infty H_2^2(\omega) S_{2, n(0)}(\omega) \omega \, d\omega}{\int_0^\infty S_{2, n(0)}(\omega) \omega \, d\omega} =$$

$$= 2\pi \int_0^\infty H_2^2(\omega) S_{2, n(0)} \omega \, d\omega. \qquad (11.12)$$

In formulas (11.11) and (11.12) function $H_2(\omega)$ is the spectral characteristic of integral transform (11.8), taking (11.9) into account:

$$H_2(\omega) = (\omega z + 1) e^{-\omega z}. \qquad (11.13)$$

The variance of the two-dimensional spectral density are, respectively,

$$\sigma_D^2(z) = \frac{4}{9} \pi^2 (J_{cld} - J_{no\,cld}) \sigma_{n(0)}^2 K_e(z) \qquad (11.14)$$

and

$$S_{2, D}(\omega) = \frac{1}{K_e(z)} H_2^2(\omega) S_{2, n(0)}(\omega). \qquad (11.15)$$

11.2.3. *Total radiation*

Using the mean fluxes of direct (11.6) and scattered (11.10) radiation, we get the mean total flux in the form

$$Q_{cld} = S_{cld} + D_{cld}. \qquad (11.16)$$

The variance of the total radiation is

$$\sigma_Q^2 = \sigma_S^2 + \sigma_D^2 + 2\sigma_S\sigma_D r_{S,\,D}\,(X=0), \qquad (11.17)$$

and the correlation function of the total radiation is written as

$$r_Q(x) = \frac{\sigma_S^2 r_S(x) + \sigma_D^2 r_D(x) + 2\sigma_S\sigma_D r_{S,\,D}(X)}{\sigma_Q^2}. \qquad (11.18)$$

Function $r_{S,D}(X)$ in formulas (11.17) and (11.18) is the normalized cross-correlation function for the fluxes of direct and scattered radiation:

$$r_{S,\,D}(X) = \pm\,\frac{2\pi}{\sqrt{K_e(z)}}\int\limits_0^\infty H_2(\omega)\,S_{2,\,n(0)}(\omega)\,J_0(\omega X)\,\omega d\omega, \quad (11.19)$$

where the distance of the cloud field

$$X = \sqrt{x^2 + z^2\,\mathrm{tg}\,\zeta - 2xz\,\mathrm{tg}\,\zeta\,\cos\psi_*}. \qquad (11.20)$$

Here ψ_* is the azimuth of the Sun relative to the x axis.

Parameters S_2, $n(0)$, $K_e(z)$, and $H_2(\omega)$ in formula (11.19) are found using formulas (3.18), (11.12), and (11.13), respectively.

In (11.19) a plus sign indicates that $J_{cld}(0) < J_{no\ cld}(0)$, and a minus sign that $J_{cld}(0) > J_{no\ cld}(0)$. Accordingly, the correlation radius of the total radiation flux is larger or smaller than the correlation radius of the direct radiation, as is also borne out by measurement data [18, 20].

11.3. Parametrization of radiation regime of cumulus field using experimental aircraft data

In order to investigate the short-wave radiation regime of a field of cumuli, the layer being studied is divided into three parts: subcloud layer, cloud layer, and layer above the clouds. The cloud layer is defined as the layer within which the cumulus clouds develop. In our studies this layer was located between the 1 and 3 km levels. The influx of short-wave radiation to the cloud layer was determined with the aid of the equation

$$R_{sw}(z_u,\ z_l) = F_u^{\downarrow}\,[(1 - A_u) - T\,(1 - A_l)].\qquad (11.21)$$

Here subscripts u and 1 denote, respectively, the upper and lower boundaries of the clouds.

For broken clouds (fractocumulus, for instance) at low albedos of the underlying surface (in the absence of snow), the radiation entering the cloud layer from below can be neglected. Then the absorptivity of the cloud layer can be found using the formula (see Chap. 10)

$$\Pi_{eff} = \frac{R_{sw}(z_u,\ z_l)}{F_u^{\downarrow}\,(1 - A_u)} = 1 - \frac{T\,(1 - A_l)}{1 - A_u}.\qquad (11.22)$$

As we see from formula (11.21), parametrization of the influx of short-wave radiation to the cloud layer reduces to parametrizations of albedos A_u, A_l and transmission coefficient T.

For the albedo parametrizations 262 determinations of A_u and 241 determination of A_l were used. Each of these involved averaging over paths 70 to 140 km long, which, as follows from [18], is sufficient for this type of cloud. Most of the flights were in the Dnepropetrovsk region, about 90% of the measurements being made above fields of Cu hum. and Cu med., the other 10% being over Cu cong.

The relative amount of cloudiness (cloud cover) n and the

Table 11.1. Frequency (%) of measured values of albedo of upper boundary $A_u(n)$ of cloud layer for various amounts of cloudiness.

A_u %	Amount of cloudiness								
	0	1—2	2—3	3—4	4—5	5—6	6—7	7—8	8—10
5.0... 13.0	15.0	10.6	3.1	3.3					
13.1... 21.0	85.0	84.2	73.3	53.3	40.9	24.5	7.3		
21.1... 29.0		10.5	23.6	41.7	50.1	56.3	61.1	44.8	2.7
29.1... 37.0				7.1	6.0	15.8	24.3	30.9	10.8
37.1... 45.0					3.0	3.4	7.3	22.6	21.7
45.1... 53.0								1.7	27.0
53.1... 61.0									21.6
61.1... 69.0									10.8
69.1... 81.0									5.4
Mean A_u %	14.5	18.8	19.1	20.7	22.9	25.0	27.2	31.9	50.0

mean cloud depth H were taken as the parameters characterizing the state of the cloud field. The relative cloudiness n was ascertained visually from the ground or from an aircraft at a height of not less than 0.5 km above or below the clouds.

Table 11.1 indicates in considerable detail the variability of $A_u(n)$, in terms of the frequency of its values for various amounts of cumulus cloudiness. The values of $A_u(n)$ for $n = 0$ in the table are assumed to correspond to A_1.

In order to gauge the effect of n on A_u, we considered the differences $\Delta A(n) = A_u(n) - A_1$ and plotted the relation between the reciprocals $(\Delta A)^{-1}$ and n^{-1}, a relation which turned out to be close to linear. It was then possible to approximate ΔA as a function of n as follows [9]:

$$\Delta A(n) = \frac{n}{0.826 - 0.060 n},\qquad (11.23)$$

where $\Delta A(n)$ is in percent and n is in points on the 10-point scale. These results are described in more detail in [7-9].

The correlation coefficient of (11.23) $r = 0.92 \pm 0.01$; the rms error is $\pm 5.3\%$; the standard deviation of the calculated ΔA values from the actual values is $\pm 4.5\%$; and the arithmetic mean of the absolute deviation of these values is -3.3%. A check of parametrization formula (11.23) indicated it to be reliable.

An attempt to take into account the influence of the cloud depth H on ΔA and A_u did not meet with positive results: for a variation of H from 0.2 to 2.5 km, this effect was practically nonexistent ($r = 0.19 \pm 0.10$).

Table 11.2 gives the mean values and frequency of values of transmission coefficient T as a function of the amount of cloudiness. On the basis of these data, an attempt was made to devise formulas for the parametrization of T [8]. This led to the following expression

$$T = T_0 - \Delta T(n).\qquad (11.24)$$

Here T_0 is the coefficient of transmission of short-wave radiation by the cloud layer when there are no clouds in it, and $\Delta T(n)$ is the variation in T due to the presence of clouds.

The averaged values of $T(n)$ were used to plot the linear relationship between $(\Delta T)^{-1}$ and n^{-1}, after which T was approximated as a function of n by the formula

$$T = T_0 - \frac{n}{a - bn}.\qquad (11.25)$$

According to data for 1971-1974, for 94 cases a mean value of $T_0 = 95.1\%$ was obtained. Table 11.2 gives an idea of the

Table 11.2. Frequency (%) of values of transmission coefficient T of solar radiation by "cloud" layer as function of amount of cumulus.

$T\%$	Amount of cloudiness								
	0	1—2	2—3	3—4	4—5	5—6	6—7	7—8	8—10
99,0... 90.1	95.7	70.6	34.2	15.7	6.9	1.9			
90,0... 81.1	4.3	24.8	57.7	59.7	41.5	16.2	5.9	2.9	
81,0... 72.1		4.6	6.9	22.1	36.2	47.5	26.4	11.7	
72,0... 63.1			1.2	1.9	12.3	21.9	20.6	8.8	
63,0... 54.1				0.6	3.1	8.7	28.0	35.7	11.2
54,0... 45.1						3.8	17.6	20.6	16.7
45,0... 36.1							1.5	5.8	16.7
36,0... 27.1								5.8	27.8
27,0... 18.1								5.8	19.4
18,0... 12.1								2.9	8.2
No. of cases	94	109	158	159	130	105	68	34	36
Mean $T\%$	95	91	88	86	82	77	70	58	33

variability of this quantity. The close correlation between T and n is shown by the coefficient value $r = 0.82 \pm 0.011$. Coefficients a and b in formula (11.25) depend on the mean cloud depth. This relationship is approximated by the formulas

$$a = 0.14 + 0.15/H, \quad b = 0.0034 + 0.0121/H, \qquad (11.26)$$

where H is in kilometers.

A check of formulas (11.25) and (11.26) using independent material indicated that they are valid over the entire range of n and for depths H up to 3 km. The values of T calculated with them have absolute errors of $\pm$ 3-8% and relative errors of 3.7-24%.

Table 11.3 gives the frequency of values of the effective absorptivity Π_{eff} for various amounts of cumulus in the cloud layer.

The parametrization of Π_{eff} should be carried out taking into account the foregoing formulas relating A and T to the amount and depth of the cumulus cloudiness. Formula (11.22) can be written somewhat differently (all quantities in percent) as

$$\Pi_{eff} = \frac{\Pi_{eff}(0) + \left[\Delta T - \dfrac{\Delta A\,(n) \cdot 100}{100 - A_1}\right]}{1 - \dfrac{\Delta A\,(n)}{100 - A_1}} \qquad (11.27)$$

Here $\Pi_{eff}(0)$ is the absorptivity of the cloud layer when no clouds are present in it. On the basis of data for 1971–1974, for 80 cases a mean value of $\Pi_{eff}(0) = 4.8\%$ was obtained.

Table 11.3. Frequency (%) of values of effective absorptivity of cloud layer as function of amount of cumulus present.

Π_{eff} %	Amount of cloudiness								
	0	1—2	2—3	3—4	4—5	5—6	6—7	7—8	8—10
0.0... 9.0	97.5	85.6	68.8	54.8	33.2	15.3	14.1	11.0	
9.1... 18.0	2.5	12.6	27.5	33.0	46.5	54.0	24.2	6.0	2.8
18.1... 27.0		1.8	3.1	11.0	16.7	18.4	25.0	17.0	25.7
27.1... 36.0			0.6	1.2	2.9	6.8	18.5	24.0	11.6
36.1... 45.0					0.7	3.6	16.8	21.0	28.5
45.1... 54.0						0.9	1.4	3.0	11.6
54.1... 63.0								6.0	11.4
63.1... 72.0								12.0	5.6
72.1... 78.0									5.5
Mean Π_{eff} %	4.5	6.4	8.0	9.8	12.1	14.8	19.2	27.6	41.0

Formula (11.27) was checked using actual values of Π_{eff}, averaged for each point on the cloudiness scale (see Table 11.4). As can be seen from Table 11.4, the suggested approach gives a quite satisfactory result. The coefficient of correlation between the actual values of $\Pi_{eff}(n)$ and the values calculated for each specific case $r = 0.78 \pm 0.02$.

Table 11.4. Actual Π_a and calculated Π_c values of coefficients $\Pi_{eff}(n)$.

Para-meter	Amount of cloudiness									
	1	2	3	4	5	6	7	8	9	10
Π_a	5.2	7.5	8.5	11.0	13.2	16.5	21.9	39.3	42.0	46.0
Π_c	5.7	6.9	8.5	10.6	13.6	17.4	21.0	32.5	41.5	49.0

11.4. Variability of fluxes of short-wave radiation for broken cloudiness

Recordings of fluxes S, D, and Q at the Earth's surface, carried out by the authors during a 10-year period at various places in the European USSR (Leningrad Region, Estonian SSR, Moscow and Dnepropetrovsk Regions), served as initial material for a study of the statistical structure of the short-wave radiation [13, 15, 18, 21-23]. The relative fluxes were determined using the following relations from [10, 16]:

$$S^* = S\left(I_0 e^{-\tau_0 m_\odot}\right)^{-1},\qquad(11.28)$$

$$D^* = D\left(0.190 m_\odot^{-0.36} - 0.033\right)^{-1},\qquad(11.29)$$

$$Q^* = Q I_0^{-1}\left[m_\odot\left(1 + f m_\odot\right)\right],\qquad(11.30)$$

where I_0 is the solar constant, τ_0 is the optical thickness of the atmosphere, $m_\odot$ is the mass of the atmosphere in the Sun's direction, and f is a parameter depending on the optical state of the atmosphere and the albedo of the underlying surface.

The variability of the relative fluxes during the recording period T = 90-240 min was considered to be a random steady-state process, which could be processed using mathematical statistics [18, 23]. The characteristics computed according to individual events were compared with the amount of cloudiness n.

The mean values of the relative fluxes of total $\overline{Q}^*$ and direct $\overline{S}^*$ radiation for cumuli are linear functions of n; they are described by regression equations of the form

$$\overline{S}^* = 1.04 - n,\qquad(11.31)$$

$$\overline{Q}^* = 1.00 - 0.6n.\qquad(11.32)$$

The absolute error in calculating the mean fluxes using formulas (11.31) and (11.32) increases with n; it does not exceed 0.18 for $\overline{S}^*$ and 0.16 for $\overline{Q}^*$. The mean fluxes of relative scattered radiation $\overline{D}^*$ increase as the amount of cumulus rises to n = 0.7-0.8, when the mean scattered flux may be 2 or 3 times the flux D for a clear sky. As n increases further, the mean flux $\overline{D}^*$ becomes lower. It is difficult to ascertain $\overline{D}^*$ as a function of n, since fluctuations in the scattered radiations are greatly affected by, in addition to the amount of cloudiness, variations in the optical depth of the clouds and the amount and nature of the atmospheric aerosol.

The variances of S^* and Q^* are the highest for average amounts of cumulus cloudiness ($n \approx 0.6$), and they decrease for both high and low n. The variability of $\overline{Q}^*$ is mainly determined by variations in the direct radiation. However, in contrast to the direct radiation, the variance of which is zero, if $n = 0$ or 1.0 then variance σ^2_{Q*} does not approach zero upon transition from variable cloudiness to a continuous cloud cover, and for $n = 1$ it is determined by the variability of the scattered radiation. For $0.2 \leq n \leq 0.9$ the variances of the direct and total radiation are, to a first approximation, described by the empirical relations

$$\sigma^2_{S*} = 0.15 \sin\left[1.2\pi\,(n - 0.15)\right], \tag{11.33}$$

$$\sigma^2_{Q*} = 0.1 \sin\left[\pi\,(n - 0.1)\right]. \tag{11.34}$$

The probability densities of S^* and Q^* for cumulus clouds are, as a rule, bimodal. For the left-hand mode the Sun is covered by clouds ($S = 0$, $Q = D$), and for the right-hand mode it is not ($Q = S \sin h_\odot + D$). The mode ratio varies monotonically with the amount of cloudiness: as n increases, the probability of the right-hand node becomes lower and that of the left-hand mode becomes higher. The probability density of scattered radiation for cumuli is monomodal, and it is more extended toward higher relative fluxes for higher n. This can apparently be attributed to the nonlinearity of the relationships between the optical depth and brightness of the clouds [18]. The probability densities of D^* and Q^* become identical in the limiting case of continuous cloudiness.

Since the variability of Q^* is mainly determined by the variability of S^*, it follows that autocorrelation functions $r_{S*}(t)$ and $r_{Q*}(t)$ are similar. These functions, which are as a rule monotonic (Fig. 11.5), can be approximated by the following formula from [21]:

$$r\,(t) = e^{-\alpha\,(n)\,t}, \tag{11.35}$$

where t is in seconds, and $\alpha(n)$ has values from 0.0215 to 0.0035. Compared to S^* and Q^*, the variability of D^* in time has a closer correlation, and the correlation radius $t_{0.5}$ (found from the relation $r(t_{0.5}) = 0.5$) for D^* is 2 to 4 times the correlation radius for Q^*. At the limit, for $n = 1.0$, we have $r_{Q*} = r_{D*}$.

The spectral densities of the relative fluxes S^*, D^*, and Q^* are determined according to $r(t)$ in the interval of linear frequencies $f = 5 \cdot 10^{-4}$ to $4 \cdot 10^{-2}$ s^{-1}. In the interval of f from $2 \cdot 10^{-3}$ to $4 \cdot 10^{-2}$ s^{-1} the spectra are approximated by a power function of the form

$$S\,(f) \sim f^{-k}. \tag{11.36}$$

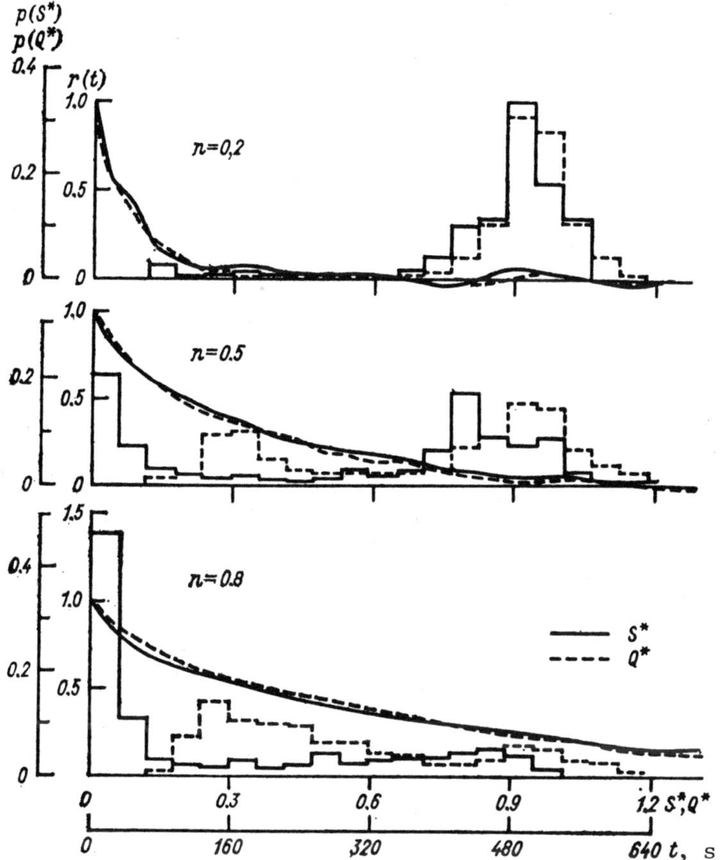

Fig. 11.5. Average correlation functions $r(t)$ and probability densities p (histograms) of fluxes S^* and Q^*.

The value of exponent k depends on the ratio between the upper and lower frequencies in the spectra. For the Q^* and S^* spectra it lies in the range $1.4 \leq k \leq 1.6$. In the spectra of scattered radiation low frequencies play a greater part, and k has values from 2.1 to 2.7.

Table 11.5 gives the mean statistical characteristics for S^*, D^*, and Q^* for various cloud conditions.

Table 11.5. Mean statistical characteristics of relative short-wave fluxes for various cloud conditions.

Radiation type	Type and amount of cloudiness		$\overline{Q}^*, \overline{S}^*, \overline{D}^*$	σ	t_{05} min	k
S^*	Cu	0.1... 0.3	0.66	0.40	0.8	1.6
	Cu	0.3... 0.6	0.54	0.46	1.3	1.6
	Cu	0.6... 0.9	0.28	0.42	1.6	1.6
D^*	Cu	0.1... 0.3	1.69	0.88	4.5	2.1
	Cu	0.3... 0.6	2.44	1.30	2.7	2.7
	Cu	0.6... 0.9	2.58	1.33	2.2	2.5
Q^*	Cu	0.1... 0.3	0.89	0.27	1.0	1.5
	Cu	0.3... 0.6	0.80	0.37	1.2	1.4
	Cu	0.6... 0.9	0.51	0.30	1.3	1.6
	Ac	0.8... 1.0	0.74	0.16	0.7	1.5
	Ci	0.6... 1.0	0.82	0.12	4.5	1.5
	Sc tr	0.4... 0.7	0.77	0.22	3.0	1.6
		0.8... 1.0	0.53	0.19	3.3	1.8
	Sc op.	1.0	0.38	0.19	3.7	1.8
	St	1.0	0.25	0.11	6.2	3.0
	Cs	1.0	0.79	0.04	6.2	1.6

11.5. A comparison of calculated and experimental radiation characteristics of a cumulus field

The optical model of a cumulus cloud described in Section 11.1 was based on calculations. The ratio of cloud height H to cloud diameter D was taken to be unity, and the cloud shape was assumed to be paraboloidal. The flux I_0 incident upon the cloud was taken as unit flux. Scattering and absorption of light in the atmosphere outside the cloud were neglected.

Two methods were used to calculate the mean fluxes and flux densities for the direct, scattered, and total solar radiation transmitted by a layer containing cumulus clouds.

11.5.1. The Monte-Carlo method

The intensities of the fluxes impinging upon a point receiver depend to a considerable extent on how it is oriented relative to the Sun and clouds, that is, whether it is in the shadow of the clouds or out of it. These possibilities are modeled, respectively, by the conditions $j \leq p_S$ and $j > p_S$, where j is a uniformly distributed number in the interval $[0, 1]$, and p_S is the probability of a clear line of sight to the Sun [18, 27].

Using this approach, the random value of the flux of direct radiation I^* equals either 1 (if $j > p_S$) or $e^{-\sigma r^*}$, where r^* is the random path length of a ray in a cloud with a random base diameter D. Accordingly, the size of the direct flux depends on the parameters of the specific cloud, if this cloud "shades" the receiver, as well as on the probability of "shading."

Unlike the direct flux, the scattered radiation can be assumed to depend just on the group properties of the clouds: the amount of cloudiness n, the size distribution of clouds $p(D)$, the scattering coefficient σ, etc. The random value of the flux of scattered radiation S^* is determined either by direct modeling (in terms of the characteristics of an individual cloud) [2, 3, 26, 29, 30] or with the aid of special modifications [1, 4, 6, 12, 17].

The random value of the total flux Q^* equals the sum $Q^* = I^* + S^*$. Multiple modeling then gives us the differential distribution functions for the fluxes of direct $p_I(I)$, scattered $p_S(S)$, and total $p_Q(Q)$ radiation. The mean fluxes are found with the aid of formulas of the form

$$\bar{I} = \int\limits_0^{I_0} I p_I(I)\, dI. \tag{11.37}$$

11.5.2. *The analytical method*

For zenith distances of the Sun $\zeta \le 45\text{--}50°$ an approximate analytical calculation method can be devised. The probability density of direct radiation at the receiver is given by the formulas

$$p_I(I) = \begin{cases} I_0(1 - p_S)\,\psi(I), & \text{if } I < I_0, \\ p_S\delta(I - I_0), & \text{if } I = I_0. \end{cases} \tag{11.38}$$

Function $\psi(I)$ can be interpreted as the probability density for transmission of the direct radiation by the clouds:

$$\psi(I) = \int\limits_{D_{nin}}^{D_{max}} \psi(I, D)\, p(D)\, dD, \qquad \text{if } I \geqslant e^{-\sigma D_{nin}/\cos\zeta},$$

$$\psi(I) = \int\limits_{-\cos\zeta I/\sigma}^{D_{max}} \varphi(I, D)\, p(D)\, dD, \qquad \text{if } I < e^{-\sigma D_{nin}/\cos\zeta}, \tag{11.39}$$

where D_{min} and D_{max} are, respectively, the minimum and maximum diamters in distribution $p(D)$.

Function $\varphi(I, D)$ is the probability density of direct radiation for an individual cloud with a base diameter D. It can be shown [3] that

$$\varphi(I,D) \approx \cos \zeta / \sigma D I, \qquad \text{for } e^{-\sigma D / \cos \zeta} \leqslant I < I_0. \qquad (11.40)$$

The flux of scattered radiation for a cloud with a diameter D is given by the formula

$$S(D) = T(D) - T_p(D), \qquad (11.41)$$

where

$$T(D) = \sqrt{\cos \zeta / 2} \left[1 + 0.0638 (\sigma D)^{2/3} \right]^{-1}, \qquad (11.42)$$

$$T_p(D) = \int_{\exp(-\sigma D / \cos \zeta)}^{I_0} I\varphi(I, D)\, dI =$$

$$= \frac{\cos \zeta}{\sigma D} \left[1 - \exp\left(\frac{-\sigma D}{\cos \zeta} \right) \right] \quad \text{for } I_0 = 1. \qquad (11.43)$$

Function $S(D)$ is nonmonotonic [3]. If D_0 is an extreme point, and if $G_1(S)$ and $G_2(S)$ are functions inverse to $S(D)$ and defined in intervals $D \leq D_0$ and $D > D_0$, respectively, then according to the theory of functions of random variables the probability density for scattered radiation can be written as

$$p_S(S) = \frac{a}{dS} \left[\int_{D_{min}}^{G_1(S)} p(D)\, dD + \int_{G_2(S)}^{D_{max}} p(D)\, dD \right]. \qquad (11.44)$$

The probability density for the total radiation, which can be represented as the sum of two random quantities with densities $p_I(I)$ and $p_S(S)$, is written as a Duhamel's integral

$$p_Q(\dot{Q}) = \int_0^Q p_S(Q - Q')\, p_I(Q')\, dQ'. \qquad (11.45)$$

Figure 11.6 compares values of Q calculated using the Monte-Carlo method and the analytical method with experimental values [18, 20, 22]. The fit is seen to be generally satisfactory.

In [25] the calculated and experimental dependences of the

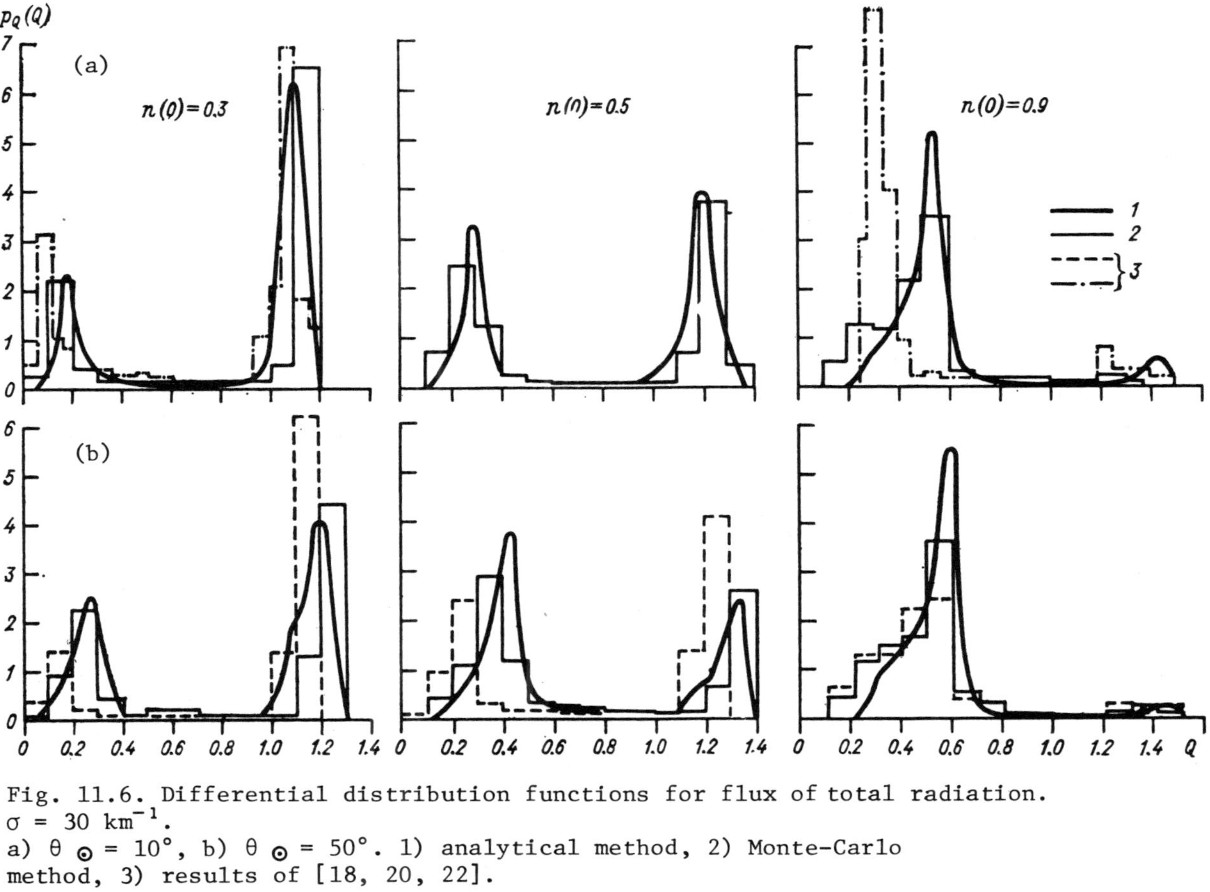

Fig. 11.6. Differential distribution functions for flux of total radiation.
$\sigma = 30$ km^{-1}.
a) $\theta_\odot = 10°$, b) $\theta_\odot = 50°$. 1) analytical method, 2) Monte-Carlo
method, 3) results of [18, 20, 22].

direct, scattered, and total radiation fluxes on the relative
amount of cloudiness are compared. The equation of linear regression
obtained in [22] agrees well with the calculation only for
mean values of n, whereas the dependence obtained in [13] is good
for any n. The experimental data of [5] fit the calculated values
if it is assumed that $\sigma \geq 30 \ km^{-1}$.

INTRODUCTION

The thermal radiation (self-radiation) of the atmosphere is
a function of the temperature and humidity fields, the amount and
distribution of cloudiness, and the contents and vertical profiles
of carbon dioxide, aerosol, ozone, and trace gaseous impurities.
It is more difficult to describe the field of thermal radiation
in a cloudy atmosphere, in comparison with cloudless conditions,
because radiation fluxes in the presence of clouds depend on a
great number of atmospheric parameters. Fluxes of thermal radiation are less affected by the optical properties of clouds than
are fluxes of solar radiation, but they are very sensitive to the
positions and temperatures of the cloud boundaries.

The cloudiness, temperature, and humidity all have a nonlinear effect on fluxes of thermal radiation, that is, the fluxes
averaged over time or space are not equal to the fluxes for the
correspondingly averaged entry parameters. The diversity and
irregularity of the cloud distribution call for a statistical approach when describing the field of thermal radiation in an atmosphere with broken clouds in it.

Available theoretical and experimental data can be used to
devise models of the thermal radiation in a cloudy atmosphere.
These models may differ greatly, however, depending on the required information about the radiation field, on the time and
space scales involved, and to what extent complete information
about the initial parameters can be used.

For example, in problems of the theory of climate scientists
sometimes restrict themselves to specifying the global balance of
radiant energy at the limit of atmosphere, and they use semiempirical relations between the main meteorological and radiation fields
(see Section 12.3). In problems of cloud formation, on the other
hand, radiation fluxes have to be calculated with a careful spatial
resolution and taking detailed account of absorption and radiation
by atmospheric gases and water droplets. All the necessary data
on the microphysical and optical properties of clouds are therefore computed during the solution of the cloud-formation problem.
Naturally, however, between these extremes there are intermediate
cases involving models of the general circulation of the atmosphere and problems of long-term weather forecasting, mesometeor-
ology, etc.

CHAPTER 12

METHODS OF CALCULATING THERMAL RADIATION

The methods used to calculate the field of thermal radiation in a cloudy or cloudless atmosphere have been described in detail in monographs [7, 11, 22, 24], and in a number of collections and original articles. In this chapter we will present only some of the initial relations and we will describe some of the approximate calculation methods used in practice.

12.1. Fundamental relations and use of integral transmission function

When describing the thermal-radiation field in problems of atmospheric physics and meteorology, it is convenient to use the integral transmission function for diffuse radiation, determined from the relation

$$D(z_1, z_2) = 2 \int_0^1 P(z_1, z_2, \mu)\, \mu\, d\mu. \tag{12.1}$$

Here $\mu = \cos \theta$, and z_1 and z_2 are the boundaries of the layer transmitting radiation and containing absorbing substances, the effective masses of which are found with the aid of formulas (5.4) or (5.5). The transmission function for directed radiation $P(z_1, z_2, \mu)$ or $P(m_1, \ldots, m_n)$ is given by formula (5.12).

More precise spectral calculation methods enable us to ascertain the subtle radiation effects, but these become unjustifiably cumbersome when it is necessary to determine the variation in the atmosphere's radiation regime in global dynamic problems with variable fields of temperature, humidity, and cloudiness. The lack of information about cloudiness and the use of the "blackbody" approximation (see Chap. 14) make spectral calculation methods for a cloudy atmosphere even less justifiable than for a cloudless atmosphere.

When the integral transmission function is used, the integral hemispherical fluxes of thermal radiation at a level z of a plane-stratified medium can be written as

$$F^{\uparrow}(z) = B_s D(0, z) + \int_0^z B(z') \frac{dD(z, z')}{dz'}\, dz', \tag{12.2}$$

$$F^{\downarrow}(z) = \int_{\infty}^{z} B(z') \frac{dD(z,\,z')}{dz'}\,dz', \qquad (12.3)$$

while the radiative cooling $R(z)$ is given in terms of the divergence of the effective radiation $F(z) = F^{\uparrow}(z) - F^{\downarrow}(z)$:

$$R(z) = -dF(z)/dz. \qquad (12.4)$$

In expression (12.2) B_g is the integral radiation of the underlying surface. Approximate calculation methods based on a simplification of expressions (12.2) and (12.4) are described in [2, 4, 23, 24, and others].

When calculating radiation fluxes within the atmosphere, it should be kept in mind that the integral transmission function is not multiplicative, that is, $D(m_1 + m_2) \neq D(m_1)D(m_2)$. This influences calculations of some absorbing substances as well as of the transmission of the totality of atmospheric layers. Because of this property of the integral transmission function, it is difficult to specify the boundary condition at the upper boundary H of the layer in question, if $F^{\downarrow}H \neq 0$. Here the contribution of the layer above H cannot be taken into account in the form $F^{\downarrow}_H D(z,\,H)$ in the counterradiation of the atmosphere at levels $z < H$. Rather, the equivalent absorbing mass in the layer $(H,\,\infty)$ has to be determined and then taken into account when calculating $F^{\downarrow}(z)$.

The comparison of different calculations of the thermal-radiation field is attended by certain difficulties, since different methods can be compared only in models having the same initial fields of temperature, humidity, cloudiness, etc. Spectral data and approximation methods are also of considerable importance when constructing the integral transmission function. Therefore, comparisons of calculations of different authors are generally just qualitative. In [5] spectral calculations of fluxes $F^{\uparrow\downarrow}(z)$ were compared with calculations employing integral transmission functions constructed on the basis of the same initial data, for identical models of the atmosphere. The discrepancies between compared unidirectional fluxes were less than 1%, and the effective fluxes differed, on the average, by 5%. The differences between the calculated influxes to layers 1 km thick could reach 10–20%.

The influence on the fluxes and influxes of differences in the transmission functions manifests itself more markedly in a cloudless atmosphere than in an atmosphere with "black" cloud layers. This is because the errors in calculating the radiative fluxes are usually greater for larger optical depths of the gas layer, while in a cloudy atmosphere the layers above, between, and below the clouds are much more transparent than the entire thickness of a cloudless atmosphere. In models taking into account absorption by water droplets, differences of the transmission function are again more important.

12.2. Thermal radiation of a cloudy atmosphere

The presence in the atmosphere of an unbroken homogeneous cloud layer does not in the blackbody approximation alter the method of calculating the radiation characteristics in the layers above and below the clouds. For the layer below the cloud this method is even more precise than in the upper part of a cloudless atmosphere. It must be recalled, however, that since the cooling of the subcloud layer is low, the relative errors in calculating it are generally increased. For low-lying cloudiness in many cases the radiative cooling of the layer under the clouds can be neglected [24].

For broken clouds the radiation fluxes have to be calculated separately for the cloudy and cloudless parts of the atmosphere, and then these are added together, with weighting factors determined by the amount of cloudiness. For instance, the counterradiation of a cloudy atmosphere in the subcloud layer can be written as

$$F^{\downarrow}(z) = f_1(n) F^{\downarrow}_{cld}(z) + f_2(n) F^{\downarrow}_{no\ cld}(z). \qquad (12.5)$$

The other characteristics of the thermal-radiation field in the presence of clouds are described similarly. The values of f_1 and f_2 depend appreciably on the statement of the problem. Let us consider three alternatives: 1) calculating the average radiation fluxes over time or over territory; 2) calculating the radiation at a fixed time; 3) studying the statistical characteristics of the thermal-radiation field as a function of the statistical cloud characteristics.

In the first case we can set $f_1 = n$ and $f_2 = 1 - n$, where n represents the absolute amount of cloudiness, obtained, for instance, with satellite photos or with models of the general circulation. For the second alternative the distribution of cloudiness over the sky, and also the vertical cloud depth, are of great significance. Then, to calculate the fluxes at any level, it would be natural to use the relative cloudiness obtained from observations at this same level (determination of relative and absolute amounts of cloudiness, see [24]). In practice this is feasible only for the departing radiation using satellite data and for the counterradiation of the atmosphere according to ground observations. Several formulas are available for converting the absolute amount of cloudiness into the relative amount, required to calculate the counterradiation (see, for instance, Chap. 3). The counterradiation of a cloudy atmosphere depends considerably on where the clouds are concentrated — at the zenith or near the horizon [16, 18].

In the presence of broken clouds, for example, fractocumulus, the fluxes of thermal radiation between the clouds are practically constant: they fluctuate little within a given cloud and vary greatly upon transition from the cloud to the clear sky. Consequently, the statistical characteristics of the thermal-radiation fluxes are determined by the statistical characteristics of the cloudiness itself and by the intensity of the thermal radiation of the clouds and the cloudless atmosphere. The variability of the luminance of the thermal radiation is close to that of the cloud cover in the viewing direction, described in Chap. 3. The statistical characteristics of the counterradiation are close to the corresponding characteristics of scattered solar radiation given in Chap. 11.

12.3. Semiempirical methods

In models of the general circulation of the atmosphere, and even more so in problems of the theory of climate, the local distribution of clouds cannot be taken into account. In such problems the information about the cloudiness often reduces to specifying the amount of cloudiness as an average over a large territory or along a latitude circle. Therefore, in climatic models and in some forecasting models use is often made of empirical and semiempirical expressions of the type of Brunt's formula for the balance of thermal radiation at the level of the underlying surface (see [14]):

$$F(0) = \delta B(0) \left(A - D\sqrt{e}\right) f(n). \qquad (12.6)$$

In these models the departing radiation is sometimes described by Budyko's formula

$$F^{\uparrow}(H) = A_1 + B_1 t - (A_2 + B_2 t) n. \qquad (12.7)$$

Here, A, A_1, A_2, B_1, B_2, and D are empirical constants, e is the partial pressure of water vapor at the surface, δ is the emissivity of the underlying surface, and $t = T - 273.2$. The cooling of the entire air column is determined by the difference between $F^{\uparrow}(H)$ and $F(0)$. In some recent works ([13], for instance) a simplified form of (12.6) is used, in which the value of e is averaged over the whole world.

Other works have recently appeared [6, 15, et al.] in which expressions like (12.6) or (12.7) were obtained from initial formulas (12.2) and (12.3), and the physical meaning of the coefficients was clarified. For example, in [15] coefficients A_i and B_i of (12.7) are represented as functions of the integral transmission function and t, it being shown that with an error of less than 2% these coefficients are constant over the latitude and also that the flux of departing radiation for the annual mean state of the climatic system is approximated satisfactorily by (12.7). Within error limits of 10%, this linear form with the same coefficients describes the seasonal variation of $F^\uparrow(H)$, and it can also be used to describe climatic states differing from the state of the moment.

A set of empirical and semiempirical formulas are available for calculating fluxes and influxes of thermal radiation; these were derived individually for specific experimental material. Some of these formulas, obtained on the basis of measurements with aircraft, will be given in Chap. 13.

Numerous measurements of the radiation balance of the ocean surface, carried out by the Main Geophysical Observatory, led to the following expression for the effective radiation at sea level (see [10] et al.):

$$F(0) = \delta\left[B_S - \left(1.63\sqrt{B(0)} - 0.775\right)(1 + kn^2)\right], \quad (12.8)$$

where coefficient k depends on the air temperature. This formula gives a valid description of the variation of $F(0)$ with a change in meteorological conditions. Since the radiation balance of the ocean surface varies appreciably with the moisture content of the atmosphere, therefore in [6] measurements made during the ATEP-74 [GATE-74] expedition were used to arrive at the following formula for $F(0)$:

$$F(0) = B_S - B(0)(1 - a/e + bn). \quad (12.9)$$

Expressions like (12.8) and (12.9) are, however, of limited applicability, since they can only be used under conditions similar to those for which they were obtained.

12.4. Algorithm for models of general circulation

In models of the general circulation various methods can be used to calculate radiative influxes of heat, depending on the number of layers and on the cloud data being considered. In the simplest of these methods, the radiation budget of the entire thickness of the atmosphere is computed with the aid of formulas like (12.6) and (12.7). Then, to each calculated layer is attributed the portion of the long-wave radiative cooling corresponding to the mass of this layer [19, 21]. In many models contemporary with general-circulation models, the method of Manabe and Strickler [12] is applied, with various modifications [3, 9].

On the basis of expressions (12.2)-(12.4) it is possible to construct a quite general algorithm for calculating radiation fluxes in an atmosphere containing cloud layers or aerosol layers, for a given amount of cloudiness or aerosol, expressed in units n_i or fractions of a unit, and for a given emissivity a_i where i is the layer number.

Let us consider a model comprising N atmospheric layers plus

an underlying layer [7]. The number of the layer corresponds to
the number of the level of its upper boundary. For each layer a
mean temperature is specified (that of the underlying layer being
the surface temperature). Then, from (12.2) and (12.3) we get

$$F_i^{\uparrow} = \sum_{j=i}^{N} B_j (D_{i,\,j} - D_{i,\,j+1}), \qquad (12.10)$$

$$F_i^{\downarrow} = \sum_{j=0}^{i-1} B_j (D_{i,\,j+1} - D_{i,\,j}). \qquad (12.11)$$

The fluxes are calculated at the ith level; B_j is the radiation function for the mean temperature of the jth layer, and $D_{i,\,j}$
is the transmission function of the layer between levels i and j.
The effective radiation at the ith level can then be expressed
as

$$F_i = \sum_{j=1}^{N} B_i (D_{i,\,j} - D_{i,\,j+1}). \qquad (12.12)$$

In a cloudy atmosphere fluxes of thermal radiation can be
written in form (12.10)–(12.12), provided that the transmission
function is formulated taking into account the amount and emissivity of the cloudiness. This function is then

$$\widetilde{D}_{i,\,j} = D_{i,\,j} \prod_{k=j}^{i-1} c_k, \qquad (12.13)$$

where coefficients $c_k = 1 - a_k n_k$. If both clouds and aerosol are
present in the layer, then $c_k = (1 - a_k n_k)(1 - a_k^a)$, where a_k is
the emissivity of the cloud or aerosol layer.
    Unidirectional radiation fluxes are practically never used
in the energetics of the atmosphere. The influx of heat to a volume
element is equal to the divergence of the effective flux, and the
influx to a layer equals the difference between the effective
fluxes at its boundaries. The principal energy characteristics,
namely the departing radiation for the natural assumption that
$F^{\downarrow}(\infty) = 0$ and the balance of radiant heat at the underlying surface,
are also effective fluxes. It is precisely the effective fluxes
which are described here using the simplest, most universal method.
The calculation procedure actually reduces to the following. The
matrix D of order $N + 1$ representing the transmission function for
the gaseous components is calculated. Matrix c is then calculated
separately, its elements $c_{i,\,j}$ being given by the formula

$$c_{i,\,j} = \prod_{k=j+1}^{i} c_k, \quad c_{i,\,i} = 1. \qquad (12.14)$$

Next the matrix $\overline{D}_{i,j}$ of order N is constructed, with the elements

$$\overline{D}_{ij} = D_{ij}c_{ij} - D_{i,\,j+1}c_{i,\,j+1}. \qquad (12.15)$$

When matrix $\overline{D}$ is multiplied by vector B with elements σT_i^4, we get
the vector of the effective fluxes

$$F = \overline{D}B. \qquad (12.16)$$

This method makes it possible to ascertain individually the
effects of changes in the temperature and the mass of absorbing
gases, as well as the role of "gray" absorbers: clouds, aerosol,
and underlying layer. For example, if only the temperature varies
with time, then the calculations can be made with a fixed matrix
$\overline{D}$. This will be the case when computing the temperature profile
of the radiative equilibrium in an atmosphere with given optical
properties.

12.5. Radiation calculations in cloud-formation models

In the previous section the blackbody approximation was
practically no longer used, since the emissivity of the clouds was
introduced, defined in terms of the transmission function for
water droplets:

$$a_{\mathbf{d}} = 1 - D_{\mathbf{d}}(m_v,\ m_w). \qquad (12.17)$$

In models of cloud formation the transfer of thermal radiation must be taken into account even more carefully, since the
optical properties of a cloud vary in time along with the microstructure. In [1, 25] the two-stream approximation was invoked
as a basis for a detailed calculation of radiation fluxes (see
Section 8.11). As the droplet spectrum broadens with the evolution
of the cloud, parameter $\alpha_{\nu,w}$ beyond the atmospheric transparency
window from 8 to 13 μm decreases from 2000 to 600 cm^2/g, while
inside the window $\alpha_{\nu,w} \approx 500$ cm^2/g, varying slightly with the
wavelength and the droplet spectrum. These estimates are based on
formulas in Section 4.2.

Different parts of the spectrum contribute differently to
the fluxes and radiative influx of heat. In the determination of
the influx, the mean coefficient $\overline{\alpha}_\nu$ over the spectrum should be
calculated with a weighing factor $(F_\nu^\uparrow + F_\nu^\downarrow - 2B_\nu)$ rather than
B_ν, since then the principal transparency window for the heat

influx inside the cloud (8–13 μm) can be distinguished. The effective absorption coefficient for water droplets, taking absorption by water vapor into account, can in this case be written as

$$\bar{\alpha}_w = \frac{\int\limits_0^\infty (\alpha_{\nu v} q_v + \alpha_{\nu w} q_w)(F_\nu^\uparrow + F_\nu^\downarrow - 2B_\nu)\, d\nu}{q_w \int\limits_0^\infty (F_\nu^\uparrow + F_\nu^\downarrow - 2B_\nu)\, d\nu}, \qquad (12.18)$$

where q_v and q_w are the specific humidity and the specific water content.

Coefficient $\bar{\alpha}_w$ determined from (12.18) turns out to be close to $\alpha_{\nu w}$ in the window [25]. Consequently, when calculating the radiation influx during the evolution of low-lying stratus and fog, it is advisable to use a value of $\bar{\alpha}_w$ between 500 and 600 cm^2/g, regardless of the cloud microstructure. The radiation influx in the two-stream approximation then becomes

$$R = r\bar{\alpha}_w (F^\uparrow + F^\downarrow - 2B). \qquad (12.19)$$

In (12.19) $F^\uparrow$ and $F^\downarrow$ are found with the aid of formulas like (12.2) and (12.3), allowing for absorption by water droplets in the transmission function, and r is the diffusivity.

It should be noted in conclusion that developing cloudiness is sensitive to variations in the radiation parameters, in view of the inverse relation between the water-content gradient and the radiative influx of heat, so that the choice $\alpha_{\nu w}$ can appreciably influence the cloud-formation process.

CHAPTER 13

EXPERIMENTAL STUDIES OF THE THERMAL RADIATION
OF A CLOUDY ATMOSPHERE

Experimental studies of the thermal radiation of the atmosphere are at present being carried out at ground level, from outer space with the aid of Earth satellites, and in the atmosphere itself with the aid of actinometric radiosondes, aircraft, and aerostats. This chapter describes studies of thermal radiation in a cloudy atmosphere, carried out directly in the atmosphere and providing the most complete information about the radiation properties of clouds and the role of cloudiness in the transfer of thermal radiation.

13.1. Actinometric radiosonde observations of atmosphere

The actinometric radiosondes which appeared in the 1960's in the USA, West Germany, Japan, and the USSR [16] led to a number of experimental studies of the relation between the field of long-wave radiation (thermal radiation) and the meteorological conditions, as well as of the variability in time and space of radiation fluxes and heat influxes in the free atmosphere up to heights of 25 or 30 km in various parts of the world [12, 32, 33].

In 1963 a Soviet network of stations for actinometric radiosonde observations of the atmosphere (ARS) was set up. The stations were situated in various climatic and geographical regions of the USSR. Data collected from 1963 to 1967 on the air temperature T and humidity u, the upward $F^\uparrow$ and downward $F^\downarrow$ fluxes of long-wave radiation, the effective radiation F, and the rate of radiation-caused variation of air temperature dT/dt are given in [10], in the form of extremal, mean, and rms deviations of the parameter values at standard levels for individual seasons and years, and for the entire observation period as a whole.

The initial results of these actinometric observations revealed a marked dependence of the field of long-wave radiation of the free atmosphere on the humidity u and the temperature T. The errors in determining fluxes of long-wave radiation in the troposphere from aerosonde data may be as high as 10% in a cloudless atmosphere, due to errors in measuring the air temperature and humidity [23]. The fluxes of long-wave radiation correlate well with the temperature of the level being considered [3, 18, 20, 26]. As higher and higher levels are observed, this correlation becomes weaker. The effect of the temperature $T(z)$ of lower-lying levels on $F^\uparrow(z^1)$ generally diminishes with an increase in $(z - z^1)$, whereas

under conditions of weak absorption (low humidity) the correlation
of $F^\uparrow$ with the temperature of the underlying surface is the highest.
The structure of the field of the upward long-wave flux in the
free atmosphere is governed mainly by the structure of the temperature field of the atmosphere. A leading role in shaping $F^\downarrow(z)$ is
also played by the temperature of the level in question, but the
correlation of $F^\downarrow$ with T diminishes for higher z levels, until at
the height of the tropopause it goes to zero. The weakest correlations in the free atmosphere are those between $F^\uparrow(z)$ and the affec-
tive mass of water vapour in the layer $(0, z)$, and the highest are
those between $F^\uparrow(z)$ and $\sigma T^4(0)$, between $F^\uparrow(z)$ and $T(z)$, and also
between $F(z)\sigma T^4(0)$ and $\Delta T = T(z) - T(0)$. The discovery of this
interrelationship suggested the following formula for the approximate description of $F^\uparrow(z)$ in the free atmosphere:

$$F^\uparrow(z) = \sigma T^4(0) \cdot 10^{-\alpha \Delta T}, \qquad (13.1)$$

where $\alpha = 0.0036$ [3]. Here there is an agreement with the mean
measured values to within 10%.

In [11, 17] actinometric radiosonde data obtained under
cloudy conditions were processed statistically. It was found that
the presence of cloudiness essentially destroys the statistical
correlation between the radiation parameters and aerological
parameters of the layers above and below the clouds, creating a
unique kind of screen. The matrices of the correlation coefficients
for these parameters under cloudy and cloudless conditions have
different structures.

The marked effect of cloudiness on the radiation field in
the free atmosphere becomes especially evident from its variation
over a 7 to 10 hour period (maximum duration of measurements 1.5-
2 hours): the appearance or disappearance of cloudiness leads to
a sharp variation in the structure of the radiation field and in
the values of the fluxes themselves [9].

Attempts to obtain mean annual values of $F^\uparrow$ and $F^\downarrow$, or mean
values over many years, for different cloud types have not been
successful. The annual variations in air temperature and humidity
and in the heights of cloud boundaries in the middle latitudes
level out the flux profiles so much that they differ little from
the mean profiles of the cloudless atmosphere.

13.2. Actinometric model of the atmosphere

The network material from the actinometric radiosonde observations of the atmosphere from 1963 to 1967 (about 4000 ARS launchings
at 15 sites in the USSR) was given a statistical analysis at the
Central Aerological Observatory. This analysis showed that the
mean many-year profiles of fluxes $F^\uparrow$ and $F^\downarrow$ in the free atmosphere
for individual stations lie within about $\pm$ 10% of the mean many-
year values for all the stations [11]. Tables 13.1 and 13.2 give

the mean many-year values of the upward and downward long-wave
fluxes in the free atmosphere, according to the ARS data. As a
first approximation these can be considered as a characteristic of
the radiation field of some "average" atmosphere over the USSR.

Table 13.1. Mean many-year values of upward fluxes (watts/m^2) in
free atmosphere for 1963 to 1967.

Station	Level, mb								
	Ground	850	700	500	400	300	200	100	50
Moscow	331	324	301	259	233	204	175	172	175
Minsk	338	331	312	270	243	213	182	179	184
Rostov-on-Don	369	360	332	284	256	224	191	184	190
Sverdlovsk	314	316	300	263	239	210	182	180	182
Vladivostok	328	325	307	271	246	218	191	180	181
Aralsk	362	356	332	284	251	223	193	184	189
Kuibyshev	351	351	325	280	253	223	189	186	187
Leningrad	329	324	302	261	237	207	177	173	179
Riga	344	337	314	270	243	212	184	184	190
Tbilisi	390	386	357	301	268	233	196	182	185
Petropavlovsk-Kamchatka	319	317	300	263	237	210	184	180	179
South Sakhalin	312	314	296	261	236	211	193	193	195
Kiev	352	346	320	274	246	216	184	177	182

Table 13.2. Mean many-year values of downward fluxes (watts/m^2) in
free atmosphere for 1963 to 1967.

Station	Level, mb								
	Ground	850	700	500	400	300	200	100	50
Moscow	325	296	252	188	152	115	83	73	67
Minsk	318	300	260	193	157	118	83	71	64
Rostov-on-Don	347	314	270	200	161	119	83	69	61
Sverdlovsk	294	279	244	182	147	110	77	66	61
Vladivostok	303	281	248	189	156	119	85	70	62
Aralsk	349	316	267	198	160	119	82	71	62
Kuibyshev	322	305	263	196	157	117	80	66	60
Leningrad	308	288	248	187	151	111	79	70	60
Riga	327	300	257	189	153	112	80	70	64
Tbilisi	364	349	305	226	183	136	91	70	61
Petropavlovsk-Kamchatka	295	281	248	189	154	115	82	71	64
South Sakhalin	288	274	237	178	144	108	80	66	64
Kiev	342	306	263	196	159	120	84	71	63

In view of the considerable similarity of the vertical profiles of the radiation fluxes and their low relative scatter, an
analytical representation of these fluxes is possible. The authors

of [13] propose a model of a radiation atmosphere which is suitable
for large-scale problems in climatology and numerical models of
the general circulation of the atmosphere. The upward and downward fluxes of long-wave radiation are represented as analytical
functions of the pressure and temperature of the level being considered, with an error of about 5%. The following formulas were
used:
for 1000 mb $\geq p \geq$ 300 mb

$$F^{\uparrow}(p) = \left(2.65 - 0.36\,\lg p - 0.061\,\lg^2 p\right)\sigma T^4(p),$$
$$F^{\downarrow}(p) = (0.158 + 0.26\,\lg p)\,\sigma T^4(p);$$

for 300 mb $\geq p \geq$ 150 mb

$$F^{\uparrow}(p) = (0.675 + 0.333\,\lg p)\,\sigma T^4(p),$$
$$F^{\downarrow}(p) = \left(-0.35 + 0.38\,\lg p + 0.03\,\lg^2 p\right)\sigma T^4(p);$$

for 150 mb $\geq p \geq$ 10 mb

$$F^{\uparrow}(p) = \left(0.62 + 0.932\,\lg p - 0.265\,\lg^2 p\right)\sigma T^4(p),$$
$$F^{\downarrow}(p) = (0.15 + 0.20\,\lg p)\,\sigma T^4(p). \tag{13.2}$$

The proposed actinometric model of the atmosphere can be used
to calculate not only the annual mean values of $F^{\uparrow}$ and $F^{\downarrow}$, but
the seasonal means as well, for points situated in middle and
polar latitudes [8]. However, when computing fluxes in specific
synoptic situations, these formulas are not applicable, even though
for some cloud types (stratocumulus, for instance) the deviations
from average conditions are insignificant.

13.3. Aircraft studies of thermal radiation of a cloudy atmosphere

As a characteristic of the regime of long-wave radiation in
an atmosphere containing clouds, the effective radiation $F(z)$ is
usually employed, together with the influx of radiation to the
cloud layer, defined as

$$R(z_{ub},\ z_{lb}) = F(z_{lb}) - F(z_{ub}) = F^{\downarrow}(z_{ub}) - F^{\uparrow}(z_{ub}) - \tag{13.3}$$
$$- F^{\downarrow}(z_{lb}) + F^{\uparrow}(z_{lb}).$$

Methods of aircraft radiation measurements and determination
of $F(z)$ have been described in [5]. Here we will present the main
data ensuing from experimental studies of the regime of effective

radiation in the presence of continuous (stratiform) and broken (cumulus) cloudiness.

Because of thermal radiation the cloud layer of the atmosphere loses some of its internal energy. Since the size of this loss $R(z_{ub}, z_{lb})$ is determined to a considerable degree by the value of $F(z_{ub})$, while the latter is a function of the state of the cloudiness above the cloud layer in question, it is advisable to consider the variation of $R(z_{ub}, z_{lb})$ in the absence of high clouds (or with a cloud cover of less than 3). Table 13.3 gives data on the radiation regime of single-layer St–Sc clouds of various depths. There is seen to be a definite dependence of $R(z_{ub}, z_{lb})$ on the cloud depth.

Table 13.3. Effective radiation of upper boundary $F(z_{ub})$, radiative influx of heat $R(z_{ub}, z_{lb})$, and rate of radiative cooling dT/dt of stratiform clouds, as functions of cloud depth H and level of upper boundary z_{ub}.

Parameter	Depth interval, m			
	140... 270	280... 380	390... 590	600... 800
H_{av} m	220	330	460	670
z_{ub} m	730	980	960	1280
$F(z_{ub})$ watts/m^2	95.6	89.3	86.6	104.0
$R(z_{ub}, z_{lb})$ watts/m^2	−71.9	−75.4	−77.5	−56.3
dT/dt K/h	−0.95	−0.69	−0.49	−0.42
No. of cases	14	20	19	18

The mean cooling due to long-wave radiation is quite significant, being much greater than the radiative heating via short-wave radiation (see Chap. 10). The rate of cooling of the entire cloud drops considerably as its thickness increases (for the same reasons as the rate of heating due to short-wave radiation). Since the amount of long-wave radiation lost by a cloud is determined by the difference between $F(z_{ub})$ and $F(z_{lb})$, it is worthwhile to consider the patterns of variation of these quantities. Figure 13.1 plots these fluxes against the boundary heights. It is seen that the effective radiation at both cloud boundaries increases with height. The dependence on the height z_{ub} in kilometers is approximated by the linear equations [6]:

$$F(z_{ub}) = 54.44 + 33.5 z_{ub} \tag{13.4}$$

$$F(z_{lb}) = 2.79 + 8.3 z_{lb}. \tag{13.5}$$

These equations provide an approximate quantitative estimate of the variation in radiative cooling of St–Sc clouds as a function

of the cloud level. For instance, a cloud 400 m in depth with its
lower boundary 250 m above the ground radiates 71 watts/m^2, on the
average; if the lower boundary of this cloud is elevated to 1000 m,
then $R(z_{ub}, z_{1b})$ increases to 90 watts/m^2, and a further elevation
to 2000 m makes $R(z_{ub}, z_{1b})$ 115 watts/m^2, or more than 1.5 times
the initial value.

By adding together the diurnal radiative heating given in
Chap. 10 and the cooling obtained by recalculating the data in
Table 13.3 for a day, we obtain the effect of the total radiative
heat transfer between cloud and atmosphere: an St-Sc layer about
0.5 km thick cools off 10-12 K per day in winter and 8-10 K in
spring. During the daylight hours in the absence of higher clouds
an St-Sc layer cools about 2-4 K in both winter and spring.

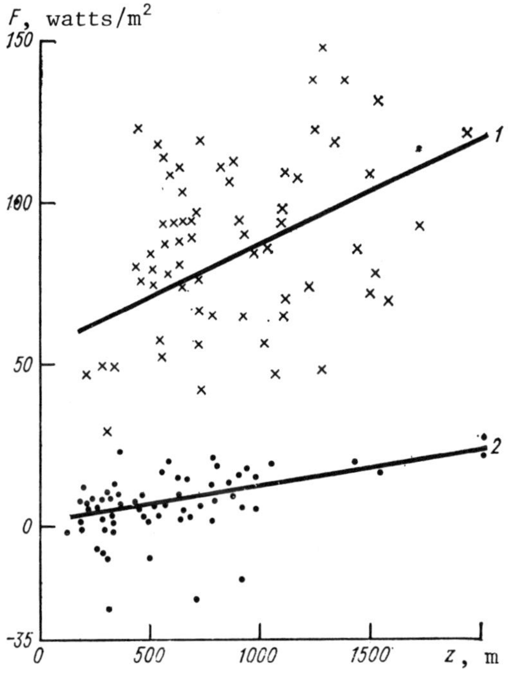

Fig. 13.1. Effective radiation F at upper (1) and lower (2) boundaries of cloud layer, as function of boundary height z.

Let us consider the behavior of $R(z_{ub}, z_{1b})$ for cumuli as a
function of the amount of cloudiness. Although direct measurement
data are scanty, the following qualitative regularities can be
established. As n increases, flux $F^{\uparrow}(z_{ub})$ diminishes and flux
$F^{\downarrow}(z_{1b})$ is enhanced, while $F^{\downarrow}(z_{ub})$ and $F^{\uparrow}(z_{1b})$ remain unchanged.
Here $F(z_{ub})$ and $F(z_{1b})$ decrease simultaneously, but the former
drops more gradually than the latter for an actual arrangement of
cumulus clouds somewhere in the layer from 1 to 2 km. Thus the
main factor affecting the variation of $R(z_{ub}, z_{1b})$ with increasing

n is the effective radiation under the cloud layer. Accordingly, the dependence of $R(z_{ub}, z_{1b})$ on n should be similar to the dependence of $F(z_{1b})$ on the amount of cumulus cloudiness. Ground-level actinometric observations can be used to evaluate the latter dependence, and Table 13.4 includes the mean values $\overline{F}(0)$ as a

Table 13.4. Influx of long-wave radiation to cloud layer and effective radiation of Earth's surface (watts/m^2).

Parameter	Amount of cloudiness n						
	0	1—2	3—4	5	6—7	8—9	10
$R(z_{ub}, z_{1b})$	—17.4	—53.7	—36.9	—45.3	—80.2	—139.6	—
R_{max}	—56.5	—90.7	—100.5	—94.2	—124.9	—216.3	—
R_{min}	+3.4	+12.5	—6.2	—18.8	—41.8	—62.8	—
No. of cases	33	21	30	18	12	9	—
$\overline{F}(0)$	117.9	114.4	107.4	100.5	84.4	64.2	17.4
$\overline{R}_t(z_{ub}, z_{1b})$	13.9	18.1	30.0	36.2	48.1	85.1	—
$R_t(z_{ub}, z_{1b})$ (calculated)	13.9	16.0	26.5	38.3	53.7	84.4	—

function of n. This relation can be approximated by a parabola:

$$F(0) = F_{no\,cld}(0)\,(1 - kn^2). \qquad (13.6)$$

where $k = 0.7$, n is in fractional units, and $F_{no\,cld}(0)$ is the effective radiation for $n = 0$. Formula (13.6) can be used to evaluate the radiation effect of cumulus cloudiness at the Earth's surface with an error of less than 30%.

The obtained relation for $F(0)$ as a function of n suggests seeking a similar relation for $R(z_{ub}, z_{1b})$. The data used to construct this relation are given in Table 13.4. This dependence is also approximated well by a parabola:

$$R(z_{ub}, z_{1b}) = R_{no\,cld}(z_{ub}, z_{1b})\,(1 + 0.085n^2). \qquad (13.7)$$

Here $R_{no\,cld}(z_{ub}, z_{1b}) = -17.4$ watts/m^2 is the mean influx of long-wave radiation to the cloud layer when there are no cumulus clouds in the layer.

The total radiative influx of heat $R_t(z_{ub}, z_{1b})$ to the cloud layer is the sum of $R_{sw}(z_{ub}, z_{1b})$ and $R_{1w}(z_{ub}, z_{1b})$. However, in view of the difficulty in determining $R_{sw}(z_{ub}, z_{1b})$ it is advisable to directly parametrize $R_t(z_{ub}, z_{1b})$. Although this approach is less accurate, it is much simpler, which is the important thing in practical calculations. For the parametrization, we used the values of $R_{sw}(z_{ub}, z_{1b})$ given in Chap. 10 and the values of $R(z_{ub}, z_{1b})$ calculated using the formula (13.7). These values were then used to find the mean total radiation $R_t(z_{ub}, z_{1b})$ for each amount of cloudiness (see Table 13.4). It was found that the total radiation influx as a function of the amount of cloudiness can also be approximated by a parabola:

$$R_t (z_{ub}, z_{lb}) = R_{no\ cld,\ t} (z_{ub}, z_{lb}) (1 + 0.070 n^2). \qquad (13.8)$$

where $R_{no\ cld\ t} (z_{ub}, z_{lb})$ = 13.9 watts/m^2. This simple approximation is suitable for use in numerical analyses. A comparison of the values obtained using this formula with empirical values of $R_t (z_{ub}, z_{lb})$ (see Table 13.4) indicates a satisfactory fit between this approximation and real conditions.

13.4. Characteristics of an "average" stratiform cloud

Data of aircraft measurements of $F(z_{ub})$ at the height of the upper boundary were averaged in order to arrive at the profile of the effective radiation in an "average" cloud. A mean value of $F(z_{ub})$ = 90.7 watts/m^2 was obtained (range of variation 70 to 115 watts/m^2). In addition, the mean influxes of long-wave radiation $R(z)$ were calculated per unit volume every 50 m in the cloud and in the layers above and below it, that is, throughout the cloud-formation layer (0 to 3 km). Table 13.5 gives mean values of $R(z)$, together with its variability and other characteristics. The variability of $R(z)$ is determined both by instability of the actual conditions and by the measurement errors.

Table 13.5. Variability of influxes of long-wave radiation.

Layers of "average" cloud, m	$R(z)$ 10^2 watts/m^2				Frequency of negative values,%	No. of cases
	mean	minimum	maximum	mode interval		
Above upper boundary	−2.1	−4.9	2.1	−1.4... −2.8	88	25
0... 50	−107.5	−178.7	−55.8	−77.5... −118.7	100	25
50... 100	−37.0	−64.2	−11.2	−34.9... −43.3	100	25
100... 150	−14.0	−50.3	−4.2	−2.8... −11.9	92	25
150... 200	−4.9	−22.3	20.9	−2.1... −9.1	70	25
200... 250	0.0	−18.1	29.3	0.0... −6.3	58	24
250... 300	0.7	−22.3	19.4	3.5... −2.8	35	20
300... 350	6.3	−6.9	36.3	13.7... 7.8	25	16
350... 400	2.8	−16.8	13.7	—	30	10
Below lower boundary	0.0	−11.2	9.8	—	44	25

A profile of the effective radiation was constructed on the basis of the values of $F(z_{ub})$ and $R(z)$ for the selected layers. The effective radiation in the above-cloud layer increases from 90.8 watts/m^2 at level z_{ub} to 104 watts/m^2 at 1.9 km and then to 132 watts/m^2 at 3.0 km, which corresponds to a negative volume heat influx (01.4·10^{-2} and -2.5·10^{-2} watts/m^3). A negative influx was observed in 90% of all cases, that is, a radiative cooling of the above-cloud layer via long-wave radiation (the natural situation for this layer).

In the top 50 m of the average cloud $F(z)$ drops to 28 watts/m^2, while $R(z_{ub}, z_{1b})$ increases in absolute value by a factor of 60 there, in comparison with the layer above the cloud, reaching $107 \cdot 10^{-2}$ watts/m^3, which corresponds to a radiative cooling of -3.6 K/h. This layer plays the most active role in the radiative transfer between the average cloud and the overlying layers of the atmosphere. The radiative cooling of the next 50-m layer is about 1/3 as great. In these two layers positive values of $R(z)$ were never observed. The appreciable radiative cooling penetrates to a depth of 150 m below z_{ub}. In the middle of the average cloud, 150 to 300 m below the upper boundary, a state close to radiative equilibrium is observed.

Close to z_{1b} the value of $R(z)$ is positive in 70% of the cases; the heating here is 0.1–0.2 K/h, on the average. In the subcloud layer the heating and cooling are about equal, and the radiation-caused temperature variation is, on the average, close to zero. For the entire average cloud, $R(z_{ub}, z_{1b})$ is -76 watts/m^3, of which the upper 100-m layer accounts for almost 95% of the total long-wave influx.

It should be noted that the radiative cooling at the upper cloud boundary obtained according to our measurements is an order of magnitude higher than the cooling measured by other authors [4, 14, 19]. This discrepancy is apparently explained by the fact that our measurement technique made it possible to investigate radiation effects close to the upper boundary of the cloud layer and inside the layer with a finer vertical resolution $(50 \text{ m} \leq \Delta z \leq 100 \text{ m})$.

The described behavior of $F(z)$ and $R(z)$ conforms closely to the qualitative conclusions of the theory developed in [27], and it corresponds reasonably well to the calculation data given there. A comparison of the calculations with averaged measurement results and with individual measurements leads to the following conclusions:

a) the integral transmission function for water vapor and carbon dioxide, and also for water droplets, given in [7] (see Chap. 5 and [27]) can be used to describe the behavior of the radiation influx for stratiform clouds both inside and outside a cloud;

b) the good fit between the calculations and the mean experimental model indicates that the parameters of this model were reliably determined;

c) for a sufficiently precise calculation of the radiation fluxes and influxes in the vicinity of cloud layers in specific cases, the real water-content gradients have to be known and information about the droplet sizes must be available; the order of magnitude of the cooling and heating peaks at the cloud boundaries can also be found from preliminary data.

The experimental results on the thermal radiation of a cloudy

cloudy atmosphere presented in this chapter prove that actinometric radiosounding of the atmosphere provides a good basis for constructing a statistically justified radiation model of a cloudy atmosphere. Aircraft data, however, in view of their complexity and accuracy, constitute suitable material for studying the physical picture of the radiation transfer in clouds and in the adjacent atmospheric layers, as well as for constructing a radiation model of the clouds themselves.

13.5. Dependence of calculation results on accuracy of measuring atmospheric parameters

The errors in the initial parameters used to calculate the fluxes of long-wave radiation may influence the calculation results. Thus it is important to know the contributions of various components in different atmospheric layers to producing the flux at the level in question [1]. Table 13.6 gives the relative errors for the upward and downward fluxes of integral thermal radiation at various heights in the atmosphere, if in the calculations we neglect the conribution of water vapor (v = 0) or carbon dioxide (u = 0) or ozone (m = 0).

Table 13.6. Relative errors in downward ($\delta F^{\downarrow}$) and upward ($\delta F^{\uparrow}$) fluxes of thermal radiation at heights z.

Con-dition	z km							
	0	5	10	25	5	10	25	50
	$\delta F^{\downarrow}$ %				$\delta F^{\uparrow}$ %			
$w = 0$	77	70	38	36	−23	−33	−34	−34
$u = 0$	10	26	53	50	−7	−10	−12	−11
$m = 0$	1	2	10	15	0	0	−1	−2

Table 13.7 shows how errors in measuring the temperature and humidity affect the accuracy of calculating the thermal radiation of the atmosphere. This relationship was considered in [23], for both integral and spectral fluxes.

Table 13.8 evaluates the sensitivity of flux $F^{\downarrow}(z)$ to variations in the atmospheric carbon-dioxide content, according to [1]. $\Delta F^{\downarrow}(z)$ is seen to depend little on the choice of the atmospheric model, but it varies substantially with height: at the ground the sensitivity of $F(z)$ to a variation in the total CO_2 content is about 1/10 of the sensitivity to a variation in the total mass of water vapor; in the stratosphere, on the other hand, carbon dioxide is every bit as important as water vapor. These evauluations show a general good fit with the results in [28–30].

Table 13.7. Absolute errors in integral downward ($F^{\downarrow}$) and upward ($F^{\uparrow}$) fluxes of thermal radiation, arising due to errors in determining air temperature and humidity.

Error	z km									
	0	2	6	10	30	0	2	6	10	30
	$F^{\downarrow}$ watts/m^2					$F^{\uparrow}$ watts/m^2				
	285	203	94,1	27.9	4.5	391	347	280	247	241
	ΔF watts/m^2					ΔF watts/m^2				
$\Delta T = 0$ K, $\Delta u = 5$ %	2.5	1.8	1.8	11.7	25,2	0,0	−0.4	−1,0	−1,4	−1,5
$\Delta T = 1$ K, $\Delta u = 0$ %	6.1	4.6	3.0	1.2	0.2	5,5	4.6	3.3	2,3	2,2

Random and systematic errors in measuring the temperature practically do not alter at all the spectral variation of the departing radiation. Random errors less than 0.5 K and systematic

Table 13.8. Variations of $F^{\downarrow}(z)$ in % for an increase in CO_2 concentration from 0.03 to 0.04%.

	z km				
	0	5	10	15	20
Tropics	0.3	1.0	3.0	5.1	7,8
Subarctic (winter)	0.6	1.7	3.6	5,8	8,2

errors less than ±1 K allow flux calculations with errors up to 3%. If, however, random errors of up to 10% in measuring the relative humidity are taken into account, then the maximum errors in calculating radiation-caused temperature variations in the troposphere are 0.25 K/day, or about 10%.

For clouds the error in calculating the radiation fluxes increases, because of the difficulty in precisely determining the height, depth, and geometrical shape of the clouds, as well as their distribution over the sky, water content, microstructure, and other characteristics.

The distribution of clouds in the cloud layer and the geometrical shape of the cloud elements have a greater effect on the radiation properties of the cloud layer as a whole than do variations in the cloud microstructure [21-24, 25, 34, 35]. The errors in calculating fluxes of thermal radiation in the atmosphere associated with the uncertainty of the cloud characteristics are lowest near the Earth's surface, because of the shielding

action of the dense lower layers of the atmosphere, and they are
highest at the boundaries of the cloud layer and inside it, especially for cumulus clouds. The term "cloud layer" is somewhat
arbitrary: actually, no sharp gradient exists, but rather a kind
of transition zone, where the meteorological and radiation characteristics are the most variable. The difficulty in determining
the location, extent, and optical properties of this region leads
in turn to difficulties in calculating the thermal radiation in
the cloud layer and its surroundings.

13.6. Comparison of measurements and calculations under specific conditions

For a number of reasons, it is not so simple to compare the
calculated and measured (by actinometric radiosounding) fluxes of
thermal radiation under cloudy conditions. First of all, it is
difficult to arrange ascents of actinometric radiosondes for which
continuous stratus cloudiness with an even upper boundary is definitely present. Second, cloud boundaries are ascertained with
a discreteness of 250 m for cloud depths ranging from 1200 to
2500 m, that is, a comparison of the fluxes at the cloud boundaries
is purely tentative. Third, it is not yet known how temperature,
humidity, and radiation sensors act in clouds and as they emerge
from clouds. However, actinometric radiosounding does not provide
us with information about the water content and microstructure of
clouds. Therefore, a model of a perfectly black cloud is used for
the calculations, and this introduces substantial errors in a
number of cases (see Chap. 14). The accuracy of calculating the
thermal radiation of clouds depends considerably on the accuracy
of the measured temperatures in the cloud layer, which in turn
depends on how reliably the height and boundaries of the layer are
determined.

Let us consider a comparison of actinometric radiosonde
measurements with calculations for single-layer cloudiness according
to the data of [2] for the atmospheric characteristics given in
Table 13.9. These calculations allowed for absorption by water
vapor and carbon dioxide, as well as absorption by ozone, aerosol,
and complexes of H_2O molecules.

Figure 13.2 gives the relative divergences $\alpha^{\uparrow\downarrow} = \Delta F^{\uparrow\downarrow}/F^{\uparrow\downarrow}$
corresponding to these calculations. The calculated (c) and measured
(e) divergences under the clouds are seen to agree well: the α
values are low. However, close to the cloud layer and inside it
the values of α are high and quite variable. For the blackbody
approximation a comparison of $F_c^{\uparrow\downarrow}$ and $F_e^{\uparrow\downarrow}$ inside the cloud layer
is, generally speaking, meaningless. Moreover, measurements show
that in a cloud $F_e^{\uparrow}/\sigma T^4 < 1$ (where T is the air temperature in
the cloud layer, decreasing with height), that is, the clouds can
by no means be considered black. At the same time, in a cloud
the ratio $F_e^{\downarrow}/\sigma T^4$ is 2 to 6% greater than 1, which is unrealistic,

Table 13.9. Atmospheric characteristics for sounding under cloudy conditions.

Parameter	Petro-pavlovsk-Kamchatka 6 XII 1966 2033 LT	Sverdlovsk 30 VIII 1966 2354 LT	Kiev 23 VIII 1967 2002 LT
Cloudiness	10/10 Ac, As	10/10 Sc, Cb	8/8 Sc
z_{lb}, km	3.8	1.9	1.8
z_{ub}, km	6.2	3.1	3.4
Total content, cm			
under clouds			
H_2O	0.340	1.220	1.790
CO_2	148.2	70.3	73.3
O_3	0.011	0.006	0.006
above clouds			
H_2O	0.021	0.343	0.524
CO_2	71.5	134.3	126.4
O_3	0.285	0.190	0.190

since above the level being considered the air temperature drops. Despite the fact that 6% of the errors are less than the errors in measuring the long-wave fluxes, the constancy of the sign of ratio $F_e^{\downarrow}/\sigma T^4 > 1$ calls for a careful analysis of the results, and such an analysis was carried out in [2]. This is mainly because of three factors: first, in the cloud the thermoresistor measuring the air temperature gets wet, making T_e slightly low; second, as the radiosonde ascends in the cloud, the upper polyethylene shield may get wet, or may even ice over at temperatures below 0°C; finally, the inertial error of the ARZ-1 instrument affects the ratio $F_e^{\downarrow}/\sigma T^4 > 1$, making $F_e^{\downarrow}$ too high. The use of a semitransparent model of a cloud layer does not eliminate simultaneously the discrepancies between F_c and F_e above and below the clouds. It should be noted that divergences $\alpha^{\uparrow\downarrow}$ in the free atmosphere above the clouds retain the features of their vertical distributions under cloudless conditions.

It is of great practical interest to compare calculations and measurements in multilayer cloudiness, especially double-layer cloudiness. To do this, we selected an ascent of an ARZ-1 radiosonde over Kiev on 23 December 1966 for 10/10 Ac, Sc with the following boundary heights: first layer 2.0-3.1 km; second layer 6.6-7.4 km. The distributions of absorbing components (in cm) under the first layer, between the clouds, and above the second layer were as follows: for H_2O 0.60, 0.36, 0.01; for CO_2 85, 72, 55; for O_3 0.0056; 0.0100; 0.253.

The behavior of the radiation fluxes under the lower cloud layer, in the layer itself, and above the second layer is the same as in the case of single-layer cloudiness (see Fig. 13.2).

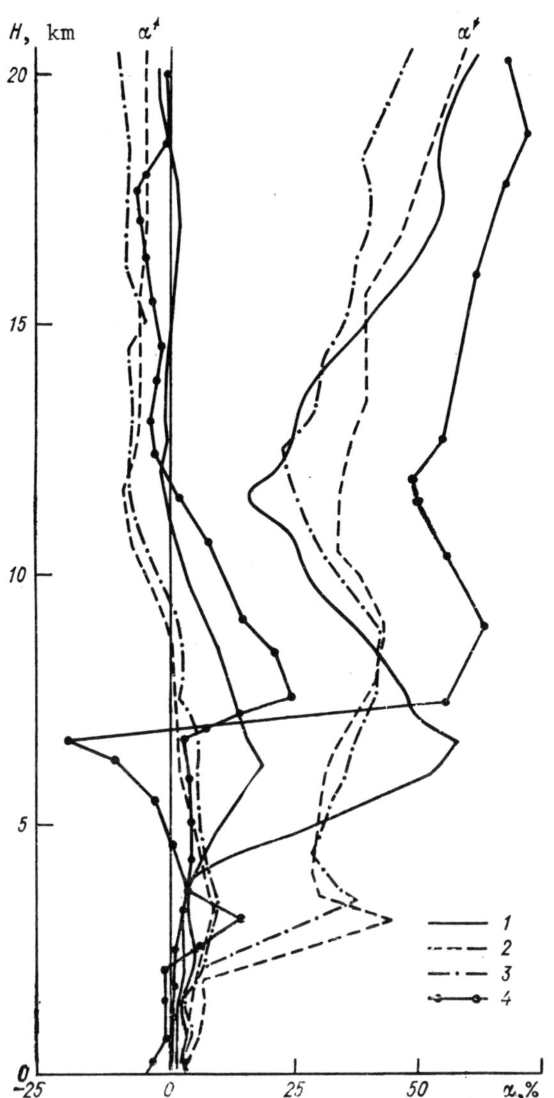

Fig. 13.2. Relative divergences of fluxes $\alpha = \Delta F^{\uparrow\downarrow}/F^{\uparrow\downarrow}$ under cloudy conditions (single-layer cloudiness).
1) Petropavlovsk-Kamchatka, 2) Sverdlovsk, 3) Kiev (double-layer cloudiness), 4) Kiev (single-layer cloudiness).

The values of $\Delta F^{\uparrow} = F_c^{\uparrow} - F_e^{\uparrow}$ in the atmosphere at the upper cloud boundary are naturally higher than for single-layer cloudiness, since thin As clouds were assumed to be black in the calculations.

In the As layer itself $F_e^{\downarrow}/\sigma T^4 < 1$. Because the air temperature at this height is about $-40°C$, the clouds can be assumed to be ice clouds, and no moisture or snow will be deposited on the poly- ethylene shield. A comparison of the behavior of $F_e^{\downarrow}$ vis-à-vis σT^4 in the lower and upper cloud layers indicates that, in the lower layer, of the three above-mentioned factors influencing the accuracy of a measurement in a cloud, wetting of the polyethylene shield is the main one. Between the clouds F_c and F_e differ little from one another ($\Delta F^{\uparrow\downarrow} < 8$ watts/m^2, $\alpha < 3\%$).

The results indicate that a critical analysis of discrepancies $\Delta F^{\uparrow\downarrow}$ and α can best be carried out on the basis of a large number of soundings, making it possible to group the data according to the synoptic situation and according to the season, and allowing the investigator to distinguish recurrent similar discrepancies $\Delta F^{\uparrow\downarrow}$ against the background of random errors. The theoretical and experimental results can be made to jibe better if spectral methods are used to calculate the radiation fluxes in the free atmosphere, if simultaneous account is taken of the continual absorption in the 8–12 μm window, and if the aerosol is considered. However, more work is necessary, in order to guarantee good agreement between these data in the free atmosphere.

At present the main objectives are: 1) to further increase the accuracy of determining the transmission function in the strato- sphere (specifying correctly the absorption as a function of temper- ature and pressure, determining the absorption by trace gaseous ad- mixtures); 2) using existing calculation methods, to take into ac- count with sufficient precision th absorption by the aerosol (numer- ous visual observations with aircraft indicate that the turbidity of the atmosphere is very high right up to the tropopause, but measure- ments are very infrequent, that is, practically no quantitative in- formation about the aerosol is available); 3) the methods of calcu- lating fluxes under conditions must be improved considerably. These problems have been well studied theoretically, but their practical solution has been sorely neglected. On the other hand, for this we must have reliable data on the position of the cloud layer and on its vertical structure (water content, microstructure). From the experimental point of view the problem is equally complicated, since special ascents guaranteeing a high measurement accuracy are needed, in order to enable a comparison of experimental and calculated data. It is not individual ascents leading to a compar- ison of $F_e^{\downarrow\uparrow}$ and $F_c^{\downarrow\uparrow}$ which are needed here, but rather whole series of soundings under different synoptic conditions.

One of the important tasks of the comparison is to determine the limits of the absolute and relative minimum divergences of the calculated and experimental long-wave radiation fluxes in the free atmosphere; this should be attempted during the physical and sta- tistical analysis of the data.

CHAPTER 14

EFFECT OF OPTICAL PROPERTIES OF CLOUDS ON THERMAL RADIATION

The scattering of thermal radiation in a cloudy medium is usually neglected. The thermal albedo of a cloud is assumed to be zero, and only the cloud emission and transmission are studied.

If the content of water droplets (or ice crystals) $m_w \geq 0.5$ g/cm^2 (see [6, 7] and Chap. 5), then clouds stop transmitting radiation and their emissivity approaches unity, the emission of a black body. However, this picture, which is valid as a first approximation, is incomplete and often incorrect, because of various factors, the main ones being the following:

1. Real clouds are spatially inhomogeneous; their boundary parts may be optically thin, and there may be semitransparent regions inside clouds.

2. Some cloud types are almost never optically dense. These are generally ice-crystal clouds of the upper (Ci) and middle (Ac, As) levels. In the liquid phase clouds are semitransparent during the initial and final stages of their existence.

3. The scattering power of clouds is, strictly speaking, not zero in the range of thermal radiation. Ice-crystal clouds are better scatterers (relative to a single particle) than droplet clouds, since the particles of the former are larger.

4. Radiation fluxes penetrating a cloud change rapidly in intensity, creating strong boundary effects (radiation sinks and heat sources). The latter are not described in the "blackbody" approximation.

The effect of these factors on the radiation regime of clouds, or the effect of their nonblackness, will be considered in this chapter.

14.1. Flux distribution inside a cloud layer. Estimates of error of "blackbody" approximation

The assumption that clouds are "black" is expressed mathematically as

$$\frac{dD\left(|z-z'|\right)}{dz} = \delta\left(|z-z'|\right) \tag{14.1}$$

for $z_{lb} \leq z' \leq z_{ub}$. Here z_{lb} and z_{ub} are the lower and upper boundaries of the cloud layer;

$$D\left(|z-z'|\right) = D\left[m_v\left(|z-z'|\right);\quad m_w\left(|z-z'|\right)\right] - \text{ITF};{}^{*)}$$

$m_v(|z-z'|)$ and $m_w(|z-z'|)$ are the contents of water vapor and water droplets (or ice crystals) in the air column (z, z') or (z', z). Assuming condition (14.1), it follows from (12.2) and (12.3) that $F^{\uparrow}(z) = F^{\downarrow}(z) = B(z)$ and $F(z) = 0$ for $z_{1b} < z < z_{ub}$. For flux $F^{\uparrow}(z)$ the additional condition $D(0, z) = 0$ for $z > z_{1b}$ must be imposed.

Formula (14.1) implies that there is a discontinuous drop in $F^{\uparrow}(z)$ for $z = z_{1b}$ ($z < z_{1b}$) and a like increase in $F^{\downarrow}(z)$ for $z = z_{ub}$ ($z < z_{ub}$). This in turn implies that

$$-\frac{dF}{dz}\bigg|_{z=z_{1b}} = \infty, \qquad -\frac{dF}{dz}\bigg|_{z=z_{ub}} = -\infty,$$

that is, infinite heat sources and sinks are created at the cloud boundaries.

Actually, even in a dense homogeneous cloud layer dD/dz is not the δ function, the extremal influxes $dF/dz\big|_{z = z_{1b}, \, z_{ub}}$ are finite and are "diffused out" over the boundary layers.

In [5, 7] a detailed study was made of the described effects for the following approximate assumption (see Sections 5.4 and 12.5):

$$D\left(m_v, \, m_w\right) = D_v\left(m_v\right) e^{-\overline{\alpha}_w^m w}.$$

Estimates were obtained of the boundary-layer thicknesses corresponding to the peaks (extreme values of cooling and heating):

$$\Delta z_i \approx \cfrac{1}{\sqrt{\overline{\alpha}_w^{(i)} \, \dfrac{d\rho_w}{dz}\bigg|_{z=z_i}}}. \tag{14.2}$$

Here $i = 1$ refers to the lower boundary region of the cloud $z_1 = z_{1b}$, and $i = 2$ refers to the upper region $z_2 = z_{ub}$. The intensity of maximum cooling R_u in the upper layer and heating R_1 in the lower layer are approximately equal to

$$R_u \approx - \sqrt{\overline{\alpha}_w^{(2)} \, \frac{d\rho_w}{dz_{ub}}}\, B(z_{ub}) \, D_v\left[m_v(z_{ub}, \, H)\right],$$

$$R_1 \approx \sqrt{\overline{\alpha}_w^{(1)} \, \frac{d\rho_w}{dz_{1b}}}\, \left[B(0) - B(z_{1b})\right] D_v\left[m_v(0, \, z_{1b})\right]. \tag{14.3}$$

By multiplying together (14.2) and (14.3), respectively, for $z = z_{1b}$ and $z = z_{ub}$ and adding, we obtain expressions for the radiative cooling (heating of entire cloud layer) in the form

---

${}^{*)}$ITF = integral transmission function

$$R\left(z_{\mathrm{lb}},\ z_{\mathrm{ub}}\right) \approx \left|B_{(0)} - B\left(z_{\mathrm{lb}}\right)\right| D_v\left[m_v\left(0,\ z_{\mathrm{lb}}\right)\right] -$$
$$- B\left(z_{\mathrm{ub}}\right) D_v\left[m_v\left(z_{\mathrm{ub}},\ H\right)\right], \qquad\qquad (14.4)$$

since $R(z_{\mathrm{lb}} + \Delta z_1,\ z_{\mathrm{ub}} - \Delta z_2) \approx 0$.

Formulas (14.2)-(14.4) can be used to investigate the dependences of R_{u}, R_1, and $R(z_{\mathrm{lb}},\ z_{\mathrm{ub}})$ on the thickness and level of the cloud layer, on the contents of water vapor and cloud particles in the atmosphere, and on the temperature distribution with height.

Detailed data, as well as complete profiles of $R(z)$ over the thickness of a cloudy atmosphere, can be found in [5, 7]. The method developed in [7] does not resort to the assumption of "blackbody" radiation of clouds, and thus can be used to estimate the error of this assumption. The blackbody approximation turns out to be practically accurate for $m_w > 0.03$ g/cm^2, and it can be used to evaluate the fluxes at the atmospheric boundaries $z = 0$ and $z = H$, as well as the influxes to the entire thickness of the atmosphere and to the part of it above the clouds. The influx to the cloud region can be evaluated with an error of about 10 or 20%.

Intense local influxes at cloud boundary layers cannot be detected using the blackbody approximation. The error in calculating small influxes above a cloud or in spaces between clouds is high, up to several tens of percents. During the 1970's spectral calculations more precise than [7] were carried out [13–15]. In these papers polydisperse homogeneous cloud layers were considered, for different variants of the model. Thus only a qualitative comparison of these with one another, or for that matter with [5, 7], is possible.

The spectral calculations completely verify all the conclusions and orders of magnitude obtained previously in [7] with the aid of simpler methods. The most interesting new finding of [13–15] is that the window from 8 to 13 μm is responsible for the main thermal effect (see Fig. 14.1; cf. Section 12.5). Outside the window the absorptivities of water vapor and water droplets are so high, and the

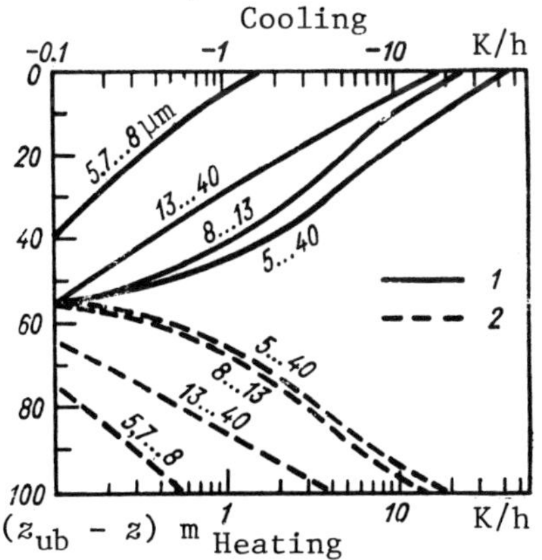

Fig. 14.1. Rate of radiative cooling °C/h (1) and heating (2) of cloud layer 100 m thick [15]. Spectral intervals indicated on curves.

regions of extremal influxes Δz_i (see (14.2)) are correspondingly so narrow, that they cannot be resolved during the calculations. Thus, with regard to thermal radiation, a cloud 100 m thick (with a droplet concentration $N = 450$ cm^{-3} and $w = 0.28$ g/m^3) is

essentially semi-infinite, with very pronounced edge effects
[14, 15].

Extremal influxes at cloud boundary layers are clearly distinguished during aircraft experiments with a detailed height resolution and with rapid ascent, in selected instances of low-level
stratus clouds with even boundaries (see Section 13.4). The large
volume of actinometric-radiosounding data gives a more smoothed
picture, because of the low height resolution and the high inertia
of the measurements, but mainly because of the actual inhomogeneity
of clouds. Nevertheless, the radiation effect of clouds is quite
pronounced (see [1] and Chap. 16).

14.2. Thermal albedo of clouds

Kirchhoff's law for the boundary of an absorbing, emitting,
scattering layer has the form*)

$$E_\lambda + T_\lambda + A_\lambda = 1. \tag{14.5}$$

where E_λ is the emissivity, T_λ is the transmittance, and A_λ is
the albedo. The radiation fluxes leaving the layer can be repre-
sented correspondingly as

$$F_\lambda^\uparrow(z_{ub}) = E_\lambda B_\lambda(z_{ub}) + T_\lambda F_\lambda^\uparrow(z_{lb}) + A_\lambda F_\lambda^\downarrow(z_{ub}), \tag{14.6}$$

$$F_\lambda^\downarrow(z_{lb}) = E_\lambda B_\lambda(z_{lb}) + T_\lambda F_\lambda^\downarrow(z_{ub}) + A_\lambda F_\lambda^\uparrow(z_{lb}). \tag{14.7}$$

Strictly speaking, expressions (14.5)-(14.7) can be used to

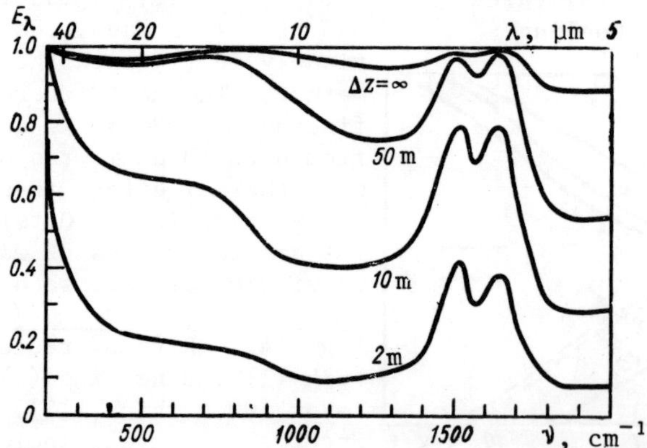

Fig. 14.2. Spectral emissivity of cloud layers of thickness Δz,
according to [14].

---

*) Here Kirchhoff's law is interpreted approximately, as ap-
plied to hemispherical radiation fluxes.

determine E_λ, T_λ, and A_λ separately for fluxes measured at the boundaries in the case of a homogeneous layer. However, attempts at such estimates have failed, due to the inhomogeneity of the clouds and also because of the measurement errors, which are approximately equal to the small quantity A_λ. In the experiments, for simplicity, optically dense clouds are often considered, and it is assumed that $T_\lambda = 0$ or, alternatively, $A_\lambda = 0$. Then E_λ and A_λ, or else E_λ and T_λ, are found using (14.5)–(14.7). Let us consider in more detail one of the more recent studies of this type [2].

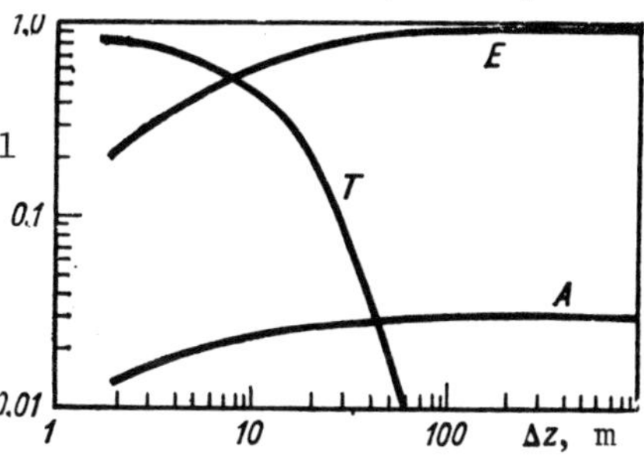

Fig. 14.3. Integral E, T, and A curves, according to [14].

The measurements were carried out in a regime of horizontal flight at distances of 5 to 50 m above the summits of dense clouds. The instrument used was a radiometer operating in the range $\Delta\lambda = 10.5$–12 µm. The instrument error in measuring $F^\uparrow$ was 3%. On the basis of 12 series of measurements, parameter $E_{\Delta\lambda}$ was found to be 0.91 to 0.99 and, correspondingly, $A_{\Delta\lambda} = 0.01$–0.09. In eight instances a temperature inversion was observed above the upper cloud boundary, a situation which may lead to excessively low values of $A_{\Delta\lambda}$. According to the data of [4], it may be that $A_\lambda = 10$–20% for $\lambda \in (8$ to 12 µm$)$. Solutions of the equation of radiation transfer in homogeneous cloud layers are quite suitable for simultaneous determinations of E_λ, T_λ, and A_λ. We will just mention the earlier works [6, 8] and take a closer look at the more recent ones [14, 15].

A model of polydisperse water clouds was considered, with a number of particles 450 cm^{-3}, a water content $w = 0.28$ g/m^3, and a humidity $\rho_v = 1.44$ g/m^3. Scattering and absorption by droplets are taken into account, as well as absorption by water vapor. The difficulties associated with the nonexponential absorption of water vapor are resolved by expanding the transmission functions in power series (see Chap. 9).

Figures 14.2 and 14.3 show the spectral values E_λ and integral values E, T, and A for clouds of various thicknesses, according to [14].

14.3. Emissivity of clouds

If scattering is neglected, expressions (14.5) and (14.7) can be used to find the emissivity E_λ and transmittance T_λ of clouds, from measurements of $F^\uparrow_\lambda(z_{ub})$, $F^\uparrow_\lambda(z_{lb})$, and $B_\lambda(z_{ub})$ or $F^\downarrow_\lambda(z_{lb})$, $F^\downarrow_\lambda(z_{ub})$, and $B_\lambda(z_{lb})$. The working formulas have the form:

$$E_\lambda^\uparrow = \frac{F_\lambda^\uparrow(z_{lb}) - F_\lambda^\uparrow(z_{ub})}{F_\lambda^\uparrow(z_{lb}) - B_\lambda(z_{ub})},$$

$$E_\lambda^\downarrow = \frac{F_\lambda^\downarrow(z_{lb}) - F_\lambda^\downarrow(z_{ub})}{B_\lambda(z_{lb}) - F_\lambda^\downarrow(z_{ub})}, \tag{14.8}$$

$$T_\lambda^{\uparrow\downarrow} = 1 - E_\lambda^{\uparrow\downarrow}.$$

These formulas are widely used to evaluate the spectral emissivity and transmittance of semitransparent clouds. As applied to integral fluxes, formulas (14.8) give useful effective parameters $E^{\uparrow\downarrow}$ [10, 12].

Values of $E^{\uparrow\downarrow}$ calculated with the aid of formulas (14.8) for layers Δp = 100 mb are given in [10]. Each value of $E^{\uparrow\downarrow}$ in [10] represents the mean over all those cases when, according to the author of [10], at the p level being considered, clouds were observed in the layer Δp = 100 mb. Similar calculations were carried out using the data of Chap. 16 (see Table 16.8). Just as for the data of [10], it was found that $E^\downarrow > E^\uparrow$ for $p < p(0)$, so that clearly the downward radiation varies more markedly as it passes through the cloud than the upward radiation does. The difference $E^\downarrow - E^\uparrow$ according to the data of Chap. 16 is seen to be much greater than the difference according to the data of [10]. Under cloudless conditions the emissivity of the atmospheric layers is lower than when clouds are present. However, in moist tropical air this difference is no longer so great.

On the whole, we are less convinced than the author of [10] that it is advisable to use the described experimental estimates of $E^{\uparrow\downarrow}$ to parametrize the profiles of the radiant fluxes. Moreover, as pointed out in [10] as well, in the lower layers of the atmosphere, and also in layers close to isothermal or inversion layers, the differences in the numerators and denominators of formulas (14.8) are of the order of the measurement errors. These differences are small, and uncertainty is introduced.

Recently, in order to provide a simple parametrization of the integral radiant fluxes for clouds, expressions like the following have often been proposed:

$$T^{\uparrow\downarrow}(z_1, z_2) = e^{-k^{\uparrow\downarrow} m_w(z_1, z_2)} = e^{-\tilde{k}^{\uparrow\downarrow} w(z_2 - z_1)}. \tag{14.9}$$

Parameters $k^{\uparrow\downarrow}$ or $\tilde{k}^{\uparrow\downarrow}$ are selected on the basis of measurement data for the fluxes and water content. Expression (14.9), unlike (14.8), is suitable for calculating radiation fluxes inside a cloud layer (see Section 14.4). In [11] it was shown that the results obtained with formulas (14.8) for the calculated fluxes and specified water content are approximated well by expressions (14.9).

Externally, formula (14.9) resembles the "graybody" variant of $D_w(m_w)$ proposed in Chap. 5. However, in contrast to the mean absorption coefficient for water droplets $\bar{\alpha}_w$ in Section 5.4, parameters $k^{\uparrow\downarrow}$ represent certain effective values depending on the temperature, humidity, cloud levels, cloud depths, etc. Formula (14.9) may prove useful if statistically reliable mean values of $k^{\uparrow\downarrow}$ are available for various conditions. So far, however, only individual values have been obtained, and these cannot always be compared with one another.

Now let us consider some more physically valid estimates of the emissivity obtained with formulas (14.8). These estimates are trustworthy for spectral radiation, but there is some doubt as to whether formula (14.5) is satisfied for nonmonochromatic fluxes.

We will briefly summarize the situation according to [1], using $E^{\downarrow}$ as an example. Assuming for simplicity that $B(z)$ = const = B_0 for $z_{1b} \le z \le z_{ub}$, from (12.3) it is easy to obtain

$$F^{\downarrow}(z_{1b}) = B_0\,[1 - D(m_0)] - \int\limits_{z_{ub}}^{H} B(z')\,\frac{dD\,[\,m(z') + m_0 - m(z_{ub})\,]}{dz'},\tag{14.10}$$

where $m_0 = m(z_{ub}) - m(z_{1b})$ is the content of absorbing substance in the specified layer. Here, for brevity, only one absorbing substance is designated. Comparing (14.10) and (14.7) for $A_\lambda = 0$, we see that $E^{\downarrow} = 1 - D(m_0)$, whereas relation $T^{\downarrow} = D(m_0)$ is satisfied only for the additional condition

$$D(m_1 + m_2) = D(m_1)\,D(m_2).\tag{14.11}$$

The integral transmission function does not satisfy this condition according to the estimates of $T^{\downarrow} < D(m)$ in [1] either. Under moist-air conditions for loose clouds one more obstacle stands in the way of evaluating the cloud emissivity. It was found in [1] that the emissivities of layers of a tropical atmosphere containing Ci clouds and devoid of these (but of the same thickness and at the same level) are close to one another. In [1] a method based on separating the absorption by water droplets from the absorption by water vapor:

$$D(m_v,\ m_w) = D_v(m_v) \cdot D_w(m_w),$$

is suggested in order to distinguish the effect of the Ci clouds proper. This corresponds to the integral transmission function in Chap. 5. Since water droplets are present only in layer $(z_{1b},\ z_{ub})$, it follows from (14.10) that

$$B_0 - F^{\downarrow}(z_{1b}) = D_w(m_w)\left\{ B_0 D\left(m_{v0}\right) + \right.$$

$$\left. + \int\limits_{z_{ub}}^{H} B(z')\,\frac{dD_v\,[\,m_v(z') - m_v(z) + m_0\,]}{dz'}\,dz' \right\}.\tag{14.12}$$

Here the expression in braces represents the quantity $B_0 - F^{\downarrow}(z_{1b})$ in a cloudless atmosphere. Hence

$$\widetilde{T}^{\downarrow} = D_w^{\downarrow}\left(m_{w_0}\right) = \frac{B_0 - F^{\downarrow}_{cld}\left(z_{1b}\right)}{B_0 - F^{\downarrow}_{no\,cld}\left(z_{1b}\right)}\,,$$

$$\widetilde{T}^{\uparrow} = D_w^{\uparrow}\left(m_w\right) = \frac{F^{\uparrow}_{cld}\left(z_{ub}\right) - B_0}{F^{\uparrow}_{no\,cld}\left(z_{ub}\right) - B_0}\,. \tag{14.13}$$

Calculations using relations (14.13) involve a number of difficulties. First of all, the question arises of how to determine $F^{\uparrow\downarrow}_{no\,cld}(z)$.

In a tropical atmosphere under cloudless conditions the measured fluxes $F^{\uparrow\downarrow}(z)$ are very stable (see Chap. 16). Therefore, the mean fluxes according to actinometric-radiosounding data in the absence of clouds (see Table 16.8) can be successfully represented in (14.13). However, Ci clouds can be quite thick and the temperature at the limits of the cloud layer may vary considerably, so what do we take as B_0? We assumed, as is often done, that $B_0 = B(z_{1b})$ when calculating $T^{\downarrow}$ and $B_0 = B(z_{ub})$ when calculating $T^{\uparrow}$. Table 14.1 gives typical calculations of $T^{\downarrow\uparrow}$ using formulas (14.8) and (14.13), for several actinometric radiosoundings from [1]. Inspection of the table reveals a systematic enhancement of the transmission, that is, a decrease in the radiation of the clouds proper as compared with the total transmission (radiation) of the layer containing the clouds. The emissivity of Ci clouds (especially tropical clouds) is a subject of great interest. These high clouds alter the thermal regime of the atmosphere appreciably [9].

Table 14.1. Transmittances of Ci clouds, calculated using formulas (14.8) and (14.13) for several soundings.

Sounding	z_{1b} km	z_{ub} km	$T^{\downarrow}$	$\widetilde{T}^{\downarrow}$	$T^{\uparrow}$	$\widetilde{T}^{\uparrow}$
1	7.9	12.1	0.36	0.56	0.41	0.59
2	7.9	10.0	0.20	0.56	0.51	0.95
3	12.1	16.5	0.47	0.87	0.63	0.52
4	10.8	13.7	0.81	1.00	0.61	0.69
5	11.3	14.5	0.52	1.00	0.73	0.80

According to the data of [1], the emissivities of tropical Ci do not exceed 0.44 for $\Delta z = z_{ub} - z_{1b} \leq 2$ km. In the other studies the emissivity of an atmospheric layer containing Ci is evaluated. According to the experimental data in [11], E increases approximately linearly with Δz.

14.4. Effect of "nonblackness" of clouds

If the cloud transmittance is described by a formula of type (14.9), then we can formulate an interesting inverse problem: to determine the flux profiles inside a cloud corresponding to this transmission function, that is, corresponding to the specified radiation fluxes coming to the cloud boundaries from outside, $F^\uparrow(z_{lb})$ and $F^\downarrow(z_{ub})$.

The fluxes inside the cloud are calculated using the formulas:

$$F^\uparrow(z) = F^\uparrow(z_{lb})\, T\,(z - z_{lb}) - \int_{z_{lb}}^{z} B\,(z')\, \frac{dT\,(z' - z_{lb})}{dz'}\, dz',$$

$$F^\downarrow(z) = F^\uparrow(z_{ub})\, T\,(z_{ub} - z) + \int_{z}^{z_{ub}} B\,(z')\, \frac{dT\,(z_{ub} - z')}{dz'}\, dz'. \qquad (14.14)$$

Here $T(z_2, z_1)$ is the transmittance of the layer (z_1, z_2).

A comparison of relations (14.14) and (12.2), (12.3) reveals that the former expressions are not completely accurate, since the fluxes inside the layer are not expressed perfectly in terms of the fluxes at the boundaries, for a nonexponential transmission function outside the layer (see Section 14.3). Nevertheless, the solution of this problem is interesting and it lies within the framework of the same ideas as formula (14.9). Formulas (14.4) were used to calculate fluxes $F^{\uparrow\downarrow}(z)$ for a cloud layer 1 km thick, situated at heights of 1-2 km, 5-6 km, and 9-10 km. Three values of the transmittance were considered: k_w = 0.046, 0.0016, and 0.0007 m^{-1}, which corresponds to $T(z_{ub} - z_{lb})$ = 0.01, 0.20, and 0.50. The fluxes at the boundaries $F^\uparrow(z_{lb})$ and $F^\downarrow(z_{ub})$, as well as the temperatures $T(z_{ub})$ and $T(z_{lb})$, were taken from [3]. All

Table 14.2. Fluxes and temperatures at boundaries of cloud layer.

	$z_{lb(ub)}$					
	1	2	5	6	9	10
$F^\downarrow(z_{ub})$ watts/m^2	—	209	—	108	—	36
$F^\uparrow(z_{lb})$ watts/m^2	372	—	311	—	272	—
$T(z_{ub})$ K	—	275.2	—	249.2	—	223.2
$T(z_{lb})$ K	281.7	—	255.7	—	229.7	—

the data used are given in Table 14.2. The temperature inside the
cloud drops linearly with height, the lapse rate being γ = 6.5 K/km.

Figure 14.4 shows the calculated fluxes $F^{\uparrow\downarrow}(z)$ with and without clouds in layer (z_{lb}, z_{ub}). Inspection of the figure indicates
how the dependence of the upward radiation on the transmission
function of the cloud layer becomes greater with an elevation of
the latter. Even for a transmittance T = 0.5 the flux profiles
$F^{\downarrow}(z)$ differ appreciably from the profile corresponding to cloudless conditions. At the same time, for low-lying cloudiness and
T = 0.5, fluxes $F^{\uparrow}(z)$ with and without clouds are close to one
another.

Some very clear results are obtained by calculating the radiant
heat influx given in Table 14.3. A cloud layer is characterized
by radiative heating at its lower boundary and cooling at its upper boundary (see Section 14.1). The data in Table 14.3 indicate
that a layer with a transmission function T = 0.01 is in this sense
a cloud at all heights considered; there is a sizable increase in
the heating and in the thickness of the heated sublayer with increasing cloud height (see [7]). An absorbing layer with a transmission function T = 0.2 at heights of 1 to 2 km can no longer be
called a cloud; here a slight increase in radiative cooling with
height is observed, as in the absence of clouds. At heights of 5
to 6 km the transition from cloud to "noncloud" takes place somewhere in the interval T = 0.4-0.5. At heights of 9 to 10 km, all
three of the absorbing layers considered have profiles of the
radiant heat influx typical of clouds, but both the heating and
cooling are more intense, the lower the transmission function.

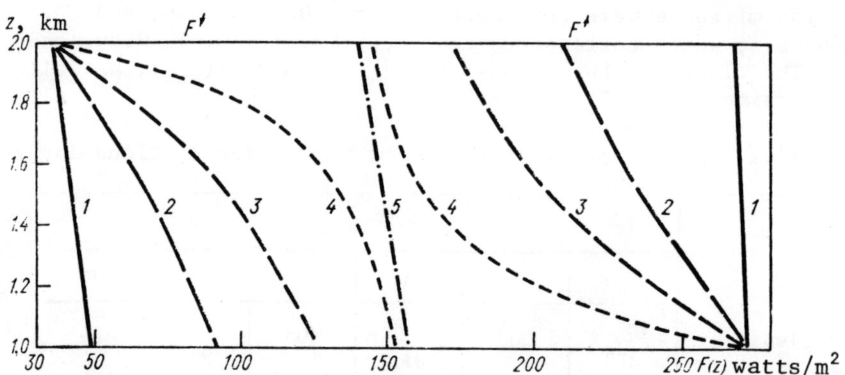

Fig. 14.4. Flux profiles $F^{\uparrow\downarrow}(z)$ in layer from 1 to 2 km.
1) cloudless conditions; 2) T = 0.5; 3) T = 0.2; 4) T = 0.01;
5) blackbody radiation at corresponding temperature.

Consequently, a semitransparent layer of absorbing substance
in the lower atmosphere may well not cause the distortions in the
radiation profiles typical of a cloud layer, since its absorptivity
differs little from that of the ambient air. If, on the other hand,

such a layer is in the upper troposphere, where it is much denser
than its surroundings, it will be capable of radiation and thermal
effects typical of cloudiness. This conclusion has a direct relationship to estimates of Ci clouds on the thermal regime of the
atmosphere.

Table 14.3. Radiant heat influx (mw/m^3) inside cloud layer (1-2,
5-6, 9-10 km) as a function of transmission function and height
of layer.

z km	T			z km	T			z km	T		
	0.01	0.20	0.50		0.01	0.20	0.50		0.01	0.20	0.50
1.0	31	—39	—46	5.0	287	55	—2	9.0	504	136	34
1.1	16	—43	—46	5.1	176	38	—6	9.1	313	106	28
1.2	2	—49	—47	5.2	105	21	—9	9.2	191	79	22
1.3	—1	—55	—47	5.3	55	3	—13	9.3	109	51	17
1,4	—25	—63	—48	5.4	16	—12	—17	9.4	52	30	10
1.5	—46	—72	—49	5.5	—19	—30	—20	9.5	5	7	5
1.6	—77	—84	—51	5.6	—58	—48	—24	9.6	—42	—15	—1
1.7	—125	—98	—52	5.7	—110	—67	—29	9.7	—97	—39	—7
1.8	—199	—114	—54	5.8	—186	—87	—33	9.8	—171	—61	—12
1.9	—317	—134	—57	5.9	—303	—111	—33	9.9	—285	—90	—19
2.0	—503	—157	—60	6.0	—484	—137	—41	10.0	—460	—119	—57

On the whole, the formalism adopted in this section did not
stand in the way of obtaining physically valid, quantitatively
correct results. This is verified by a comparison with the aggregate of data presented above.

INTRODUCTION

When describing global heat-circulation processes, it is
important to give special attention to tropical regions, in which
solar energy accumulates, and polar regions, which are regions of
energy "runoff." It will be shown in Chaps. 15 and 16 that clouds
in these regions possess special regional features with regard to
their spatial structure and physical characteristics. These special
features, together with astronomical factors, tend to make the
radiation regimes of the tropics and the polar regions somewhat
unique. Unfortunately, however, these regions are insufficiently
covered by the network of observation stations. In the tropical
seas no regularly operating network of actinometric stations exists
at all. In this part of the book an attempt will be made to generalize the data of Soviet observations in the Arctic and Antarctic,
as well as data obtained aboard Soviet vessels in the tropical
Atlantic in summer 1974, during the international ATEP expedition.*)

*) Usually referred to in English as GATE (Translator).

CHAPTER 15

THE POLAR REGIONS

Radiation conditions in the polar regions exhibit a number of special features, because of the way the solar radiation arrives there, because of the irregularity of the underlying surface (especially in the Arctic), with its varying reflectivity, and also because of the structure of the polar atmosphere and cloudiness.

The present polar network of actinometric stations comprises 16 stations on the mainland or on islands, two stations on drifting arctic ice, and six stations in the Antarctic. For periods ranging from 15 to 30 years these stations have regularly recorded all fluxes of short-wave radiation, and until 1965 they also measured the long-wave radiation.

However, radiation measurements in the free atmosphere were unfortunately only sporadic. From 1953 to 1965 an Il-14 aircraft operated quite regularly in the Arctic as a "flying laboratory," aboard which comprehensive studies of the radiation fluxes and cloud microphysics were carried out. After a long interruption these observations in the Arctic were resumed in 1976 aboard Il-18 aircraft, within the framework of the POLEKS program.

In the Antarctic actinometric radiosonde observations with aircraft were carried out only from 1958 to 1961 at Mirny [13]. The data obtained gave an idea of the profiles of thermal-radiation fluxes and they revealed the uniqueness of the radiation-transfer processes in the antarctic atmosphere.

In 1966 actinometric radiosonde observations were organized in the Antarctic, for the three winter seasons 1966–1969 at Molodezhnaya Station, and then from 1969 to 1972 at Bellingshausen [4]. In the Arctic actinometric radiosounding was carried out for several years at Murmansk Station [3]. Unfortunately, however, this region is not sufficiently representative of the Central Arctic.

15.1. Structural features of the atmosphere

Let us now consider the special features of the polar regions which must be taken into account in order to analyze and calculate the radiation characteristics of the atmosphere.

1. The low moisture content of the atmosphere, the annual mean of which is only 2.5 mm for the Antarctic and 6 mm for the entire Arctic (the analogous mean value for the whole world is 25 mm). Table 15.1 gives the seasonal variation of the water-vapor content. To a considerably lesser degree than in the Arctic, the antarctic

Table 15.1. Atmospheric moisture content in polar regions (mm) during middle months of seasons [1, 2].

Latitude	I	III	IV	VII	IX	X
90°N	1.4	2.0		11.2		5.5
80	1.6	2.6		11.9		3.6
70	2.7	3.1		14.8		4.5
90°S	1.1		0.5	0.2	0.5	
80	1.6		0.8	0.5	1.1	
70	4.0		2.9	1.9	2.9	

seasonal difference in moisture content is due to the special features of the atmospheric circulation there. During the cold season the meridional circulation in the Antarctic is the most developed, as is the associated influx of moist air from the sea. Since the local source of moisture, evaporation, is minor, therefore in contrast to the Arctic no sharp summer increase in water content is observed in the antarctic atmosphere. The Antarctic is the region with the lowest atmospheric moisture content in the world.

2. In their spatial structure and physical characteristics, clouds in the polar regions differ from clouds everywhere else in the world. Information about the polar clouds was included in Chaps. 1 through 3.

3. The stratification of the atmosphere turns out to have an appreciable effect on the regime of long-wave radiation in the polar regions. Its main idiosyncrasy there is a high frequency of inversions of temperature and humidity. Especially deep inversions, with thicknesses of as much as 1 to 5 km, are observed in the

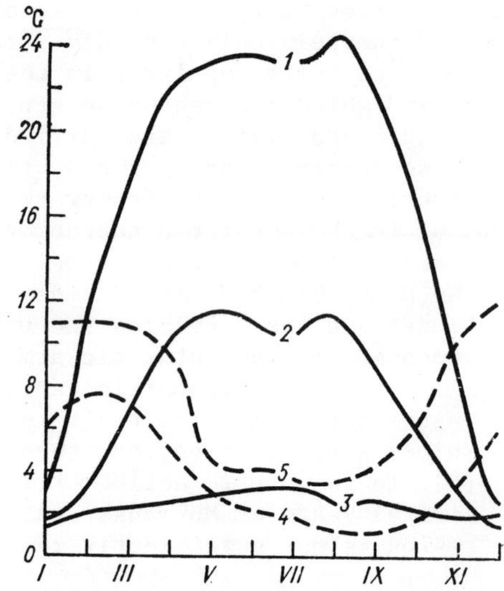

Fig. 15.1. Annual variation in intensity of surface inversions. 1) Central Antarctica (at level of 3.5 km), 2) antarctic coast (regions without converging winds), 3) antarctic coast (regions of converging winds), 4) arctic coast, 5) Arctic Basin.

central regions of the Arctic and Antarctic, the majority of them
beginning at the ground. High inversion intensities are also typical of these regions, the intensity being defined as the difference
between the maximum temperature and the temperature at the base
of the inversion (Fig. 15.1). Inversions are most frequent in winter; in spring and autumn they have a lower frequency, and in summer over the melting ice the frequency is higher again (see also
Chap. 1).

On the antarctic coast, especially in regions of converging
winds, temperature inversions are observed more rarely (frequencies
of 70 to 90%), and their intensities are lower. There inversions
at heights of 1 to 2 km ("high" inversions of subsidence) often
occur, at frequencies of more than 30%.

In winter surface inversions may have depths of 1.5 to 2 km,
while in summer the depths are less than 0.5 km. High inversions
are not as deep (0.5-0.9 km), but in rare instances, when a surface radiation inversion merges with a high inversion, these systems may reach to a height of 4 km, with intensities of 25°C.

15.2. Mean data on radiation regime of cloudless atmosphere

Because of the low moisture content and low pollution of the
polar atmosphere, its transparency is typically high: for sec ζ =
2 the transmission coefficient is 0.75-0.82 in the Arctic and
0.80-0.88 in the Antarctic. Thus the intensity of direct solar
radiation is very high, especially in mountainous regions such as
those of Greenland and Antarctica (Fig. 15.2a). In the latter the
maximum amounts of direct solar radiation are close to those typical of an ideal atmosphere, being in individual instances almost
1200 watts/m^2.

Table 15.2 gives the individual components of the attenuation
of the integral direct solar radiation S in the atmosphere, calculated using the method of [5], for several regions in the Arctic
and Antarctic. Data for Pavlovsk are presented, so as to provide
a comparison with the middle latitudes. Here ΔS_{tot} is the total
attenuation, ΔS_{H_2O} is the absorption by water vapor, and ΔS_{aer} is
the aerosol attenuation, determined as the residual effect in
Rayleigh attenuation and equal to 251 watts/m^2 at sea level and
154 watts/m^2 at a height of 3.5 km. Only during the warmest months
does the attenuation by water vapor exceed the aerosol attenuation.
At the low air temperatures that characterize most of the year,
the main factor is the aerosol attenuation created by products of
condensation and sublimation.

The annual variation of the total radiation attenuation in
the polar regions is not very evident, due to the opposite annual
variation in the water vapor and aerosol content of the atmosphere
in these regions. The total attenuation of the radiation during
the winter season is often greater than in summer.

On the antarctic coast, especially in regions of converging

winds, the aerosol-caused attenuation is less than in the Arctic, because of the dryness of the air. On the antarctic continent, at very low temperatures a large number of ice crystals form in the air, so that the aerosol component there is greater than on the coast. This property of the antarctic atmosphere explains the very low vertical gradient of S, which is on the average 2-3% per 1000 m; for mountainous regions in the middle latitudes, on the other hand, according to the data of [10] this gradient is 5-9%.

For fluctuations of the transparency of the atmosphere, the total radiation varies less than the direct radiation (Fig. 15.2b), since a partial compensation of the direct and scattered radiation takes place. The compensation effect is especially marked for an increased aerosol turbidity and a high albedo of the underlying surface. For instance, during the period of anomalous turbidity of the antarctic atmosphere after the eruption of the volcano Agung, when the transparency of the atmosphere dropped by 15 to 20%, the total radiation almost did not vary at all, in comparison with the mean data.

Calculation results had to be used to study the distribution of thermal radiation in the atmosphere, because of the insufficiency, or complete lack, of experimental data. The initial material, calculation methods, and volume of obtained information were different for the Arctic and Antarctic. For the Arctic the nomogram of F.N. Shekhter was used, together with

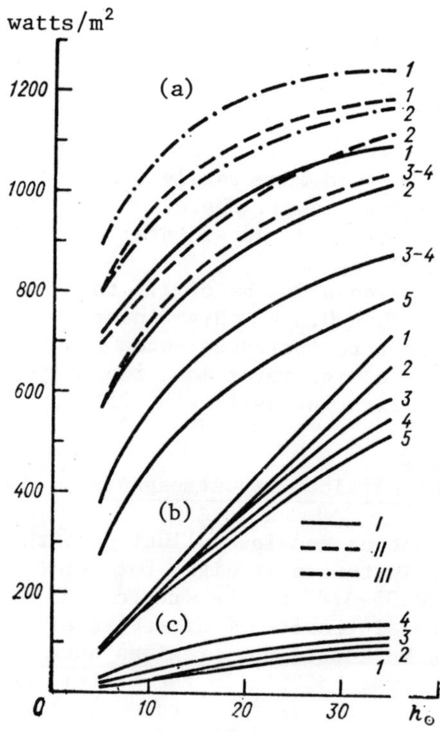

Fig. 15.2. Fluxes of direct (a), total (b), and scattered (c) radiation as functions of Sun's height.
I) mean, II) maximum, III) in ideal atmosphere.
1) Central Antarctica (at level of 3.5 km), 2) antarctic coast, 3) Arctic, winter, 4) Arctic, summer (underlying surface without snow), 5) middle latitudes (summer) [10].

mean monthly values of the aerological data; radiation fluxes from the ground up to a height of 7 km were obtained [7, 8]. In the calculations for Antarctica daily data of aerological soundings were used and fluxes of thermal radiation were computed for the principal isobaric surfaces up to 200 mb [9].

Figure 15.3 gives some typical profiles of temperature, up-

ward and downward radiation fluxes, and rate of radiative cooling
in summer and winter, for two arctic regions differing in their
conditions of atmospheric stratification.

Table 15.2. Components of attenuation of solar radiation (watts/m^2)
for Sun height of 30°.

Region	Winter			Summer		
	ΔS_{tot}	ΔS_{H_2O}	ΔS_{aer}	ΔS_{tot}	ΔS_{H_2O}	ΔS_{tot}
Arctic coast	560	113	196	551	175	125
Arctic Basin	567	106	210	560	169	140
Antarctic coast	426	105	70	440	140	49
Central Antarctic (3.5 km)	330	50	125	315	91	70
Pavlovsk	483	112	120	608	196	161

Figure 15.4 gives us an idea of the distribution of fluxes
of thermal radiation over the coastland and central regions of
Antarctica. Here and in the following, Central Antarctica will refer
to the region around the Vostok Station, which is about 3.5 km
above sea level (surface pressure 620–630 mb).

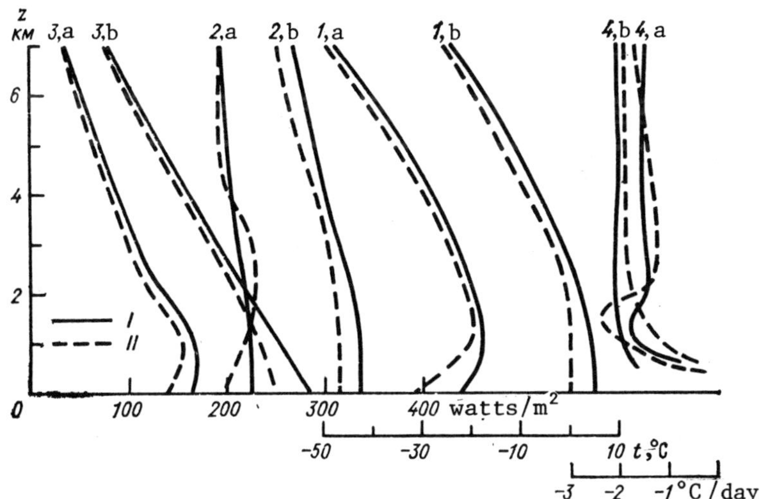

Fig. 15.3. Mean profiles of air temperature (1), ascending (2)
and descending (3) fluxes of thermal radiation, and rate of radiative cooling (4). Curves a refer to January and curves b to July.
I) Arctic coast, II) Arctic Basin.

The special features of the temperature stratification in each of the regions considered manifest themselves quite clearly in the distribution of the long-wave fluxes. For example, attenuation of $F^\uparrow$ with height occurs only in the absence of a surface inversion or for a very slight inversion. In summer on the arctic and antarctic coasts at a level of 200 mb, $F^\uparrow$ comprises 80 to 100% of the radiation of the underlying surface (Table 15.3). In the presence of an inversion $F^\uparrow$ increases with height, reaching a maximum in the layer above the inversion. The spatial variability of $F^\uparrow$ over Antarctica is much greater than over the Arctic, throughout the atmosphere, since the effect of the marked inhomogeneity of the field of the surface temperature manifests itself up to very high levels.

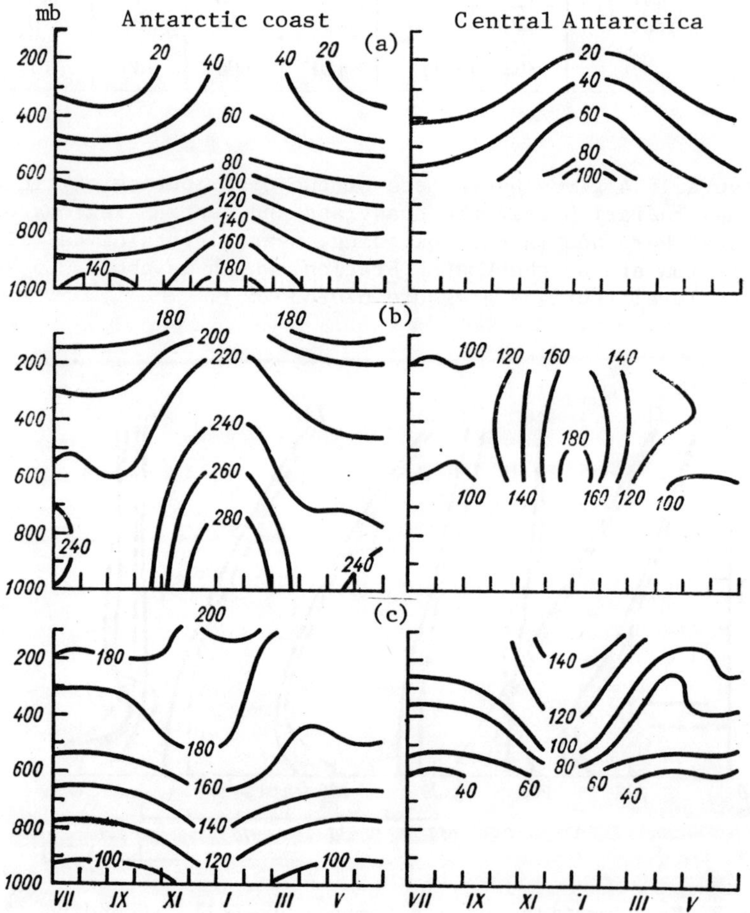

Fig. 15.4. Annual variation of long-wave radiation in troposphere over Antarctica (watts/m^2).
a) downward flux $F^\downarrow$, b) upward flux $F^\uparrow$, c) effective flux F.

Table 15.3. Values of parameters $\dfrac{F^{\uparrow\downarrow}(z)}{\sigma T^4(0)}$ for various levels of troposphere.

Region	Season	Level, mb and km					
		ground	850 1	700 3	500 5	300 7	200
$F\uparrow/\sigma T^4(0)$							
Antarctic coast	Winter	1.0	1.03	1.06	1.00	0.92	0.90
	Summer	1.0	0.91	0.91	0.88	0.80	0.80
Central Antarctica (3.5 km)	Winter	1.0	—	—	1.02	1.04	1.00
	Summer	1.0	—	—	0.95	0.90	0.88
Arctic coast	Winter	1.0	1.0	0.97	0.86	0.84	—
	Summer	1.0	0.96	0.93	0.84	0.78	—
Arctic Basin	Winter	1.0	1.08	1.12	1.0	0.95	—
	Summer	1.0	1.02	0.98	0.87	0.79	—
Middle latitudes, dry winter [11]	Winter	1.0	0.96	0.92	0.85	0.79	0.77
$F\downarrow/\sigma T^4(0)$							
Antarctic coast	Winter	0.60	0.50	0.40	0.20	0.10	—
	Summer	0.65	0.60	0.48	0.30	0.20	—
Central Antarctica	Winter	0.45	—	—	0.32	0.10	—
	Summer	0.55	—	—	0.40	0.20	—
Arctic coast	Winter	0.71	0.71	0.58	0.28	0.06	—
	Summer	0.80	0.72	0.63	0.38	0.21	—
Arctic Basin	Winter	0.71	0.71	0.58	0.28	0.06	—
	Summer	0.80	0.72	0.63	0.38	0.21	—
Middle latitudes, dry winter [11]	Winter	0.54	0.44	0.34	0.23	0.13	—

The amount of thermal radiation absorbed by the atmosphere is evaluated in terms of the quantity $\Delta F^\uparrow = F^\uparrow(0) - F^\uparrow(H)$. It varies considerably from season to season as a function of the moisture content of the atmosphere. The heat loss $\Delta F^\uparrow$ in the troposphere over the arctic and over the antarctic coast in summer ranges from 100 to 125 watts/m^2, which corresponds to a cooling of 1°C/day. In winter the heat loss drops to 10–50 watts/m^2 (or 0.1–0.5°C/day). In Central Antarctica for most of the year the atmosphere barely alters the flux $F^\uparrow$ at all. A slight amount of heat is absorbed by it only in summer (10 to 15 watts/m^2), which is equivalent to a cooling of the atmospheric layer between 620 and 100 mb by 0.2–0.5°C/day.

Flux $F^\downarrow$ has the same patterns of vertical variation as $F^\uparrow$, but its vertical gradients are considerably higher. A spatial

Table 15.4. Ratio $\dfrac{F^{\downarrow}(0)}{\widetilde{F}^{\downarrow}(0)}$ as function of temperature gradient in atmospheric layer from 0 to 1 km.

$\gamma°C/100$ m	Air temperature at ground, °C		
	-20	-10	0
0.6	1.00	1.00	1.00
0.4	1.04	1.03	1.02
0.2	1.08	1.06	1.05
0	1.12	1.10	1.07
-0.2	1.19	1.15	1.12
-0.4	1.25	1.20	1.16
-0.6	1.28	1.22	1.19

variation of $F^{\downarrow}$, in contrast to $F^{\uparrow}$, is observed only in the lower troposphere, up to a level of 500 mb.

The gradual formation of the downward radiation flux from the upper atmosphere to sea level is shown in the second half of Table 15.3. The emissivity of the atmosphere $F^{\downarrow}(0)/\sigma T^{4}(0)$ over Antarctica is on the whole much less than over the Arctic, because of the lower atmospheric moisture content. However, even for Antarctica (coast) this quantity is higher than in the "dry winter" model for the midlatitudes [11].

A temperature inversion in the lower layers of the atmosphere raises $F^{\downarrow}(0)$ appreciably. This is evident from Table 15.4 in terms of the quantity $F^{\downarrow}(0)/\widetilde{F}^{\downarrow}(0)$, where $\widetilde{F}^{\downarrow}(0)$ is the flux in the absence of an inversion for $\gamma = 0.6°C/100$ m.

Table 15.5. Matrix of correlation coefficients $[r(F_{i}^{\uparrow}, T_{j})]\cdot 10^{2}$. Winter, antarctic coast.

P_{ij} mb	1000	850	700	500	300
1000	98	97	91	64	61 $F_i^{\uparrow}$
850	84	87	93	63	65
700	66	69	84	66	66
500	58	59	68	51	66
300	51	51	57	41	67
T_j					

A study of the fluxes $F^{\uparrow\downarrow}(z)$ obtained with actinometric radiosondes on the antarctic coast in winter indicated a marked temperature dependence of the thermal-radiation field of the free atmosphere. $F^{\uparrow}$ was found to correlate best with the temperature of the underlying surface, and $F^{\downarrow}$ with the temperature at the level above the one being considered (Tables 15.5 and 15.6).

Table 15.6. Matrix of correlation coefficients $[r(F^{\downarrow}_i, T_j)]\cdot 10^2$. Winter, antarctic coast.

P_{ij} mb	1000	850	700	500	300
1000	86	84	76	65	42 $F^{\downarrow}_i$
850	74	74	76	64	44
700	59	79	70	69	53
500	62	60	79	79	68
300	59	59	60	61	63
T_j					

When all the foregoing results were compared with corresponding data for the middle latitudes, it became evident that the vertical profiles of the thermal-radiation fluxes in the polar atmosphere cannot be described using models worked out for other parts of the world.

15.3. Effect of cloudiness on regime of solar radiation

Because of their lower water content, their thickness, and their phase makeup (see Chaps. 1 and 2), polar clouds have a higher transparency than clouds elsewhere in the world.

According to observations in the polar regions, for unbroken cloudiness of any type the total radiation is almost always less than under cloudless conditions (Table 15.7). For unbroken lower-level cloudiness, and in the Arctic middle-level cloudiness as well, the attenuation of the radiation differs greatly from season to season. This is because the cloud properties themselves vary, and also because the properties of the underlying surface in the observation region are different.

Lower-level clouds in Antarctica attenuate the radiation more than in the Arctic (conditions over a snowy surface are compared). The extensive single-layer cloud fields typical of the Arctic apparently attenuate radiation less than do the multilayer clouds over the antarctic coast.

The transmission T of radiation by clouds in the polar regions increases substantially with an increase in the Sun's height. This characteristic is more sharply defined in the Arctic than in the Antarctic. For example, in the Arctic an increase in the height of the Sun $h_{\odot}$ from 10 to 40° almost doubles the relative total

Table 15.7. Mean amounts of total radiation for unbroken cloudiness (as percent of radiation under cloudless conditions).

Region	Season	Sun's height, °			
		10	20	30	40
		Upper level			
Arctic coast	Winter	88	92	97	—
	Summer	80	86	92	94
Arctic Basin	Winter	88	92	97	—
	Summer	86	90	96	98
Antarctic coast	Year	82	88	95	97
Central Antarctica	Year	96	98	94	94
		Middle level			
Arctic coast	Winter	72	74	85	—
	Summer	38	42	50	64
Arctic Basin	Winter	72	74	87	—
	Summer	52	58	68	78
Antarctic coast	Year	58	66	74	80
		Lower level			
Arctic coast	Winter	45	55	66	—
	Summer	26	33	40	52
Arctic Basin	Winter	59	64	73	—
	Summer	40	46	57	70
Antarctic coast	Winter	48	55	61	—
	Summer	36	44	49	54
Middle latitudes [10]	Summer	23	20	21	23

radiation. In the midlatitudes variations in the attenuation of radiation with the height of the Sun are also observed, by about 15% according to the data of [10]. The high values of $dT/dh_\odot$ in polar regions in comparison with the middle latitudes are due to the lower optical depths of polar clouds and to the special features of the scattering function for polar ice clouds (see Chap. 8).

In [6, 12] the attenuation of the total radiation by a cloud layer was determined as a function of the layer thickness on the basis of observations with a "flying laboratory" (Fig. 15.5).

Here $P = \dfrac{Q_1 - Q_2}{Q_1} \cdot 100\%$, where Q_2 and Q_1 are the amounts of total radiation at the upper and lower cloud boundaries, respectively. The curves in Fig. 15.5 are approximated by the exponential

$Q_1 = Q_2 e^{-kH}$, where H is the cloud depth in meters, and k is the attenuation factor in m^{-1}.

Attenuation factor k depends not only on the cloud properties but also on the nature of the underlying surface, since for a snow-ice surface Q_1 increases due to multiple reflection, For clouds over ice $k = 0.11 \cdot 10^{-2}$, and for clouds over water $k = 0.20 \cdot 10^{-2}$ to $0.35 \cdot 10^{-2} m^{-1}$.

Observations with the "flying laboratory" showed that the albedos of St, Sc, and Ns varied from 20 to 85%. The albedos of clouds less than

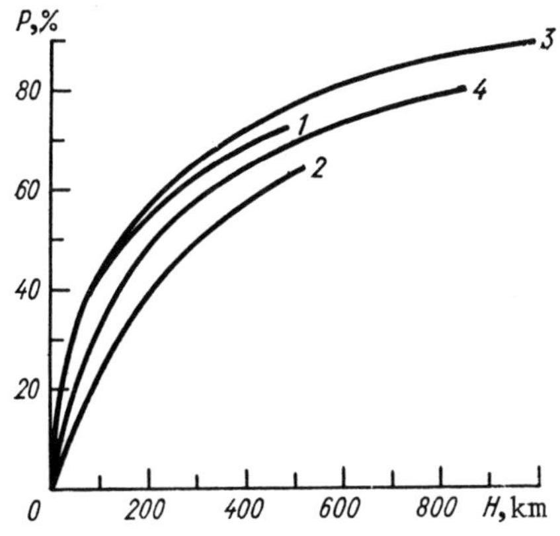

Fig. 15.5. Attenuation of solar radiation as function of cloud depth. 1) according to data of Koptev [6], 2) according to data of Timerev [12], 3) POLEKS-Yug (lower-level cloudiness), 4) POLEKS-Yug (middle-level cloudiness).

500 m thick depend on the nature of the underlying surface. On the average, the albedos of clouds over water are from 30 to 35%, and those of clouds over ice are from 60 to 70%.

15.4. Effect of cloudiness on regime of thermal radiation

The low humidity of the atmosphere in the polar regions enhances the effect of cloudiness on fluxes of thermal radiation. Observations have indicated that the counterradiation of the atmosphere for an unbroken cloud cover increases on the average by 30% in winter and by 20% in summer. The variations of these quantities are considerable, and they depend on the height and thickness of the cloud layers. A quantitative evaluation of the effect of cloudiness on the counterradiation of the atmosphere in the Arctic [7] showed that the emissivity of the lower cloud boundary for air temperatures of less than -6°C at its level is reduced, relative to the radiation of a perfect blackbody (Table 15.8).

Table 15.8. Emissivity of cloudiness as function of temperature at lower boundary.

t,°C	—6	—10	—15	—20	—25
$\dfrac{F\uparrow}{B(T)}$	0.97	0.90	0.88	0.87	0.85

The effect of the cloudiness on the outgoing radiation was calculated assuming the upper boundary of the lower-level cloudiness

to be at a height of 1000 m, and that of the middle-level cloudiness to be at 3000 m. Calculations for the Arctic showed that
lower-level clouds located between 200 and 1000 m enhance flux
$F^\uparrow(H)$ by 10%, on the average for a year. In individual instances,
for warm subinversion clouds this increase may be as high as 20%.
Middle-level cloudiness situated between 2000 and 3000 m always
reduces $F^\uparrow(H)$.

    Data of actinometric sounding were used to evaluate the role
of cloudiness in the antarctic atmosphere (Fig. 15.6). Profiles
1 and 2 pertain to clear-sky conditions; they characterize the
vertical variation of F in the presence of a surface inversion.

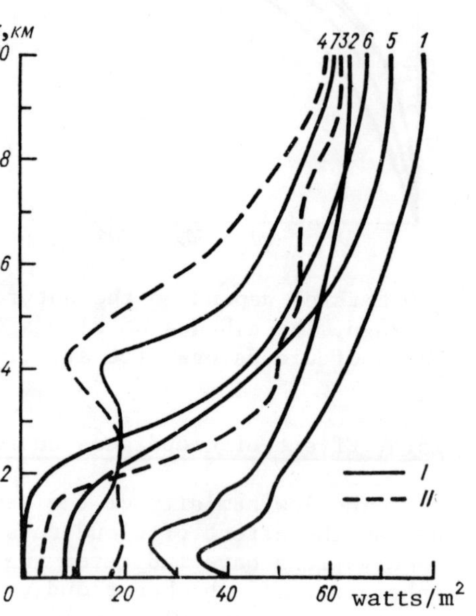

Fig. 15.6. Vertical profiles of
effective flux of long-wave radiation for different cloud conditions.
I) mean data for various numbers
of cases, II) individual cases.
1) clear, surface temperature
-10°C, 2) clear, surface temperature -25°C, 3) 10/10 St, Sc
in layer from 600 to 200 m,
4) 10/10 As in layer from 3000
to 4500 m, 5) 10/10 St, Sc in
layer from 600 to 1000 m,
6) 10/10 St, Sc in layer from
100 to 200 m, 7) 10/10 As.

    Most of the measurement for unbroken (10/10) cloudiness were
in the temperature range from -8 to -15°C. The upper cloud boundary
for St and As in each individual case (curves 3, 4) and even for
averaging (curves 5-7) is clearly shown by the discontinuity on
the profile of flux F. The lower boundary is evident only for As
clouds (curves 4, 7). The lower boundary of St and Sc does not
show up on the curves, and the height indicated by the observer
is used.

    Flux F forms mainly up to a height of 7 km, above which it
increases only very gradually. The maximum values of F are observed
for clear skies and a comparatively high temperature of the underlying surface (curve 1). In contrast to the Arctic, unbroken lower-
level cloudiness in the Antarctic does not enhance appreciably the
flux in the above-cloud layer. Only if cloud and cloudless conditions at a low air temperature are compared, is a slight increase
in F above 5 km observed for lower-layer cloudiness.

CHAPTER 16

RADIATION REGIME OF THE TROPICAL CENTRAL ATLANTIC

The radiation regime over the tropical seas has been measured only sporadically during recent years, within the framework of expeditions with scientific-research vessels. In 1974 the Atlantic Tropical Experiment (ATEP)*) was successfully carried out in the tropical Atlantic. This was one of the first major experiments of the international Global Atmospheric Research Program (GARP), and 13 Soviet research vessels (R/V) and research weather vessels (RW/V) participated in it.

The GATE-74 project, its goals, and the Soviet findings ensuing from it have been described in several special collections (for instance, TROPEKS-74 [22]) and in numerous papers. This chapter is based on an analysis of actinometric material gathered aboard all the Soviet vessels in GATE as well as of data from actinometric radiosonde observations made with 10 R/V.

16.1. Fluxes of solar radiation in water-adjacent layer under cloudy and cloudless conditions

When studying the tropics, we have to consider separately the tradewind region and the intertropical convergence zone (ITC), since these regions are characterized by completely different conditions of cloud genesis and development.

An analysis of charts of the distribution of daily sums of the total solar radiation (Q) for the GATE region clearly reveals an area of minimum Q, which is a consequence of the cloud cover of the ITC. In the ITC the presence of a denser cloud cover makes the amount of solar heat reaching the ground about 20% less than in the tradewind region. There are sizable day-to-day fluctuations in the total solar fluxes in the convergence zone. The standard deviation of Q there ranges from 60 to 80 watts/m^2, while in the tradewind region it is only 25 watts/m^2.

The small-scale structure of the solar-radiation field was investigated via a harmonic analysis of the values of the total solar radiation measured aboard R/V *Akademik Kurchatov* during the midday hours with a resolution of 36 seconds. Fluctuations with a period of about 20 min were detected, originating against a background of diurnal periodicity and as a result of banks of convective clouds.

With an increase in the frequency f, the spectral density s

---

*) GATE in English (Translator).

can be approximated as $s = f^{-k}$, where $k = 0.82$ [26]. For multi-layered cloudiness of the ITC in the Indian Ocean, parameter k was assigned a value of 2.4, on the basis of data from the Musson-77 experiment.

In contrast to the tradewind region, in the ITC low-frequency fluctuations with a period of several hours predominate, since the main role in the variability of the radiation fluxes near the surface is played not by individual clouds but rather by cloud clusters.

An important characteristic of the regime of solar radiation in the tropics is an increase in the flux of total solar radiation, in comparison with cloudless conditions, observed for large Sun heights and for clouds located close to the solar disk. To describe this phenomenon, Table 16.1 gives the frequencies p of the ratios Q/Q_0, where Q represents the observed hourly sums of the total solar radiation, and Q_0 represents the possible sums of the total solar radiation in the absence of clouds and for the transparency of the cloudless parts of the sky during the observations [19].

Table 16.1. Frequency of ratios of actual hourly sums of total solar radiation (Q) to corresponding possible sums for clear sky (Q_0).

Q/Q_0	$p, \%$	Q/Q_0	$p, \%$
0.50... 0.59	2	1.00... 1.09	28
0.60... 0.69	4	1.10... 1.19	2
0.70... 0.79	6	1.20... 1.29	1
0.80... 0.89	14	1.30... 1.39	1
0.90... 0.99	42		

It is seen from the table that 32% of the cases lie in the range $Q/Q_0 \geq 1$. On the average, the clouds in the tradewind region reduce the hourly sums of Q by 6%. However, in individual instances, due to the reflection of solar highlights by the cloud edges, Q may be higher than Q_0.*)

On the basis of a statistical processing of simultaneous observations of the total radiation and the amounts of cloudiness of various types during the period of GATE-74, the following formula was obtained for the total radiation, taking the type of cloudiness into account:

$$Q_n = Q_0 - a (\sin h)^b n^z, \qquad (16.1)$$

---

*) It should also be kept in mind that the values assumed for Q_0 may not correspond accurately enough to the state of the cloudless sky at the time of measurement of Q (Editor).

where Q_n and Q_0 are, respectively, the intensities of the total radiation of a cloudy and cloudless sky, n is the total amount of cloudiness in fractional units, h is the Sun's height, and a and b are coefficients depending on the cloud type. The fit between the measured and calculated values of Q_n was best for $\alpha = 3$. Table 16.2 gives the values of a and b in formula (16.1).

Table 16.2. Values of coefficients a and b in formula (16.1) for various cloud types.

Cloud type . . .	Ci	Ac	Cu	Sc	Cb
a	4.9	24.4	34.9	48.2	65.6
b	0.70	0.70	1.15	1.10	1.15

Table 16.3 gives values of the total radiation over the ocean, calculated with formula (16.1) for a cloud cover of 10 (different cloud types) at various heights of the Sun. A comparison of these values with similar data for continental conditions led the authors of [5] to conclude that over the ocean clouds of the lower and middle levels are considerably more transparent to the total radiation than over the continent. For Ci clouds the differences are negligible: at Sun heights less than 50° the radiation over the oceans is somewhat less than over the continents, while for greater Sun heights the opposite is true.

Table 16.3. Total radiation (watts/m^2) over ocean for cloud cover of 10 (various cloud types).

Cloud type	Sun height								
	10	20	30	40	50	60	70	80	90
Ci	105	246	395	539	669	778	860	911	928
Ac	48	155	275	396	507	601	673	718	733
Sc	49	122	200	278	350	411	455	485	495
Cb	32	79	130	180	227	266	296	315	321

Aerosol turbidity of the atmosphere, caused by the arrival of dust particles from arid regions, is also typical of the tropical latitudes of the eastern Atlantic, as well as some other parts of the World Ocean [10, 15]. In such cases in the absence of clouds the lower transparency makes the daily sums of the total solar radiation 10 to 15% less than those possible under normal transparency conditions.

Table 16.4. Statistical characteristics of relative fluxes of solar radiation in tropical Atlantic (mean values S^*, Q^*; variances $\sigma^2_{S^*}$, $\sigma^2_{Q^*}$, etc.)

Amount of cloudiness	Flux	$\overline{S}^*$ or $\overline{Q}^*$	$\sigma^2_{S^*}$ or $\sigma^2_{Q^*}$	Main parameters of distributions $p(S^*)$ and $p(Q^*)$						
				min S^*,Q^*	max S^*,Q^*	mode		probability of mode		median
						1	2	1	2	
0.2	S^*	0,82... 0,86	0,042	0	1,0	0... 0,06	0,85... 0,90	0.02... 0.04	0,53... 0.83	0.85... 0,90
0,3	S^*	0.80... 0.83	0,074	0	1,1	0... 0.03	0.80... 0,85	0.10... 0,15	0,44... 0.50	0.80... 0,85
0.4	S^*	0.65... 0.70	0.10	0	0.95	0	0,85	0.20	0.50	0,80
0.4	Q^*	0.63... 0.70	0,028	0.25	1,0	0,35... 0.45	0.70... 0.85	0,03... 0.04	0,23... 0.37	0.70... 0.75
0.5	Q^*	0.43... 0.73	0,021... 0,042	0.05... 0.35	1.20	0,30... 0,50	0,10... 0.75	0,35... 0.50	0.10... 0,12	0,45... 0.65
0.6	Q^*	0.55	0,024... 0.025	0.25	1.20	0,40... 0.50	0.80	0.13... 0,15	0.06	0.55
0.7	Q^*	0,44... 0.49	0,043... 0.066	0.10	1.20	0.30	0,80... 0.90	0.08... 0,12	0.04... 0,08	0,40... 0,45
0.8	Q^*	0.33... 0.40	0,026... 0.027	0,15	0.95	0.25... 0.30	0,70... 0.90	0,17... 0,33	0.02	0,30... 0.40
0.9	Q^*	0.32... 0.46	0,019... 0,044	0.15	1,05	0,20... 0.25	0.80	0.17... 0.21	0,04... 0.06	0,2)... 0.35

A number of authors [9, 13, 16] have analyzed the statistical structure of radiation fluxes in the tropical Atlantic. They found that, on the whole, the statistical characteristics of the relative fluxes of direct $S*$ and total $Q*$ radiation for cumulus clouds are qualitatively similar to those obtained for the midlatitudes of the European USSR (see Section 11.4 and [20, 21]). The relations between fluxes $S*$ and $Q*$ and the amount of cloud cover n for the GATE region turned out to be nonlinear. They can be approximated by the expression

$$Q* = 0.75 - 0.005n^2 \qquad\qquad (16.2)$$

for $0.60 \le p_2 \le 0.75$. Here p_2 is the transmission coefficient for mass 2, and n is in fractional units. The error in calculating $Q*$ with the aid of expression (16.2) does not exceed ±0.25.

Allowing for the slight differences in the statistical characteristics of the radiation fluxes for the middle latitudes, as a first approximation the structure of the radiation fields in the tropical Atlantic can be described by the approximations given in Section 11.4 and in [16, 20, 21].

Table 16.4 summarizes the statistical characteristics of fluxes $S*$ and $Q*$ on the basis of data obtained aboard two GATE vessels: R/V *Akademik Korolev* (12°N, 23°30'W) and RW/V *Ernst Krenkel'* (6°30'N, 20°W).

16.2. Classification of fluxes of thermal radiation*)

Within the framework of GATE-74 [22] fluxes of thermal radiation were measured on ten Soviet ships with the aid of actinometric radiosondes of the Central Aerological Observatory; 565 soundings were made [7, 8]. Along with the integral hemispherical fluxes of upward $F^\uparrow(z)$ and downward $F^\downarrow(z)$ thermal radiation, the temperature $t(z)$, relative air humidity $u(z)$, and pressure $p(z)$ were measured as well.

Of the ten Soviet ships equipped with actinometric radiosondes, seven were sited on the periphery and at the center of the principal GATE test area (A/B) from 5 to 12°N [18]. R/V *Akademik Kurchatov* and RW/V *Passat* operated at the equator, and RW/B *Volna* in the region from 8 to 12°N and 30 to 44°W. This section presents the results of a systematic analysis of actinometric radiosoundings for the following categories: 1) for the test area A/B cloudy and cloudless situations are considered separately; 2) all data are considered together. An equatorial model was constructed, but cloudy and cloudless situations were not distinguished in it, since the

---

*) Sections 16.2 and 16.3 are a concise version of [18].

loose trade cloudiness prevailing at the equator affects the radiation fluxes only negligibly [8,11]. Altogether, four main models were constructed: equatorial, tropical cloudless, tropical cloudy, and general tropical. In individual cases other essembles of data were invoked as well.

The method proposed in [8] was used, in order to distinguish cloud conditions objectively and obtain information about clouds to supplement the visual information. Let us consider the relation

$$f(p) = F(p)/\overline{F}_{no\ cld}(p). \qquad (16.3)$$

where $F(p) = F^{\uparrow}(p) - F^{\downarrow}(p)$ is the effective or resultant radiation flux, and $\overline{F}_{no\ cld}(p)$ is the mean effective flux according to measurements under cloudless conditions.

If $f(p) \geq f^*$ for 970 mb $\geq p \geq$ 100 mb, where f^* is a specified number less than unity, then the atmosphere was cloudless along the radiosonde path at the time of the sounding. If $f(p) < f^*$ in some region (p_1, p_2), then clouds can be expected there.

The center of the cloud, or its densest part, is located at the level p_0 corresponding to the minimum of $f(p)$, namely min $f(p) = f_0 = f(p_0)$ and $p_1 > p_0 > p_2$. The cloud boundaries are difficult to establish, but in an case they do not lie outside the region (p_1, p_2); thus p_1 and p_2 were taken arbitarily to be the levels of the cloud boundaries (Table 16.5).

Table 16.5. Levels of cloud boundaries (p_1, p_2) and cloud centers (p_0).

$p_1 - p_2$ mb	No. of cases %	p_0 mb	No. of cases %
970... 150	47.7	900... 800	26.5
970... 300	16.4	800... 700	15.5
970... 500	9.9	700... 600	8.7
970... 700	15.3	600... 500	11.1
700... 400	3.3	500... 400	14.9
500... 150	2.7	400... 300	7.9
400... 150	5.0	300	15.4

As applied to 477 soundings, this method proved to be very fruitful. All the predictions of [8] were confirmed and made more precise.

A total of 135 cases satisfying the inequalities $0.8 \leq f(p) \leq 1.3$ for 970 mb $\geq p \geq$ 150 mb were distinguished. In 95% of the cases the limits of the inequality were reduced to $0.8 \leq f(p) \leq 1.1$, and in 70% to $0.9 \leq f(p) \leq 1.1$. Here, as $\overline{F}_{no\ cld}$ in (16.3) for test area A/B, the mean flux calculated from actinometric

radiosounding data of three vessels on a principal meridian under
cloudless or slightly cloudy conditions was used. The latter were
identified on the cloud legend ($n_{tot} \leq 2$) by a careful check of
the $F(p)$ profiles and a rejection of those which revealed the
influence of clouds [8].

Clearly, the 135 soundings which lay within the limits of the
inequalities represent the radiation regime of the cloudless tropical sky itself. The narrow limits of these inequalities attest to
the uniformity of the radiation field of the cloudless tropical
atmosphere. Table 16.6 gives the distribution of cloud layers and
their centers according to the proposed evaluation, referred to
below as the "radiation evaluation."

Table 16.6. Distribution (percent) of radiation evaluations of
cloudiness according to amount of cloud cover.

Radiation evaluation	Amount of cloud cover and predominant cloud type				No. of cases
	$\leqslant 2$ Cu	3–5 Cu, Ac	6–8 Cu, Ac Sc, Ci	9–10 Cb,Cu,Sc Ac, Ci	
Cloudless	38.7	30.4	21.5	10.4	135
Slightly cloudy	7.0	23.3	30.2	39.5	43
Moderately cloudy	8.2	23.5	16.5	51.8	85
Dense cloudiness	0.5	4.2	14.0	81.3	214

As the table shows, clouds extending over the entire thickness of the troposphere, or almost all of it, are predominant,
that is, dense convective cloudiness. In a number of cases two
values of min $f(p)$ were observed, indicating two-layer clouds.
In all, 69% of the clouds were single-layered, and 31% were double-
layered. For convective clouds there was probably often a third,
upper layer of cirrus present as well, but it could not be detected.
At the same time, as Table 16.5 shows, in some cases upper clouds
were observed in the absence of lower-lying clouds.

In [8] the thickness of the clouds was estimated on the basis
of the quantity $f_0 = f(p_0)$. The clouds turned out to be dense
($f_0 < 0.4$) in 63.9% of the cases, moderate ($0.4 \leq \bar{f}_0 \leq 0.6$) in
23.6%, and slight ($0.6 \leq \bar{f}_0 \leq 0.8$) in 12.5%. Table 16.6 shows the
degree of correlation of the cloud legend of these radiation
evaluations of the cloudiness, together with an evaluation of the
cloudless state on the basis of the inequality $f(p) \geq 0.8$ for
100 mb $\leq p \leq 970$ mb.

There is evidently a good agreement between the radiation
evaluations and the meteorological evaluations for dense cloudiness.
In the case of moderate and, in particular, slight cloudiness, the
radiation evaluation is distributed with almost equal probability

over the cloud amounts. These intermediate situations apparently correspond to a passage of the radiosonde through the peripheral regions of the clouds. Finally, at first glance it might seem surprising that the radiation evaluation gives a large number (32%) of cases where $n_{tot} \geq 6$ for a cloudless sky. Inspection of the initial material indicated that in these instances Ac and Ci clouds predominated. Evidently, the variations in the radiation fluxes caused by these clouds often lie within the range of natural flux variability for a clear sky. Only quite dense Ac and Ci clouds can be definitely distinguished with a radiosonde.

Below, "cloudless" and "cloudy" models will be constructed on the basis of a radiation evaluation. Thus it should be kept in mind that the former also includes instances of nondense (loose) clouds of the middle and upper levels, when a radiosonde actually responds weakly to clouds. The cloud model unites all the situations in which the clouds in fact affect the radiation field. Consequently, these models are quite suitable for illustrating the influence of clouds on the radiative heat transfer.

16.3. Fluxes of thermal radiation and radiative cooling of a cloudy or cloudless atmosphere

The flux of effective radiation $F(p)$ is, strictly speaking, the only characteristic of the thermal radiation used in the energetics of the atmosphere. The radiative cooling is calculated with the aid of the formula $R(p) = -\text{div } F(p)$. Only fluxes $F(p^*)$ and $F(0)$ enter into the boundary conditions for the equation of heat transfer at the lower ($p = p^*$) and upper ($p = 0$) limits of atmosphere; for flux $F(0)$ the natural condition $\lim_{p \to 0} F^\downarrow(p) = 0$ obtains.

Table 16.7 gives the mean fluxes $\overline{F}(p)$ for the designated models, together with the variability coefficients $v_F(p) =$
$\sigma_F(p)/\overline{F}(p)$; $\sigma_F(p) = \sqrt{\overline{\sigma_F^2(p)}}$; here $\sigma_F^2(p)$ is the variance of flux $F(p)$. This table verifies the assumed classification of the data.

The equatorial zone clearly exhibits the highest values of the mean flux and a low variability. Here the effect of the clouds is negligible, on the average, just a slight increase in variability . In the tropics proper (test area A/B) the cloudless radiation regime is quite close to the equatorial regime for lower $F(p)$, because of the high humidity [8].

In the absence of clouds the radiation field is isotropic and uniform: the mean data and variability coefficients according to three vessels of the principal meridian (23.5°W) and four northern vessels (10°N) are practically identical. The first set of data are also close to the data for the test area as a whole. Clouds reduce flux $F(p)$ and make its variability higher.

For four models Table 16.8 gives the mean profiles of the unidirectional fluxes $F^{\uparrow\downarrow}(p)$, the temperature $t(p)$, and the relative

Table 16.7. Mean effective radiation $\overline{F}(p)$ in watts/m^2 and variability coefficients v_F.

p mb	Equatorial general		General tropical		Cloudless tropical		Cloudy tropical	
	$\overline{F}$	v_F	$\overline{F}$	v_F	$\overline{F}$	v_F	$\overline{F}$	v_F
970	49	0.21	33	0.81	45	0.22	26	0.45
900	63	0.22	39	0.90	56	0.16	29	0.50
850	71	0.20	42	1.00	63	0.15	32	0.50
800	81	0.17	49	0.61	70	0.17	36	0.51
750	89	0.16	55	0.54	75	0.16	41	0.47
700	97	0.17	61	0.49	82	0.14	47	0.44
650	103	0.18	67	0.47	88	0.14	52	0.45
600	112	0.20	72	0.46	97	0.12	56	0.47
550	121	0.20	77	0.49	104	0.12	59	0.53
500	132	0.19	81	0.47	114	0.12	66	0.51
450	141	0.16	91	0.47	123	0.13	75	0.45
400	151	0.13	102	0.41	134	0.12	84	0.42
350	156	0.12	112	0.39	140	0.11	93	0.39
300	164	0.12	119	0.36	147	0.11	101	0.37
250	173	0.10	126	0.34	153	0.11	109	0.38
200	176	0.09	130	0.31	153	0.13	114	0.31
150	176	0.09	132	0.29	152	0.16	117	0.27
100	183	0.09	141	0.26	155	0.13	127	0.23
70	191	0.10	152	0.22	164	0.13	137	0.20
50	194	0.10	158	0.17	169	0.12	143	0.20
30	198	0.09	162	0.16	183	0.07	150	0.17
20	201	0.10	166	0.16	190	0.07	155	0.16
10	203	0.10	163	0.17	170	0.13	157	0.20

Note. The equatorial model was constructed using data gathered aboard R/V *Akademik Kurchatov*, and the cloudless tropical model using data of vessels *Akademik Korolev*, *Professor Zubov*, and *Professor Vize* (23.5°W); the general tropical and cloudy tropical models were constructed using data of all seven vessels of test area A/B.

humidity $u(p)$; the standard deviations $\sigma(p)$ are also given.

The temperature field in the tropics is stable, and the temperature of the troposphere at the equator is practically the same as the temperature at tropical latitudes. The variance of the temperature increases in the vicinity of the tropopause, as is well known [6].

The humidity and its variance are seen to increase in the tropics proper, in comparison with the equator, and under cloudy conditions they differ appreciably from the corresponding characteristics in clear weather.

The flux of upward radiation $F^\uparrow(p)$ for $p \geq 500$ mb is stable because of the stability of the temperature, and its is unaffected by the appearance of clouds, due to the high humidity of the lower

Table 16.8. Mean values and standard deviations of temperature t, relative humidity u, and radiation fluxes $F^{\uparrow\downarrow}$.

p mb	$t°C$		u %		$F^{\uparrow}$ watts/m^2		$F^{\downarrow}$ watts/m^2	
	$\bar{t}$	σ_t	$\bar{u}$	σ_u	$\bar{F}^{\uparrow}$	$\sigma_{F^{\uparrow}}$	$\bar{F}^{\downarrow}$	$\sigma_{F^{\downarrow}}$
1) Equatorial model								
970	21.4	0.9	82	3	439	7	390	7
900	17.8	1.0	80	9	432	7	369	14
800	13.9	1.2	54	14	411	7	334	14
700	9.1	1.0	42	15	397	7	299	14
600	2.2	1.2	46	18	376	7	265	21
500	−5.4	1.0	41	21	348	14	216	14
400	−15.2	1.3	27	16	328	14	174	14
300	−30.6	1.6	32	17	299	14	132	7
200	−51.3	1.5	36	15	265	14	84	7
150	−65.3	2.0	35	13	244	14	63	7
100	−77.7	2.3	36	13	223	14	42	7
70	−73.1	2.6	35	12	223	14	28	7
50	−68.1	1.7	35	11	230	14	35	7
30	−59.8	2.4	34	9	237	14	35	7
20	−53.1	2.6	31	7	237	21	35	7
10	−41.0	3.0	26	5	244	14	42	14
2) General tropical model								
970	22.6	0.9	85	6	453	7	418	14
900	18.7	1.0	86	8	439	7	397	14
800	14.2	1.4	79	16	411	7	362	21
700	8.8	1.3	72	17	390	7	328	21
600	1.4	1.1	73	17	369	7	293	28
500	−6.3	1.2	68	20	334	14	251	28
400	−16.5	1.4	56	21	307	14	202	28
300	−31.8	1.6	56	19	272	21	146	21
200	−53.7	2.1	58	15	223	28	91	14
150	−67.0	2.5	58	13	202	28	70	14
100	−77.2	2.7	57	12	188	28	49	7
70	−72.4	2.4	55	11	188	21	42	7
50	−66.9	5.0	52	11	195	21	42	7
30	−57.8	2.9	46	10	209	21	42	7
20	−52.5	2.6	38	8	216	21	49	14
10	−42.2	5.7	32	8	223	21	63	28
3) Tropical cloudy model								
970	22.3	0.9	86	6	453	7	425	14
900	18.7	1.1	87	8	439	7	404	14
800	14.0	1.4	83	14	411	7	376	14
700	8.6	1.4	76	16	390	7	342	21
600	1.4	1.2	76	16	362	7	307	21
500	−6.3	1.2	72	20	328	7	265	28

p mb	$t°$C		u %		$F^\uparrow$ watts/m^2		$F^\downarrow$ watts/m^2	
	$\bar{t}$	σ_t	$\bar{u}$	σ_u	$\bar{F}^\uparrow$	$\sigma_{F\uparrow}$	$\bar{F}^\downarrow$	$\sigma_{F\downarrow}$
400	−16,6	1.4	61	21	300	14	216	28
300	−31,7	1.6	60	19	258	21	160	21
200	−53,7	2.2	61	15	216	21	98	14
150	−67,5	2.4	60	13	195	28	77	14
100	−77,1	2.9	58	12	174	21	49	14
70	−72,6	2.7	56	11	181	21	41	7
50	−67,3	2.4	53	11	188	21	42	7
30	−58,1	3.1	47	10	195	21	49	7
20	−52,6	2.8	40	6	202	21	49	7
10	−44,5	4.0	31	8	216	21	56	21

4) Tropical cloudless model

p mb	$\bar{t}$	σ_t	$\bar{u}$	σ_u	$\bar{F}^\uparrow$	$\sigma_{F\uparrow}$	$\bar{F}^\downarrow$	$\sigma_{F\downarrow}$
970	22,3	1.0	83	2	453	7	404	14
900	18,5	0.9	81	2	439	7	383	14
800	14,5	1.5	70	16	418	7	348	14
700	8,0	1.2	63	16	397	7	321	14
600	1,4	1,1	67	14	376	7	279	7
500	−6,2	1.4	60	19	348	7	230	14
400	−16,3	1.4	42	13	320	7	188	14
300	−31,4	1.2	46	14	286	7	139	14
200	−52,7	2.0	55	13	251	14	98	14
100	−65,9	1.9	58	10	223	21	70	7
70	−77,0	2.8	57	10	202	21	49	7
50	−72,4	2.2	55	9	202	14	42	7
30	−58,4	2.8	46	8	223	14	42	7
20	−52,9	2.3	37	6	237	7	42	7
10	−38,3	3,6	36	9	244	14	57	28

troposphere. At levels of $p < 500$ mb under cloudy conditions, $F^\uparrow(p)$ decreases, while $\sigma_{F\uparrow}(p)$ increases. The rise in $\sigma_{F\uparrow}$ in the vicinity of the tropopause is associated with an increase in σ_t in this layer.

Parameter $\sigma_{F\downarrow}$ definitely increases with a rise in p for $p = 200$–300 mb. The upper cloud boundaries are evidently often lo- cated there, and also σ_u increases. Note, too, that for $p \geq 400$ mb in all the models $\sigma_{F\uparrow} \ll \sigma_{F\downarrow}$. This means that the variability of the effective flux $\tilde{F}(p)$ is mainly a consequence of the behavior of flux $F^\downarrow(p)$.

If the values of σ_F (where $f = t, u, F^{\uparrow\downarrow}$) for the tropics are compared with the analogous characteristics for the midlatitudes [6], it is seen that in the former case the atmosphere is very stable.

The variation of $F^{\uparrow\downarrow}(p)$ with height is seen to be more

pronounced in the tropics, as
compared with the mean profiles
of fluxes $F^{\uparrow\downarrow}(p)$ in the middle
latitudes (see Table 13.1 and
13.2). In all the cases con-
sidered, $F^{\uparrow\downarrow}_{\text{trop}}(p^*) > F^{\uparrow\downarrow}_{\text{midl}}(p^*)$,

while $F^{\uparrow\downarrow}_{\text{trop}}(100 \text{ mb}) < F^{\uparrow\downarrow}_{\text{midl}}(100 \text{ mb})$.

The spatial structure, or lati-
tudinal variation, of the effective
flux $F(p)$ at various meridians is
portrayed in Fig. 16.1. In the
intertropical convergence zone
(ITC) the values of $F(p)$ are
seen to reach extremal values.
It is also clear that in the
lower troposphere the field can
be considered zonal. Parameters

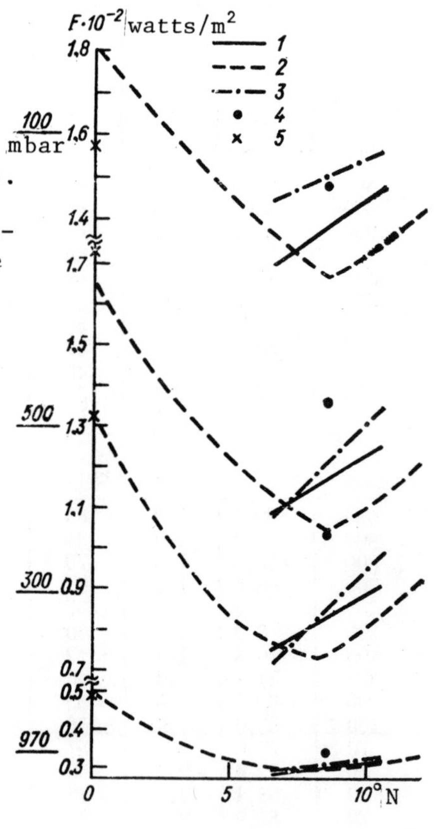

Fig. 16.1. Latitude distribution
of effective flux $F(p)$ at various
meridians and isobaric levels.
Lines join data of following ships
(from left to right): 1) $\lambda = 20°\text{W}$
(*Krenkel'*, *Poryv*); 2) $\lambda = 23°30'\text{W}$
(*Akademik Kurchatov*, *Professor
Zubov*, *Professor Vize*, *Akademik
Korolev*); 3) $\lambda = 27°\text{W}$ (*Okean
Priboi*), 4) $\lambda = 30\text{-}33°\text{W}$ (*Volna*);
5) $\lambda = 10°\text{W}$ (*Passat*).

$t(p)$, $u(p)$, and $F^{\uparrow\downarrow}(p)$ also have extrema in the ITC. For $p \leq 500$ mb
the zonal nature of the field disappears.

Let us consider the flux of departing radiation $F^{\uparrow}(0) =$
$\lim_{p\to 0} F^{\uparrow}(p)$. This flux is one of the main parameters in the theory
of climate and the general circulation of the atmosphere. At first
glance it is natural to assume that $F^{\uparrow}(0)$ is equal to the flux at
the upper level of measurement $p = 10$ mb, that is, to set $F^{\uparrow}(0) =$
$F^{\uparrow}(10)$. However, with a decrease in p for $p < 100$ mb the number
of measurements is reduced and the error in determining the mean
value of $F^{\uparrow\downarrow}(p)$ is greater. If we set $F^{\uparrow}(0) = F^{\uparrow}(100)$, then flux
$F^{\uparrow}(0)$ decreases by about 10% (see Table 16.8). The mean values
of $F^{\uparrow}(0)$ for clouds diminish by 12 to 25% (cf. sections 3 and 4 of
Table 16.8), while the difference between the equator and the tropics proper due to the humidity differences is from 10 to 17% (cf.
sections 1 and 4 of Table 16.8); the first figure in both cases
pertains to flux $F^{\uparrow}(10)$ and the second to $F^{\uparrow}(100)$.

The temperature of the underlying surface t_S has a negligible effect on flux $F^\uparrow(0)$. The total moisture content of the equatorial atmosphere is, on the average, 3.5 g/cm^2, while that of the tropical atmosphere is 5 g/cm^2. The integral transmission function of water vapor, for these amounts of it, is 0.16 in the former case and 0.13 in the latter (see Chap. 5). Thus only 13 to 16% of the radiation of the ocean traverses the entire thickness of the atmosphere. Variations in the temperature and humidity of the lower troposphere, due to the high humidity of the tropical atmosphere and the short range of action of the thermal radiation, do not have much effect on $F^\uparrow(0)$ either. Table 16.9 gives the mean effective level at which flux $F^\uparrow(0)$ forms, determined from the condition $\sigma T^4(p_{\text{eff}}) = F^\uparrow(0)$.

Table 16.9. Effective radiation levels (p_{eff}, mb) of fluxes $F^\uparrow(100)$ and $F^\uparrow(10)$.

Flux	Model			
	equatorial	general tropical	cloudy tropical	cloudless tropical
$F^\uparrow(100)$	354	294	268	318
$F^\uparrow(10)$	358	354	342	358

The level of formation of the upward flux for $p = 10$ mb is seen to be practically constant, being distorted as a result of the rising temperature of the stratosphere. When studying processes in the troposphere, it is apparently more correct to assume $p = 100$ mb as the upper limit of the atmosphere for radiation in the tropics. The level of formation of flux $F^\uparrow(100)$ lies from 270 to 350 mb and it varies by about the same amount upon transition from cloudless to cloudy conditions or from a less humid (equatorial) atmosphere to a humid (tropical) atmosphere.

As mentioned above, upper cloud boundaries are often located in the vicinity of $p = 300$ mb, and there the humidity becomes more variable (as compared with a layer $p < 300$ mb). Clearly, the humidity of the upper troposphere and the levels of the cloud boundaries, together with the number of these, are the main factors determining the size and variability of the flux of outgoing radiation. The distributions of temperature and humidity in the lower layers of the atmosphere do not affect flux $F^\uparrow(0)$ very much. For comparison, we note that in the middle latitudes, on the avarege over a year, flux $F^\uparrow(0)$ is formed at $p \approx 400$ mb [6] and it is influenced by the distribution of temperature and humidity throughout the entire atmosphere. In particular, from 20 to 25% of the radiation of the Earth's surface participates in the formation of flux $F^\uparrow(0)$.

Let us consider flux $F^\downarrow(970)$. This parameter, like $F^\uparrow(0)$, is also very important for the energetics of the atmosphere. More precisely, it is parameter $F(p*)$ which is important, the effective radiation at the level of the ocean surface, this being one of the main components of the heat balance there. Since the lowest level of actinometric radiosonde observations is 970 mb, flux $F(p*)$ cannot be measured directly with actinometric radiosondes. In [3] it is proposed to determine $F(p*)$ by extrapolating downward flux $F^\downarrow(p)$ according to radiosonde data in the lower troposphere and then calculating $F^\uparrow(p*) = \sigma T_s^4$ according to the known temperature t_s of the ocean surface. The variability of $F^\downarrow(p*)$ is close to that of $F^\downarrow(970)$, since the lower cloud boundaries are located mainly above the 970 mb level.

Table 16.10 gives the mean amounts of cooling in various layers, calculated using the data of Table 16.7. When examining this table, recall that for a 1 or 2% error in measuring a unidirectional flux [12] the error in the effective radiation $F(p)$ is about 10%, while the error in $F(p_1) - F(p_2)$ is already as much as about 25% (for $p_1 - p_2 = 200$ mb in the middle troposphere). This may in part account for the nonuniformity of the vertical distribution of the cooling. However, to an even greater extent, the irregular vertical structure of the cooling is governed by the positions of cloud boundaries, aerosol layers and their boundaries, inversions of temperature and humidity, or even layers with an accelerated (retarded) drop of the latter with increasing height.

Table 16.10. Radiative cooling of various layers of atmosphere ($^\circ$C/day).

Δp mb	Model				
	Equatorial	cloudless tropical	general tropical	cloudy tropical	cloudless northern ($\varphi \geq 10^\circ$N)
970... 800	—1.56	—1.24	—0.76	—0.49	—1.21
800... 600	—1.33	—1.15	—0.97	—0.86	—1.24
600... 400	—1.65	—1.56	—1.30	—1.21	—1.45
400... 200	—1.06	—0.83	—1.15	—1.24	—1.06
970... 150	—1.30	—1.10	—1.01	—0.94	—1.14
900... 500	—1.46	—1.21	—0.96	—1.17	—1.22

Table 16.10 reveals the following indisputably real effects:
1) increased cooling of the middle troposphere;
2) less cooling of a cloudless tropical atmosphere in comparison with an equatorial atmosphere;
3) less cooling of the lower layers with clouds than without clouds; from 600 to 400 mb the cooling is also less in the former

case, whereas from 400 to 200 mb it is greater than in the absence of clouds;

4) when clouds are present, the entire troposphere is warmer than under cloudless conditions.

The first of these effects is due to a simultaneous monotonic rise in $F(p)$ with height an drop in air density, while the second is due to the higher humidity of the tropical latitudes proper. The third and fourth effects have been described in detail in [23] (see also Chap. 14). Under clouds, close to their lower boundaries, and inside clouds, the cooling is low, but close to the upper boundaries it increases sharply, and on the whole a cloudy atmosphere with high upper cloud boundaries will experience more radiative heating than a cloudless atmosphere.

Judging by Table 16.10, the upper cloud boundaries lie mostly between 600 and 200 mb. The region from 900 to 500 mb and the last column in the table are given in an attempt to detect the effect of the aerosol. According to [24, 25], outbreaks of the Saharan aerosol appear at $\varphi \geq 10°N$ and 500 mb $\leq p \leq 900$ mb. However, the mean data do not reveal any additional effect of the aerosol in the cooling. At the same time, this effect does apparently manifest itself in individual cases (see Table 10 in [18]).

In conclusion, let us consider the radiation effects of various types of tropical clouds, as contrasted with the corresponding effects in the midlatitudes. Recall that when clouds are present fluxes $F(p*)$ and $F(0)$ become lower, the resultant effect depending on the positions of the lower (z_{lb}) and upper (z_{ub}) cloud boundaries [23].

Let us assume that $z_{lb} \approx 0.5$ km and $z_{ub} - z_{lb} \approx 0.5$ km, which is typical for a stratiform cloud of the lower level. In this case $F_{cld}(p*) \approx 0 < F_{no\ cld}(p*)$, while $F_{cld}(10) \approx F_{no\ cld}(0)$, and as a result $|R_{cld}| > R_{no\ cld}$, that is, a cloudy atmosphere is cooler than a cloudless atmosphere. Here R_{cld} and $R_{no\ cld}$ are the radiative cooling of the entire cloudy and cloudless atmosphere. As z_{ub} increases, flux $F(z_{ub})$ becomes less, and for $z_{lb} = \text{const}$ the atmosphere heats up. If, moreover, both z_{lb} and $F(p*)$ increase, then the cooling drops off more intensively and may be replaced by heating.

Low-lying stratus clouds are rarely observed in the tropical atmosphere. The effect of middle-level clouds is difficult to foresee. Thick convective clouds of the ITC with very high z_{ub} should reduce the cooling appreciably, as compared with cloudless conditions. Dense Ci clouds also cause a considerable relative heating up of the atmosphere, but since Ci clouds are usually not dense (see Chap. 14) their effect is not so substantial for the whole thickness of the atmosphere.

Judged by a small number of "pure" cases, the radiative cooling of the depth of the tropical troposphere was, on the average, found to be: $R_{no\ cld} = -1.28°C/day$ (50 cases); $R_{Ci} = -1.05°C/day$ (10 cases); $R_{ITC} = -0.68°C/day$ (20 cases). Here R_{Ci} and R_{ITC} represent radiative cooling of the layer (970 to 100 mb) for Ci clouds and for clouds of the ITC. When clouds of the latter two kinds were observed simultaneously, the cooling dropped off to -0.50°C/day.

A marked weakening of the radiative cooling of the troposphere as a whole in the presence of ITC clouds is one of the radiation features of the tropical atmosphere. At the same time, the upper parts of these clouds, which reach to heights of 400-100 mb, that is, the upper troposphere, enhance the cooling there to -1.4°C/day, for a mean amount of -0.9°C/day under cloudless conditions. Ci clouds in the tropics also lie between 400 and 100 mb, where they can reduce the cooling until it is replaced by heating. On the average, the cooling of the layer between 400 and 100 mb for Ci is -0.44°C/day.

Sharp radiation anomalies of the upper troposphere are also typical of the tropics. In particular, the radiation effects of cirriform clouds are important; tropical cirrus clouds are higher, more frequent, and denser than cirrus in the midlatitudes [1]. Table 16.11 gives the mean fluxes $F^{\uparrow}(100)$ and $F(970)$ in the specified "pure" cases.

Table 16.11. Mean fluxes $F^{\uparrow}(100)$ and $F(970)$ in watts/m^2.

Situation	$F^{\uparrow}(100)$	$F(970)$
Cloudless	230.3	52.4
ITC	139.6	22.3
Ci	188.5	44.7

This leads us to one more radiation feature of the tropical cloudy atmosphere: only thick ITC cloudiness is able so surely to reduce simultaneously $F(970)$ and $F^{\uparrow}(100)$. Finally, flux $F(970)$, and all the more $F(1000)$, are in the tropics reduced to half or even less in comparison with the effective flux at the ground in the middle and high latitudes (see [6]). This is because of the high air humidity and the small difference between the temperatures of the ocean surface and the air in this layer, as well as due to the low level of the lower boundary of ITC clouds.

REFERENCES*)

Introduction

1.   Dubrovina, L. S. (ed.). Aeroclimatological atlas-handbook
of the USSR, No. 3. Statistical characteristics of the spatial and
microphysical structure of clouds. Gidrometeoizdat, Moscow, 1975.

2.   Ginzburg, A. S., Kostyanoi, G. N., and Mullamaa, Yu.-A. R.
A tentative radiation model of a cloudy atmosphere (thermal radia-
tion and cumulus clouds). Preprint, Inst. Fiz. Atmos., Akad Nauk
SSSR, Moscow, 1977.

3.   Feigel'son, E. M. (ed.). A tentative radiation model of a
cloudy atmosphere. Preprint, Inst. Fiz. Atmos., Akad. Nauk SSSR,
1976.

4.   Mazin, I. P. (ed.). *Trudy TsAO*, No. 124. Gidrometeoizdat,
Leningrad, 1976.

5.   Feigel'son, E. M. Radiant heat transfer in a cloudy atmosphere. Israel Program for Scientific Translations, Jerusalem,
1973 (English translation).

6.   Feigel'son, E. M., and Krasnokutskaya, L. D. Fluxes of
solar radiation and clouds. Gidrometeoizdat, Leningrad, 1978.

7.   Borovikov, A. M., et al. Cloud physics. Israel Program
for Scientific Translations, Jerusalem, 1963 (English translation).

Chapter 1

1.   Dubrovina, L. S. (ed.). Aeroclimatological atlas-handbook
of the USSR, No. 3. Statistical characteristics of the spatial and
microphysical structure of clouds. Gidrometeoizdat, Moscow, 1975.

2.   Baranov, A. M. High-level clouds and flying conditions in
them. LKVVIA, Leningrad, 1960.

3.   Baranov, A. M. Heights and vertical extents of low-level
and intermediate-level clouds over the Laptev Sea. *Trudy AANII*,

---

        *) For each chapter Soviet publications are given first,
more or less in Cyrillic-alphabetical order, and these are followed
by non-Soviet works, in Roman-alphabetical order (Translator).

239 (2): 85, 1962.

4.   Baranov, A.M. High-level clouds over the Arctic. *Trudy AANII*, 239 (2): 111, 1962.

5.   Baranov, A. M. Frontal clouds and flying conditions in them. Gidrometeoizdat, Leningrad, 1964.

6.   Bol'shakov, V. N. The probability density of optical depths of low-level and intermediate-level stratiform clouds. Izv. Akad. Nauk SSSR, *Fiz. Atmos. Okeana*, 11 (6): 590, 1975.

7.   Borovikov, A. M., Lavrent'ev, E. V., and Mazin, I. P. Morphological characteristics of clouds of the eastern tropical Atlantic. In: *TROPEKS-74*. Gidrometeoizdat, Leningrad, 1976.

8.   Dolgin, I. M. A study of clouds in the Arctic. *Trudy AANII*, 239 (2): 5, 1962.

9.   Dubrovina, L. S. Characteristics of the cloud cover over the USSR according to aircraft-sounding data. *Trudy NIIAK*, No. 27, p. 3, 1968.

10.   Dubrovina, L. S. Relationship between vertical thickness of clouds and heights of cloud boundaries. *Trudy VNIIGMI MTsD*, No. 7, p. 3, 1974.

11.   Zavarina, M. V., and Romasheva, M. K. Heights of lower cloud boundaries over the Arctic. *Trudy AANII*, 217: 99, 1959.

12.   Zak, E. G. The vertical distribution of high-level clouds. *Trudy TsAO*, No. 39, p. 13, 1962.

13.   Kolenkova, S. I., and Litvinov, I. V. On the horizontal sizes of stratiform clouds. *Meteorologiya i Gidrologiya*, No. 9, p. 42, 1976.

14.   Lobanova, V. Ya. The structure of a climatic cloud field along a specified direction. *Trudy NIIAK*, No. 27, p. 21, 1968.

15.   Newton, C. W. (ed.). Meteorology of the Southern Hemisphere. Amer. Met. Soc., Boston, 1972.

16.   Marine atlas, 2: 50, 1953.

17.   Agafonova, S. I. (ed.). Handbook of the climate of the USSR, Part 5. Cloudiness and atmospheric phenomena. Nos. 1-34. Gidrometeoizdat, Leningrad, 1968-1970.

18.   Kosarev, A. L., et al. A comparison of some microphysical

cloud characteristics for various geographical regions. In: Problems in cloud physics, p. 113. Gidrometeoizdat, Leningrad, 1978.

19.   Strokina, L. A., and Beeva, I. M. On the distribution of clouds over oceans. *Trudy GGO*, No. 307, p. 94, 1974.

20.   Titov, L. V. Aerological characteristics of clouds in Northern Kazakhstan. *Trudy KazNIGMI*, No. 37, p. 97, 1969.

21.   Mazin, I. P. (ed.). *Trudy TsAO*, No. 134. Gidrometeoizdat, Moscow, 1977.

22.   Fedorova, A. A. Spatial characteristics of cirrus fields. *Trudy TsAO*, No. 39, p. 3, 1962.

23.   Borovikov, A. M., et al. Cloud physics. Israel Program for Scientific Translations, Jerusalem, 1963 (English translation).

24.   Furman, A. I. Some spatial characteristics of cumulus fields and the reliability of ground-level determinations of these. *Meteorologiya i Gidrologiya*, No. 7, p. 23, 1974.

25.   Khrgian, A. Kh. The structure of stratiform clouds. Izv. Akad. Nauk SSSR, *Fiz. Atmos. Okeana*, 13(11): 1150, 1977.

26.   Atlas international des nuages. Vol. 1. WMO, 1956.

27.   Gottwald, A. Space Research XVI.–Berlin: Akademie-Verlag, 1976.

28.   Malberg H. Mittlere, jahreszeitliche Bewölkungsverteilung und Häfigkeit von heitarem, wolkigem und stark bewölktem Wetter im europaischatlantischen Bereich nach Satellitenaufnahmen. Meteorol. Rdsch., 1973, Vol. 26, No. 6, S. 187–192.

29.   Marine climate atlas of the world. Washington, 1955–1959, 1969, Vol. 1–5, 8.

Chapter 2.

1.   Aleksandrov, E. L., and Yudin, K. B. The vertical profile of microstructure parameters of stratus clouds. *Meteorologiya i Gidrologiya*, No. 12, p. 57, 1979.

2.   Atlas of climatic characteristics of air temperature, density, and pressure, wind, and geopotential in the troposphere and lower stratosphere of the Northern Hemisphere. No. 3. Many-year average air temperatures at principal isobaric surfaces, Tables, Part 2. Gidrometeoizdat, Moscow, 1975.

3.   Borovikov, A. M., and Mazin, I. P. Microphysical characteristics of clouds. In: Aeroclimatological atlas-handbook of the
USSR, No. 3, Vol. 1, Part 2, p. 127. Gidrometeoizdat, Moscow, 1975.

4.   Veikman, Kh. K. Types of growth of atmospheric ice crystals.
In: Problems in cloud physics, p. 97, Gidrometeoizdat, Leningrad,
1978.

5.   Kosarev, A. L., and Mazin, I. P. A formula for calculating
the adiabatic value of the cloud water content. *Meteorologiya i
Gidrologiya*, No. 3, p. 107, 1972.

6.   Newton, C. W. (ed.). Meteorology of the Southern Hemisphere.
Amer. Met. Soc., Boston, 1972.

7.   Rodzhers, R. R. A short course in cloud physics. Gidrometeoizdat, Leningrad, 1979.

8.   Kosarev, A. L., et al. A comparison of some microphysical
cloud characteristics for various geographical regions. In: Problems in cloud physics, p. 113. Gidrometeoizdat, Leningrad, 1978.

9.   Mazin, I. P. (ed.). *Trudy TsAO*, No. 134, Gidrometeoizdat,
Moscow, 1977.

10.   Khrgian, A. Kh., and Mazin, I. P. Droplet-size distributions
in clouds. *Trudy TsAO*, No. 7, p. 56, 1952.

11.   Yudin, K. B. Some results of measuring local characteristics
of stratiform clouds. *Meteorologiya i Gidrologiya*, No. 12, p. 44,
1976.

12.   Magono, C., and Lee, C. W. Meteorological classification
of natural snow-crystals. *J. Faculty Sci.*, Hokkaido Univ., Ser. 7
(Geophysics), 1966, Vol. 2, No. 4, p. 321–335.

13.   Warner, J. The microstructure of cumulus clouds. P. 1.
General features of the droplet spectrum. *J. Atm. Sci.*, 1969,
Vol. 96, No. 5, p. 1049–1059.

Chapter 3

1.   Vul'fson, N. I. A study of convective motions in the free
atmosphere. Izd. Akad. Nauk SSSR, Moscow, 1961.

2.   Garger, E. K., Lukoyanov, N. F., and Pakhomov, V. I. A
study of the fine structure of trade cloudiness using the photo-
gram method. In: *TROPEKS-72*, p. 418. Gidrometeoizdat, Leningrad,
1974.

3.   Dmitrieva, L. R., Goisa, N. I., and Gorb, A. S. On the parametrization of fields of cumulus clouds and radiation. *Trudy GGO*,
No. 353, p. 13, 1975.

4.   Kendall, M. G., and Moran, P. A. P. Geometrical probability.
Griffin's statistical monographs, No. 10. Hafner Pub. Co., 1963.

5.   Kuusk, A. E. Horizontal sizes of cumulus clouds in the
tradewind regions. In: The variability of clouds and radiation
fields, p. 70. Tartu, 1978.

6.   Kasatkina, O. I., et al. On the objective determination of
cloud characteristics. *Meteorologiya i Gidrologiya*, No. 8, p. 23,
1972.

7.   Gudimenko, A. V., et al. Linear sizes of cumulus clouds
according to ground-based measurements of radiation fluxes. *Trudy
GGO*, No. 388, p. 114, 1977.

8.   Okhvril', Kh. A. Reduction of information on cover of
almucantars by cumulus clouds. In: Clouds and radiation, p. 51.
Tartu, 1975.

9.   Okhvril', Kh. A., and Epik, R. S. Describing the variability
of azimuth-averaged parameters of cumulus clouds and long-wave
radiation by means of eigenvectors. In: The variability of clouds
and radiation fields, p. 5. Tartu, 1978.

10.   Furman, A. I., et al. Space-time characteristics of cumulus
fields. Izv. Akad. Nauk SSSR, *Fiz. Atmos. Okeana*, <u>13</u> (9): 949, 1977.

11.   Rudneva, L. B. Determination of cloud characteristics from
measurements of self-radiation of clouds in 8-12 μm transmission
window. *Trudy GGO*, No. 363, p. 44, 1976.

12.   Mullamaa, Yu.-A. R., et al. The stochastic structure of
cloudiness and radiation fields. Tartu, 1972.

13.   Timanovskaya, R. G. The optical density of cumuli in the
middle latitudes of the European USSR and the tropical Atlantic.
Meteorologiya i Gidrologiya, No. 6, p. 52, 1979.

14.   Timanovskaya, R. G., and Timanovskii, D. F. Determination
of amount of cloudiness by continuous recording of direct radiation. *Trudy GGO*, No. 345, p. 8, 1975.

15.   Timanovskaya, R. G., and Feigel'son, E. M. A means of studying the statistical structure of surface fluxes of radiation under
cloudy conditions. In: Heat transfer in the atmosphere, p. 97.
Nauka, Moscow, 1972.

16.   Timanovskaya, R. G., and Feigel'son, E. M. Some cumulus parameters obtained from sky photos and surface actinometric observations. In: Heat transfer in the atmosphere, p. 107, Nauka, Moscow, 1972.

17.   Gifford, M. D., and McKee, T. B. Characteristic size spectra of cumulus fields observed from satellites. Atmos. Sci. Pap., Dep. Atmos. Sci. Colo. State Univ., 1977, No. 280, 79 p.

18.   Lopez, R. E. The log-normal distribution and cumulus cloud populations. Mon. Weather Rev., 1977, 105, No. 7, p. 865–872.

19.   Plank, V. G. The size distribution of cumulus clouds in representative Florida populations. *J. Meteorol.*, 1969, **V**ol. 8, No. 1, p. 46–67.

Chapter 4

1.   Buikov, M. V., and Khvorost'yanov, V. I. The formation and evolution of radiation fog and stratus clouds in the atmospheric boundary layer. Izv. Akad. Nauk SSSR, *Fiz. Atmos. Okeana,* 13 (4): 356, 1977.

2.   van de Hulst, H. C. Light scattering by small particles. Wiley, New York, 1957.

3.   Volkovitskii, O. A., Pavlova, L. N., and Snykov, V. P. The asymmetry of the scattering properties of a crystalline cloud medium. Izv. Akad. Nauk SSSR, *Fiz. Atmos. Okeana,* 11 (7): 757, 1975.

4.   Volkovistkii, O. A., Pavlova, L. N., and Petrushin, A. G. Light scattering by ice crystals. Izv. Akad. Nauk SSSR, *Fiz. Atmos. Okeana,* 16 (2), 1980.

5.   Georgievskii, Yu. S., and Shukurov, A. Kh. Some results of measuring the transmission of thin clouds in the visible and infrared ranges. Izv. Akad. Nauk SSSR, *Fiz. Atmos. Okeana,* 10 (1), 1974.

6.   Goody, R. M. Atmospheric radiation. Clarendon Press, 1964.

7.   Deirmendjian, D. Electromagnetic scattering on spherical polydispersions. Elsevier, New York, 1968.

8.   Dugin, V. P., et al. Anisotropy of light scattering by artificial crystalline cloud formations. Izv. Akad. Nauk SSSR, *Fiz. Atmos. Okeana,* 13 (1): 35, 1977.

9.   Zel'manovich, I. L., and Shifrin, K. S. Light scattering

tables, Vol. 4, Scattering by polydisperse systems. Gidrometeoizdat, Leningrad, 1971.

10.   Zuev, V. E. Propagation of visible and infrared radiation in the atmosphere. Wiley, New York, 1974.

11.   Kuusk, A. E. The structure of cloud fields and the radiation regime at the underlying surface. Author's Abstract of Candidate's Dissertation. Leningrad, 1979.

12.   Levin, L. M. A study in the physics of coarsely dispersed aerosols. Izd. Akad. Nauk SSSR, 1961.

13.   Volkovitskii, O. A., et al. Model studies of the optical and aerodynamic properties of ice crystals. In: Problems in cloud physics, p. 102. Gidrometeoizdat, Leningrad, 1978.

14.   Nevzorov, A. N., and Shugaev, V. F. The use of integral parameters to study the microstructure of droplet clouds. *Trudy TsAO*, No. 101, p. 32, 1972.

15.   Volkovitskii, O. A., et al. Some data on the propagation of laser radiation at CO_2 in a crystalline cloudy medium. Izv. Akad. Nauk SSSR, *Fiz. Atmos. Okeana*, 10 (11): 1157, 1974.

16.   Rozenberg, G. V., et al. Determining the optical character- istics of clouds from measurements of reflected solar radiation with *Cosmos 320*. Izv. Akad. Nauk SSSR, *Fiz. Atmos. Okeana*, 10 (1), 1974.

17.   Kosarev, A. L., et al. The optical density of clouds. Gidrometeoizdat, Moscow, 1976 (*Trudy TsAO*, No. 124).

18.   Rozenberg, G. V., et al. On the brightness of clouds (re- sults of a comprehensive study). Izv. Akad. Nauk SSSR, *Fiz. Atmos. Okeana*, 6 (5), 1970.

19.   Pavlova, L. N. Measurement of coefficients of directional light scattering in droplet and crystalline cloud media. Izv. Akad. Nauk SSSR, *Fiz. Atmos. Okeana*, 14 (9): 990, 1978.

20.   Rozenberg, G. V. Optical properties of thick layers of a homogeneous scattering medium. In: The Spectroscopy of light-scat- tering media, p. 5. Izd. Akad. Nauk BSSR, Minsk, 1963.

21.   Rozenberg, G. V. Estimating the absorptivity of haze from cloud brightness. Izv. Akad. Nauk SSSR, *Fiz. Atmos. Okeana*, 9 (5), 1973.

22.   Sobolev, V. V. The transfer of radiant energy in stellar

and planetary atmospheres. Gostekhizdat, Moscow, 1956.

23.   Kosarev, A. L., et al. A comparison of some microphysical cloud characteristics for various geographical regions. In: Problems in cloud physics, p. 113. Gidrometeoizdat, Leningrad, 1978.

24.   Mullamaa, Yu.-A. R., et al. The stochastic structure of cloud and radiation feilds. IFA Akad. Nauk ESSR, Tartu, 1972.

25.   Feigel'son, E. M., and Krasnokutskaya, L. D. Fluxes of solar radiation and clouds. Gidrometeoizdat, Leningrad, 1978.

26.   Filippov, V. L., Ivanov, V. P., and Makarov, A. S. Variations in the áerosol attenuation of radiation under ice-fog conditions. In IV All-Union Symposium on the Propagation of Laser Radiation in the Atmosphere, p. 174. Tomsk, 1977.

27.   Khvorost'yanov, V. I. Calculating the coefficients of scattering and absorption of short-wave radiation by clouds. *Trudy UkrNIGMI*, No. 178, 1979.

28.   Shifrin, K. S. Calculating the radiation properties of clouds. *Trudy GGO*, No. 46 (108): 5, 1955.

29.   Dugin, V. P., et al. Experimental studies of the optical characteristics of artificial ice clouds. Izv. Akad. Nauk SSSR, *Fiz. Atmos. Okeana,* 7 (8): 871, 1971.

30.   Blau, H. H., Espinola, R. P., and Reifenstein, E. Near infrared scattering by sunlit terrestrial clouds. Appl. Opt. 1966, Vol. 5, No. 4, p. 555-564.

31.   Danielson, R. E., Moore, D. R., and van de Hulst, H. C. The transfer of visible radiation through clouds. *J. Atm. Sci.*, 1969, Vol. 26, No. 5, p. 2.

32.   Drummond, A. Y., and Hickey, Y. R. Large scale reflection and absorption of solar radiation by clouds as influencing earth radiation budget: new aircraft measurement. Preprints of papers Intern. Conf. Weather Modification. Canberra, Australia Acad. Sci. and Amer. Met. Soc., 1971.

33.   Grassl, H. Determination of cloud drop size distribution from spectral transmission measurement. Beitrage Physik Atmosphäre, 1970, 43.

34.   Hansen, J. E., and Cheyney, H. Theoretical spectral scattering of ice clouds in the near infrared. *J. Geophys. Res.*, 1969, Vol. 74, No. 13, p. 3337-3346.

35.   Huffman, P. J., and Thursby, W. R. Light scattering by ice crystals. *J. Atm. Sci.*, 1969, Vol. 26, No. 5, p. 1073-1077.

36.   Huffman, P. J. Polarization of light scattering by ice crystals. *J. Atm. Sci.*, 1970, Vol. 27, No. 8, p. 1207-1208.

37.   Irvine, W. W., and Pollack, J. B. Infrared optical properties of water and ice spheres. Icarus, 1968, Vol. 8, p. 324-360.

38.   Liou, K. N. Light scattering by ice clouds in the visible and infrared. *J. Atm. Sci.*, 1972, Vol. 29, No. 3, p. 524-526.

39.   Morita, Y. Scattering cross-section of freely suspended ice crystals for visible light. Tenki, 1973, Vol. 20, No. 3, p. 13-22.

40.   Ohtake, Y., and Huffman, P. J. Visual range in ice fog. *J. Appl. Meteorol.*, 1969, Vol. 8, p. 499-501.

41.   Pembrook, J. D., Gryvnak, D. A., and Burch, D. B. Attenuation of solar radiation by natural clouds at several wavelengths in the visible and infrared. *J. Opt. Soc. Amer.*, 1971, Vol. 61, No. 4, p. 580.

42.   Robinson, G. D., and Drummond, A. Y. Some recent aircraft measurements of absorption and scattering of solar radiation by atmospheric aerosol. Australia Acad. Sci., Amer. Meteorol. Soc., 1971, 288p.

43.   Schaaf, J. W., and Williams, D. Optical constants of ice in the infrared. *J. Opt. Soc. Amer.*, 1973, Vol. 63, No. 6, p. 726-732

44.   Twomey, S. On the possible absorption of visible light by clouds. *J. Atm. Sci.*, 1970, Vol. 27, No. 3.

Chapter 5

1.   Kondrat'ev, K. Ya. (ed.). The atmospheric aerosol and its effect on radiation transfer. Gidrometeoizdat, Leningrad, 1978.

2.   Ginzburg, A. S., and Feigel'son, E. M. Parametrization of radiative heat transfer in models of the general circulation of the atmosphere. In: Physics of the atmosphere and the problem of climate. Nauka, Moscow, 1980.

3.   Golubitskii, B. M., and Moskalenko, M. I. Spectral transmission functions in H_2O and CO_2 vapor bands. Izv. Akad. Nauk SSSR, *Fiz. Atmos. Okeana,* 3 (3): 346, 1968.

4.   Gradus, L. M., Niilisk, Kh. Yu., and Feigel'son, E. M.
An integral transmission function for cloudy conditions. Izv. Akad.
Nauk SSSR, *Fiz. Atmos. Okeana*, 4 (4): 397, 1968.

5.   Goody, R. M. Atmospheric radiation. Clarendon Press, 1964.

6.   Deirmendjian, D. Electromagnetic scattering on spherical
polydispersions. Elsevier, New York, 1968.

7.   Dovgladyuk, Yu. A., and Ivlev, L. S. Physics of water
aerosols and other atmospheric aerosols. Izd. LGU, Leningrad,
1977.

8.   Evseeva, M. G., and Podol'skaya, E. L. The integral transmission function of solar radiation in the near infrared region.
Trudy LGMI, No. 40, p. 215, 1970.

9.   Zel'manovich, I. L., and Shifrin, K. S. Light scattering
tables, Vol. 4, Scattering by polydisperse systems. Gidrometeoizdat,
Leningrad, 1971.

10.   Krasnokutskaya, L. D., and Sushkevich, T. A. Analytical
representation of the integral transmission function of clouds.
Izv. Akad. Nauk SSSR, *Fiz. Atmos. Okeana*, No. 5, p. 505, 1977.

11.   Livshits, G. Scattered light of the daytime sky. Nauka,
Alma-Ata, 1973.

12.   Mokhov, I. I., and Petukhov, V. K. Parametrization of outgoing long-wave radiation for climatic models. Preprint, IFA Akad.
Nauk SSSR, Moscow, 1978.

13.   Niilisk, Kh. Yu., and Sammel, L. E. The integral transmission function of the atmosphere for calculations of the thermal-
radiation field in the troposphere. In: Tables of radiation
characteristics of the atmosphere. Tartu, 1969.

14.   Evseeva, M. G., et al. Calculation of the integral transmission function for water vapor and its dependence on pressure
and temperature. *Trudy LGMI*, No. 49, p. 143, 1974.

15.   Zuev, V. E., et al. A stratified model of the atmospheric
aerosol for optical-sounding wavelengths. Izv. Vuzov SSSR, *Fizika*,
No. 11, p. 30, 1974.

16.   Sivkov, S. I. Methods of calculating solar-radiation characteristics. Gidrometeoizdat, Leningrad, 1968.

17.   Sushkevich, T. A., and Khokhlov, V. F. Approximation of
functions of an exponential sum. Preprint, IPM Akad. Nauk SSSR,

No. 80, 1975.

18.   Feigel'son, E. M. Radiant heat transfer in a cloudy atmosphere. Israel Program for Scientific Translations, Jerusalem,
1973 (English translation).

19.   Feigel'son, E. M., and Krasnokutskaya, L. D. Fluxes of
solar radiation and clouds. Gidrometeoizdat, Leningrad, 1978.

20.   Shekhter, F. N. Some problems of radiative heat transfer
in a cloudy atmosphere. *Trudy GGO*, No. 187, p. 82, 1966.

21.   Davis, P. A., and Viezee, W. A model for computing infrared transmission through atmospheric water vapor and carbon dioxide. *J. Geophys. Res.*, 1964, Vol. 69, No. 18, p. 3785-3794.

22.   Elterman, L. UV, Visible and IR attenuation for altitudes
up to 50 km. Env. Res. Pap. AFCRL. 1968, No. 258.

23.   Elterman, L. Vertical attenuation model with eight surface
meteorological ranges 2 to 13 kilometres. Env. Res. Pap. AFCRL,
1970, No. 310.

24.   Howard, J. H., Burch, D. E., and Williams, D. Infrared
transmission of synthetic atmospheres. *J. Opt. Soc. Am.*, 1956,
Vol. 46, No. 3, p. 186-190.

25.   Katuyama, A. A simplified scheme for computation of radiative transfer in the atmosphere. Tech. Rep. No. 6, Dept. Met.
UCLA, 1972.

26.   Krueger, A. Y., and Minzner, R. A. A mid-latitude ozone
model for the 1976 US Standard Atmosphere. *J. Geophys. Res.*, 1976,
Vol. 81, No. 24, p. 4477-4481.

27.   Linke, F. Meteorologisches taschenbuch. 11. Leipzig, 1939,
354 S.

28.   McClatchey, R. A. Optical properties of the atmosphere.
AFCRL Environmental Research Papers, 1972, No. 411, 108 p.

29.   Rodgers, C. D., and Walshaw, C. D. The computation of infrared cooling rate in planetary atmospheres. *Quart. J. Roy. Meteorol.
Soc.*, 1966, Vol. 92, No. 391, p. 67-92.

30.   Rodgers, C. D. The radiative heat budget of troposphere
and lower stratosphere. Mass. Inst. of Technology, Planetary
Circulation Project, Rep., 1967, No. A2.

31.   Rodgers, C. D. The use of emissivity in atmospheric

radiation calculations. *Quart. J. Roy. Meteorol. Soc.*, 1967,
Vol. 93, No. 396, p. 43.

32.   Stull, V. R., Wyatt, P. J., and Plass, G. N. The infrared
absorption of carbon dioxide. Infrared Transmission Studies. Final
Report. Vol. 3, SSD-TRD-62-127, 1963.

33.   Vigroux, E. Contribution a l'étude experimentale de
l'absorption de l'osone. Annal. Phys., 1953, Vol. 8, p. 709-762.

34.   Wyatt, P. J., Stull, V. R., and Plass, G. N. The infrared
absorption of water vapor. Infrared Transmission Studies. Final
Report, Vol. 2, SSD-TDR-62-127, 1962.

35.   Yamamoto, G. Direct absorption of solar radiation by
atmospheric water vapor, carbon dioxide and molecular oxygen.
J. Atm. Sci., 1962, Vol. 19, No. 2, p. 182-188.

Chapters 6 and 7

1.   Busygin, V. P., Evstratov, N. A., and Feigel'son, E. M.
Optical properties of cumulus clouds and radiation fluxes in the
presence of cumuli. Izv. Akad. Nauk SSSR, *Fiz. Atmos. Okeana*, 9
(11): 1142, 1973.

2.   Busygin, V. P., Evstratov, N. A., and Feigel'son, E. M.
Calculations of sizes and distributions of fluxes of direct, scat-
tered, and total solar radiation in the presence of cumulus clouds.
Izv. Akad. Nauk SSSR, *Fiz. Atmos. Okeana*, 13 (3): 264, 1977.

3.   Grechko, E. I., Dianov-Klokov, V. I., and Malkov, I. P.
Aircraft measurements of effective photon path lengths for reflection and transmission of light by clouds in 0.76 μm oxygen
band. Izv. Akad. Nauk SSSR, *Fiz. Atmos. Okeana*, 9 (5): 471, 1973.

4.   Grechko, E. I., Dianov-Klokov, V. I., and Malkov, I. P.
Measurement of effective photon path lengths in real cloud systems. Izv. Akad. Nauk SSSR, *Fiz. Atmos. Okeana*, 11 (2): 125, 1975.

5.   Grechko, E. I., et al. Calculation of mean and effective
photon path lengths for model of two-layer cloudiness allowing
for reflection from underlying surface. Izv. Akad. Nauk SSSR,
Fiz. Atmos. Okeana, 12 (1): 38, 1976.

6.   Dianov-Klokov, V.I. Photon-path distribution and coorela-
tion of transmission and reflection of disperse samples with true
absorption of substance. *Optika i Spektroskopiya*, 38 (1): 124, 1975.

7.   Dianov-Klokov, V. I., and Krasnokutskaya, L. D. A comparison

of experimental and calculated effective photon path lengths in a cloud layer. Izv. Akad. Nauk SSR, *Fiz. Atmos. Okeana,* 8 (8): 843, 1972.

8.   Dianov–Klokov, V. I., et al. Calculation of photon-path distributions and effective photon path lengths for some models of two-layer cloudiness. Izv. Akad. Nauk. SSSR, *Fiz. Atmos. Okeana,* 10 (7): 728, 1974.

9.   Dianov–Klokov, V. I., Evstratov, N. A., and Ozerenskii, A. P. Calculation of radiant-energy density and equivalent photon trajectories for some cloud models. Izv. Akad. Nauk, *Fiz. Atmos. Okeana,* 13 (3): 315, 1977.

10.   Kargin, B. A., Krasnokutskaya, L. D., and Feigel'son, E. M. Reflection and absorption of radiant energy from the Sun by cloud layers. Izv. Akad. Nauk SSSR, *Fiz. Atmos. Okeana,* 8 (5): 505, 1972.

11.   Kondrat'ev, K. Ya. Radiative heat transfer in the atmosphere. Gidrometeoizdat, Leningrad, 1956.

12.   Malkevich, M. S. Optical studies of the atmosphere using satellites. Nauka, Moscow, 1973.

13.   Malkov, I. P. Twin–wave "phase" spectrophotometer for studying the transmission spectrum of clouds. Izv. Akad. Nauk SSSR, *Fiz. Atmos. Okeana,* 6 (8): 850, 1970.

14.   Minin, I. N. Light absorption in the atmosphere and at the surface of the planet for multiple scattering and reflection. Izv. Akad. Nauk SSSR, *Fiz. Atmos. Okeana,* 10 (3): 228, 1974.

15.   Romanova, L. I., and Ustinov, E. A. A generalized transfer equation for the photon-path distribution and the problem of vertically nonuniform gaseous absorption in the atmosphere. Izv. Akad. Nauk, *Fiz. Atmos. Okeana,* 18 (3): 240, 1982.

16.   Titov, G. A. Modeling of solar-radiation transport for cumulus cloudiness. Izv. Akad. Nauk, *Fiz. Atmos. Okeana,* 15 (6): 633, 1979.

17.   Feigel'son, E. M., and Krasnokutskaya, L. D. Fluxes of solar radiation and clouds. Gidrometeoizdat, Leningrad, 1978.

18.   Irvine, W. M. The distribution of photon optical paths in a scattering atmosphere. *Astroph. J.,* 1966, Vol. 144, No. 3, p. 1140–1147.

Chapter 8

1.   Germogenova, T. A. Solution of the transfer equation for a plane layer. *Zhurn. Vych. Mat. Mat. Fiz.*, No. 6, 1961.

2.   Germogenova, T. A., and Konovalov, N. V. Asymptotic characteristics of the solution of the transfer equation in the problem of inhomogeneous layer. *Zhurn. Vych. Mat. Mat. Fiz.*, 14 (4), 1974

3.   Konovalov, N. V. Asymptotic characteristics of fields of monochromatic radiation in problems of an inhomogeneous plane layer of large optical size. II. Calculation of main parameters and functions. Prepirnt, *IPM Akad. Nauk SSSR*, No. 14, 1974.

4.   Konovalov, N. V. The accuracy of the solution of the transfer equation for a plane layer of large optical thickness with highly anisotropic scattering. Preprint, *IPM Akad. Nauk SSSR*, No. 94, 1975.

5.   Ogibin, V. N. The use of "splitting" and "roulette" when calculating particle transport with the Monte-Carlo method. In: The Monte-Carlo method as applied to radiation transfer, G. I. Marchuk (ed.). Atomizdat, Moscow, 1967.

6.   Prishivalko, A. P., and Astaf'eva, L. G. Absorption, scattering, and attenuation of light by water-coated atmospheric particles. Izv. Akad. Nauk SSSR, *Fiz. Atmos. Okeana*, 10 (12): 1322, 1974.

7.   Marchuk, G. I., et al. Solution of direct and some inverse problems of atmospheric optics using the Monte-Carlo method. Nauka, Novosibirsk, 1968.

8.   Rozenberg, G. V. The luminous characteristics of thick layers of a scattering medium with a low specific absorption. *Doklady Akad. Nauk SSSR*, 145 (4), 1962.

9.   Ryym, R.I. Solution of the radiation-transfer equation for an expanded scattering function. Preprint A-2 (1979), Inst. Astrofiz. Fiz. Atmos, Akad Nauk ESSR, Tartu, 1979.

10.   Ryym, R. I. Fluxes of reflected and transmitted (by stratus clouds) radiation in the 0.4–2.2 μm wavelength interval. Inst. Astrofiz. Fiz. Atmos. Akad. Nauk ESSR, Tartu, 1979. MS dep. at *VINITI* on 12 Nov 79, No. 3841.

11.   Sobolev, V. V. Diffusion of radiation in a medium of large optical thickness for nonisotropic scattering. *Doklady Akad. Nauk SSSR*, 179 (1), 1968.

12.   Sobolev, V. V. Light scattering in planetary atmospheres.
Pergamon, Oxford, 1975 (English translation).

13.   Sushkevich, T. A. Allowing for high scattering anisotropy
in problems of radiation propagation from a unidirectional source.
Preprint, *IPM Akad. Nauk SSSR*, No. 132, 1979.

14.   Tarasova, T. A., and Feigel'son, E. M. Allowing for aerosol
effect in radiative heat transfer. Izv. Akad. Nauk SSSR, *Fiz.
Atmos. Okeana*, 17 (1): 18, 1981.

15.   Tarasova, T. A., and Feigel'son, E. M. Effect of tropospheric aerosol on integral albedo of system consisting of atmosphere and underlying surface. Izv. Akad. Nauk SSSR, *Fiz. Atmos.
Okeana*, 18 (11), 1982.

16.   Feigel'son, E. M., and Krasnokutskaya, L. D. Fluxes of
solar radiation and clouds. Gidrometeoizdat, Leningrad, 1978.

17.   Ackerman, T., and Baker, M. Short-wave radiative effects
of unactivated aerosol particles in clouds. *J. Appl. Meteorol.*,
1977, Vol. 16, No. 1, p. 63-69.

18.   Grassl, H. Albedo reduction and radiative heating of clouds
by absorbing aerosol particles. Contributions Atm. Phys., 1975,
Vol. 48, No. 3, p. 199-210.

19.   Joseph, J. H., Wyscombe, W. J., and Weinman, W. A. The
Delta-Eddington approximation for radiative flux transfer. *J. Atm.
Sci.*, 1976, Vol. 33, No. 12.

20.   Plass, G. H., and Kattawar, G. W. Radiative transfer in
water and ice clouds. *J. Appl. Opt.*, 1971, Vol. 10, No. 4, p. 738-
748.

21.   Shettle, E. P., and Fenn, R. W. Models of the atmospheric
aerosols and their optical properties. AGARD Conf. Proc. No. 183,
Optical Propagation in the Atmosphere, 1975, p. 738-748.

22.   Twomey, S. The influence of pollution on the short-wave
albedo of clouds. *J. Atm. Sci.*, 1977, Vol. 34, No. 7, p. 1149-1152.

Chapter 9

1.   Dianov-Klokov, V. I., Evstratov, N. A., and Ozerenskii,
A. I. Calculation of radiant-energy density and equivalent photon
trajectories for some cloud models. Izv. Akad. Nauk SSSR, *Fiz.
Atmos. Okeana*, 13 (3): 315, 1977.

2.   Dmitrieva, L. R., and Samoilova, L. V. Calculation of the field of short-wave radiation in the scheme of the general circulation of the atmosphere. Izv. Akad. Nauk SSSR, *Fiz. Atmos. Okeana,* <u>6</u> (1): 29, 1970.

3.   Dmitrieva-Arrago, L. R., Parshina, G. V., and Samoilova, L. V. Calculation of fluxes of short-wave radiation under cloudy conditions. *Trudy GGO,* No. 272, p. 70, 1972.

4.   Ogibin, V. N. The use of "splitting" and "roulette" when calculating particle transport with the Monte-Carlo method. In: The Monte-Carlo method as applied to radiation transfer, G. I. Marchuk (ed.). Atomizdat, Moscow, 1967.

5.   Petukhov, V. K., and Feigel'son, E. M. A model of long-period heat transfer and moisture circulation for broken cloudiness. Izv. Akad. Nauk SSSR, *Fiz. Atmos. Okeana,* <u>9</u> (4), 1973.

6.   Marchuk, G. I., et al. Solution of direct and some inverse problems of atmospheric optics using the Monte-Carlo method. Nauka, Novosibirsk, 1968.

7.   Feigel'son, E. M., and Krasnokutskaya, L. D. Fluxes of solar radiation and clouds. Gidrometeoizdat, Leningrad, 1978.

8.   Shifrin, K. S., and Minin, I. N. Toward a theory of non-horizontal viewing. *Trudy GGO,* No. 68, p. 5, 1957.

9.   Liou, K. N., and Wattan, D. Parametrization of the radiative properties of clouds. *J. Atm. Sci.,* 1979, Vol. 36, No. 7, p. 1261-1273.

10.   Twomey, S. Computations of the absorption of solar radiation by clouds. *J. Atm. Sci.,* 1976, Vol. 33, No. 6, 1087-1091.

11.   van de Hulst, M. C., and Irvine, W. M. Scattering in model planetary atmospheres.   Meteorol. Soc. Roy. Sci., Liege, 1963, Ser. 5, No. 1, p. 78-86.

Chapter 10

1.   Belov, V. F. Albedo of underlying surface and some cloud types in the vicinity of the antarctic slope and the Davis Sea. *Trudy TsAo,* No. 37, 1960.

2.   Binenko, V. I., Grishechkin, V. S., and Kondrat'ev, K. Ya. A comprehensive radiation experiment in a cloudy atmosphere. *Trudy GGO,* No. 332, p. 57, 1973.

3.   Binenko, V. I., and Kondrat'ev, K. Ya. Vertical profiles of radiation characteristics of typical cloud formations. *Trudy GGO*, No. 331, p. 3, 1975.

4.   Binenko, V. I. Determination of effective refraction indexes of the atmospheric aerosol. *Trudy GGO*, No. 434, p. 33, 1980.

5.   Goisa, N. I. Effect of depth and water content of stratiform clouds on reflection, transmission, and absorption of short-wave radiation. *Trudy UkrNIGMI*, No. 70, p. 112, 1968.

6.   Goisa, N. I. Actinometric measurements with the IL-14 aircraft. *Trudy UkrNIGMI*, No. 70, p. 112, 1968.

7.   Goisa, N. I., and Shoshin, V. M. Experimental model of radiation regions of "average" stratiform cloud. *Trudy UkrNIGMI*, No. 82, p. 28, 1969.

8.   Goisa, N. I., and Shoshin, V. M. Vertical profiles of short-wave albedo in lower troposphere. *Trudy UkrNIGMI*, No. 82, p. 42, 1969.

9.   Goisa, N. I., and Shoshin, V. M. Radiation balance of cloud layers. *Trudy UkrNIGMI*, No. 86, p. 96, 1970.

10.   Goisa, N. I., and Shoshin, V. M. Experimental studies of solar-radiation fluxes in the lower troposphere with clouds. In: Heat transfer in the atmosphere, p. 60. Nauka, Moscow, 1972.

11.   Kondrat'ev, K. Ya., et al. The global aerosol-radiation experiment (GAREKS). *Meteorologiya i Gidrologiya*, No. 9, p. 5, 1980.

12.   Binenko, V. I., et al. Measurement of spectral characteristics of reflection, transmission, and absorption of cloudiness. *Trudy GGO*, No. 322, p. 77, 1973.

13.   Kastrov, V. G. Measurement of absorption of solar radiation in cloudy atmosphere up to 3-5 km. *Trudy TsAO*, No. 8, 1952.

14.   Kastrov, V. G. Some errors in the pyranometric determination of absorption of solar radiation in the atmosphere. *Trudy TsAO*, No. 22, p. 3, 1959.

15.   Koptev, A. P. Actinometric studies with aircraft in the Arctic. Nauchn. Soobshch. Inst. Geol. Goegr. Akad Nauk LitSSR, p. 13, 1962.

16.   Koptev, A. P., and Voskresenskii, A. I. On the radiation properties of cloudiness. *Trudy AANII*, <u>239</u> (2), 1962.

17.   Kosarev, A. L., et al. Optical densities of clouds. *Trudy TsAO*, No. 124, 1976.

18.   Petrenchuk, O. P. Experimental studies of the atmospheric aerosol. Gidrometeoizdat, Leningrad, 1979.

19.   Kondrat'ev, K. Ya., and Ter-Markaryants, N. E. (eds.). A comprehensive radiation experiment. Gidrometeoizdat, Leningrad, 1976.

20.   Rozenberg, G. V. Estimating the absorptivity of atmospheric haze from cloud brightness. Izv. Akad. Nauk SSSR, *Fiz. Atmos. Okeana*, 9 (5): 460, 1973.

21.   Feigel'son, E. M., and Krasnokutskaya, L. D. Fluxes of solar radiation and clouds. Gidrometeoizdat, Leningrad, 1978.

22.   Chel'tsov, N. I. A study of reflection, transmission, and absorption of solar radiation by clouds of various types. *Trudy TsAO*, No. 8, p. 36, 1952.

23.   Fritz, S. Measurement of albedo of clouds. *Bull. Amer. Meterol. Soc.*, 1950, No. 1.

24.   Griggs, M. Aircraft measurements of albedo and absorption of stratus and surface albedos. *J. Appl. Meteorol.*, 1968, Vol. 7, No. 6, p. 1012-1017.

25.   Grassl, H. Radiation effects of absorbing aerosol particles inside clouds. In: Proc. Int. Rad. Symp. JAMAP, Garmisch-Partenkirchen, Aug. 1976, Science Press, 1977, p. 171-175.

26.   Luckiesh, M. Aerial photometry. *Astrophys. J.*, 1919, Vol. 49, No. 2, p. 108-130.

27.   Neiburger, M. The reflection, absorption and transmission of insolation by stratus clouds. *J. Meteorol.*, 1949, Vol. 6, No. 2.

28.   Robinson, L. D. Some observations from airfract of surface albedo and absorption of cloud. Arch. Meteorol. Geophys. u. Bioklimatol., 1958, Bd. 9, No. 1.

29.   Twomey, S. The influence of pollution on the short-wave albedo of clouds. *J. Atm. Sci.*, 1977, Vol. 34, No. 7, p. 1149--1152.

30.   Twomey, S. The effect of cloud scattering on the absorption of solar radiation by the atmosphere. *J. Atm. Sci.*, 1972, Vol. 29, p. 1156-1159.

31.   Welch, R., and Cox, S. A treatise on the absorption of
solar radiation in clouds. *J. Atm. Sci.*, 1979.

Chapter 11

1.   Avaste, O. A., and Vainikko, G. M. Transfer of solar radiation in broken cloudiness. Izv. Akad. Nauk SSSR, *Fiz. Atmos.
Okeana*, 10 (10): 1054, 1974.

2.   Busygin, V. P., Evstratov, N. A., and Feigel'son, E. M.
Optical properties of cumulus clouds and radiation fluxes in the
presence of cumuli. Izv. Akad. Nauk SSSR, *Fiz. Atmos. Okeana*, 9
(11): 1142, 1973.

3.   Busygin, V. P., Evstratov, N. A., and Feigel'son, E. M.
Calculation of sizes and distributions of fluxes of direct, scat-
tered, and total solar radiation in the presence of cumulus clouds.
Izv. Akad. Nauk SSSR, *Fiz. Atmos. Okeana*, 13 (3): 264, 1977.

4.   Vainikko, G.M. Equations for the mean radiation intensity
in broken cloudiness. In: Statistical studies of broken cloudiness,
A series of meteorological studies. No. 21, p. 28, Nauka, Moscow,
1973.

5.   Goisa, N. I., et al. The interrelation and diurnal variation of parameters of cumulus fields. *Trudy UkrNIGMI*, No. 161,
p. 120, 1978.

6.   Glazov, G. N., and Titov, G. A. The propagation of in-
coherent visible radiation in stochastic macrononuniform cloudiness. In: Material from the All-Union Conference on the Propagation of Visible Radiation in a Disperse Medium, p. 51. Gidrometeoizdat, Moscow, 1978.

7.   Goisa, N. I., and Gorb, A. S. A study of short-wave radiation fluxes and influxes in the atmosphere for cumulus cloudiness.
Trudy UkrNIGMI, No. 130, p. 92, 1974.

8.   Goisa, N. I., Gorb, A. S., and Dmitrieva-Arrago, L. R. The
possibility of parametrizing the radiation characteristics of
broken cumulus cloudiness. *Trudy GGO*, No. 353, p. 13, 1975.

9.   Goisa, N. I., Gorb, A. S., and Dmitrieva-Arrago, L. R.
Albedos of cumulus fields. *Trudy UkrNIGMI*, No. 146, p. 58, 1976.

10.   Kondrat'ev, K. Ya. Actinometry. Gidrometeoizdat, Leningrad,
1965.

11.   Kuusk, A. Diurnal variation of total-radiation flux in a

cloudless sky. In: The variability of cloud and radiation fields, p. 39. Tartu, 1978.

12. Marchuk, G. I. (ed.). The Monte-Carlo method in atmospheric optics. Nauka, SO, Novosibirsk, 1976.

13. Ross, Yu. (ed.). Cloudiness and radiation. Tartu, 1975.

14. Kosarev, A. L., et al. Optical densities of clouds. Gidrometeoizdat, Moscow, 1976 (*Trudy TsAO*, No. 124).

15. Pyldmaa, V. K., and Timanovskaya, R. G. Total radiation at the Earth's surface under different cloudiness conditions. In: Heat transfer in the atmosphere, p. 101. Nauka, Leningrad, 1972.

16. Sivkov, S. I. Methods of calculating solar-radiation characteristics. Gidrometeoizdat, Leningrad, 1968.

17. Avaste, O. A., et al. Statistical modeling of the transfer of short-wave radiation in broken cloudiness. In: Monte-Carlo methods in computational mathematics and mathematical physics, p. 232, Novosibirsk, 1974.

18. Mullamaa, Yu.-A. R., et al. The stochastic structure of fields of cloudiness and radiation. Tartu, 1972.

19. Feigel'son, E. M., and Tsvang, L. R. (eds.). Heat transfer in the atmosphere. A collection of papers. Nauka, Moscow, 1972.

20. Timanovskaya, R. G. The statistical structure of fluxes of direct and total radiation for cumulus cloudiness. *Trudy GGO*, No. 297, p. 142, 1973.

21. Timanovskaya, R. G. The space-time structure of fluxes of short-wave radiation for cumulus clouds. *Trudy GGO*, No. 326, p. 152, 1975.

22. Timanovskaya, R. G., and Feigel'son, E. M. Fluxes of solar radiation at the Earth's surface for cumulus cloudiness. *Meteorologiya i Gidrologiya*, No. 11, p. 44, 1970.

23. Timanovskaya, R. G., and Feigel'son, E. M. A means of studying the statistical structure of surface fluxes of solar radiation under cloudy conditions. In: Heat transfer in the atmosphere, p. 97. Nauka, Leningrad, 1972.

24. Mazin, I. P. (ed.). *Trudy TsAO*, No. 134. Gidrometeoizdat, Moscow, 1977.

25. Feigel'son, E. M., and Krasnokutskaya, L. D. Fluxes of

solar radiation and clouds. Gidrometeoizdat, Leningrad, 1978.

26.   Davis, J. M., Cox, I. K., and McKee, T. B. Solar absorption
in clouds of finite horizontal extent. Atm. Sci. Paper, 1978,
No. 282, p. 94.

27.   Kauth, R. J., and Penquite, J. L. The probability of clear
lines of sight through a cloudy atmosphere. Sci. report, 1966,
Univ. Michigan 33.

28.   Masary, Aida. Scattering of solar radiation as a function
of cloud dimensions and orientation. *J. Quant. Spectrosc. Radiat.
Transfer*, 1977, Vol. 17, p. 303-310.

29.   Masary, Aida. Reflection of Solar Radiation from an Array
of Cumulis. *Repr. J. Meteorol. Soc.* Japan, 1977, Vol. 55, No. 2,
p. 174-181.

30.   McKee, I.B., and Cox, S.K. Scattering of visible radiation
by finite clouds. *J. Atm. Sci.*, 1974, Vol. 31, No. 7, p. 1885-
-1892.

Chapter 12

1.   Buikov, M. V., and Khvorost'yanov, V. I. The formation
and evolution of radiation fog and stratus clouds in the atmospheric boundary layer. Izv. Akad. Nauk SSSR, *Fiz. Atmos. Okeana,*
<u>13</u> (4): 356, 1977.

2.   Vasil'eva, T. I., Nasper, T. M., and Podolśkaya, E. L.
Parametrization of fluxes of long-wave radiation in a cloudless
atmosphere. Izv. Akad. Nauk SSSR, *Fiz. Atmos. Okeana,* <u>11</u> (3):
239, 1975.

3.   Marchuk, G. I. (ed.). A hydrodynamic model of the general
circulation of the atmosphere and ocean (application methods).
Preprint, *VTs SO Akad. Nauk SSSR,* Novosibirsk, 1975.

4.   Ginzburg, A. S., and Feigel'son, E.M. Approximate methods
of calculating heat fluxes and influxes in the atmosphere. Izv.
Akad. Nauk SSSR, *Fiz. Atmos. Okeana,* <u>6</u> (5), 1970.

5.   Ginzburg, A. S., and Niilisk, Kh. Yu. A comparison of different methods of calculating a field of long-wave radiation. In:
Heat transfer in the atmosphere. Nauka, Moscow, 1972.

6.   Ginzburg, A. S. Calculation of thermal-radiation fluxes
using material of the ATEP expedition. In: Meteorological studies
according to the program of the international tropical experiment.

Nauka, Moscow, 1977.

7.   Ginzburg, A. S. A method of calculating the thermal radiation of a cloudy atmosphere. Izv. Akad. Nauk SSSR, *Fiz. Atmos. Okeana,* 15 (3), 1979.

8.   Goody, R. M. Atmospheric radiation. Clarendon Press, 1964.

9.   Dmitrieva-Arrago, L. R. Calculation of fluxes and influxes of long-wave radiation under cloudy conditions. *Trudy GGO,* No. 197, 1968.

10.   Egorov, B. I. The dependence of atmospheric long-wave radiation on cloudiness in the tropical Atlantic. In: *TROPEKS-72,* p. 319. Gidrometeoizdat, Leningrad, 1974.

11.   Kondrat'ev, K. Ya. Actinometry. Gidrometeoizdat, Leningrad, 1965.

12.   Manabe, S., and Strickler, R. Thermal equilibrium of the atmosphere with a convective adjustment. *J. Atm. Sci.,* 21 (4), 1964.

13.   Marchuk, G. I., and Skiba, Yu. N. A model for predicting mean temperature anomalies. Preprint, *VTs SO Akad. Nauk SSSR,* No. 120, Novosibirsk, 1978.

14.   Matveev, L. T. Meteorology. Physics of the atmosphere. Gidrometeoizdat, Leningrad, 1975.

15.   Mokhov, I. I., and Petukhov, V. K. Parametrization of outgoing long-wave radiation for climatic models. Preprint, IFA Akad. Nauk SSSR, Moscow, 1978.

16.   Niilisk, Kh. Yu. Calculating fluxes of counterradiation of the atmosphere, averaged over large territories. Izv. Akad. Nauk SSSR, *Fiz. Atmos. Okeana,* 4 (5): 518, 1968.

17.   Niilisk, Kh. Yu., and Sammel, L. E. Integral transmission function of the atmosphere for calculating thermal-radiation field in the troposphere. In: A table of radiation characteristics. Tartu, 1969.

18.   Niilisk, Kh. Yu. Cloudiness characteristics in problems of radiation energetics in the Earth's atmosphere. Izv. Akad. Nauk SSSR, *Fiz. Atmos. Okeana,* 8 (3): 270, 1972.

19.   Rubinshtein, K. G. The tropospheric heat balance at low latitudes in a model of the zonal atmospheric circulation. *Meteorologiya i Gidrologiya,* No. 7, p. 20, 1977.

20.   Mullamaa, Yu.-A. R. (ed.). The stochastic structure of cloudiness and radiation fields. Tartu, 1972.

21.   Trosnikov, I. V., and Egorova, E. N. The use of empirical formulas for calculating radiative energy influxes when modeling the general circulation of the atmosphere. *Trudy GMTs*, No. 160, 1975.

22.   Feigel'son, E. M. Radiation processes in stratiform clouds. Nauka, Moscow, 1964.

23.   Feigel'son, E. M. The radiative influx of heat in the atmosphere. Izv. Akad. Nauk SSSR, *Ser. Geofiz.*, No. 10, p. 1539, 1964.

24.   Feigel'son, E. M. Radiant heat transfer in a cloudy atmosphere. Israel Program for Scientific Translations, Jerusalem, 1973 (English translation).

25.   Khvorost'yanov, V. I. Calculation of long-wave radiation fluxes and influxes for cloudiness, with the aid of spectral transmission functions and transfer equations. *Trudy UkrNIGMI*, No. 178, 1979.

Chapter 13

1.   Arst, Kh. Yu. Calculation of the thermal radiation of the Earth's atmosphere for an incomplete set of initial data. *Izv. Akad. Nauk ESSR, Fizika, Matematika*, 26 (2): 152, 1977.

2.   Arst, Kh. Yu., and Kostyanoi, G. N. A comparison of theoretical and experimental fluxes of long-wave radiation in the free atmosphere under cloudy and cloudless conditions. Izv. Akad. Nauk SSSR, *Fiz. Atmos. Okeana*, 16 (1): 111, 1980.

3.   Barashkova, E. P. Some regularities of the vertical distribution of the upward long-wave radiation. *Trudy GGO*, No. 221, p. 167, 1968.

4.   Gaevskii, V. L. A profile of fluxes of long-wave radiation in clouds. In: Studies of clouds, precipitation, and thunderstorm electricity, p. 114. Izd. Akad. Nauk SSSR, Moscow, 1961.

5.   Goisa, N. I. Actinometric measurements with the IL-14 aircraft. *Trudy UkrNIGMI*, No. 55, p. 3, 1966.

6.   Goisa, N. I., and Shoshin, V. M. The effective radiation of stratiform clouds. *Trudy UkrNIGMI*, No. 92, P. 49, 1970.

7.   Gradus, L. M., Niilisk, Kh. Yu., and Feigel'son, E. M.

An integral transmission function for cloudy conditions. Izv.
Akad. Nauk SSSR, *Fiz. Atmos. Okeana*, <u>4</u> (4): 397, 1968.

8.   Zaitseva, N. A. Application domain of model of standard
actinometric atmosphere. In: Radiation processes in the atmosphere
and at the Earth's surface, p. 62. Gidrometeoizdat, Leningrad,
1974.

9.   Zaitseva, N. A., and Kostyanoi, G. N. Variation of field
of long-wave radiation in the free atmosphere over a 7-10 hr
period. *Trudy TsAO*, No. 70, p. 41, 1966.

10.   Zaitseva, N. A., and Kostyanoi, G. N. The field of long-
-wave radiation in the free atmosphere (reference data). Gidrometeoizdat, Moscow, 1974.

11.   Zaitseva, N. A., and Kostyanoi, G. N. Time variation of
field of long-wave radiation during summer. *Trudy TsAO*, No. 83,
p. 47, 1969.

12.   Zaitseva, N. A., Kostyanoi, G. N., and Shlyakhov, V. I.
Mean many-year characteristics of the field of long-wave radiation
in the free atmosphere (according to actinometric-radiosounding
data). *Meteorologiya i Gidrologiya*, No. 7, p. 35, 1971.

13.   Zaitseva, N. A., Kostyanoi, G. N., and Shlyakhov, V. I.
A model of the standard radiation atmosphere (long-wave radiation).
Meteorologiya i Gidrologiya, No. 12, p. 24, 1973.

14.   Kostyanoi, G. N., and Kurilova, Yu. V. Radiation properties
of clouds. *Trudy TsAO*, No. 70, p. 31, 1966.

15.   Kostyanoi, G. N., and Niilisk, Kh. Yu. Comparison of measured and calculated fluxes of long-wave radiation in the atmosphere. *Trudy TsAO*, No. 83, p. 56, 1969.

16.   Kostyanoi, G. N., and Shlyakhov, V. I. International comparisons of actinometric radiosondes. *Meteorologiya i Gidrologiya*,
No. 5, p. 99, 1967.

17.   Kurilova, Yu. V. Structural features of fluxes of long-wave
radiation. *Trudy MMTs*, No. 11, p. 128, 1966.

18.   Kurilova, Yu. V., and Federov, Yu. K. The vertical structure
of fields of long-wave radiation. *Trudy GMTs*, No. 11, p. 148, 1967.

19.   Lopukhin, E. A. Radiation fluxes in the atmosphere over
Tashkent for a cloudy sky. *Trudy Tashk. Geofiz. Observ.*, No. 13
p. 35, 1957.

20.   Malkevich, M. S. Correlations between characteristics of
the vertical structure of long-wave radiation field and fields

of temperature and humidity. Izv. Akad. Nauk SSSR, *Fiz. Atmos. Okeana,* <u>1</u> (10): 1039, 1965.

21.   Niilisk, Kh. Yu. Calculating fluxes of counterradiation of the atmosphere, averaged over large territories. Izv. Akad. Nauk SSSR, *Fiz. Atmos. Okeana,* <u>4</u> (2): 182, 1968.

22.   Niilisk, Kh. Yu. Calculations of the thermal radiation of the atmosphere under partly cloudy conditions. Izv. Akad. Nauk SSSR, *Fiz. Atmos. Okeana,* <u>4</u> (4): 383, 1968.

23.   Niilisk, Kh. Yu. The effect on calculated fluxes of atmospheric thermal radiation of errors in the initial meteorological data. In: Radiation in the atmosphere, p. 60. Tartu, 1969.

24.   Niilisk, Kh. Yu. Some cloudiness characteristics. In: Radiation and clouds, p. 72. Tartu, 1969.

25.   Niilisk, Kh. Yu. Cloudiness characteristics in problems of radiation energetics in the Earth's atmosphere. Izv. Akad. Nauk SSSR, *Fiz. Atmos. Okeana,* <u>8</u> (3): 270, 1972.

26.   Popov, S. M. Some statistical characteristics of the vertical structure of temperature and humidity fields. Izv. Akad. Nauk SSSR, *Fiz. Atmos. Okeana,* <u>1</u> (1): 18, 1965.

27.   Feigel'son, E. M. Radiant heat transfer in a cloudy atmosphere. Israel Program for Scientific Translations, Jerusalem, 1973 (English translation).

28.   Ellingson, R. G., and Gille, J. C. An infrared radiative transfer model. Part 1. Model Description and Comparison of Observations with Calculations. *J. Atmos. Sci.,* 1978, Vol. 35, No. 3, p. 523-547.

29.   Kano, M., and Miyauchi, M. On the comparison between the observed vertical profiles of long-wave radiative fluxes and the computed ones in Japan. Pap. Meteorol. a. Geophys., 1977, Vol. 28, No. 1, p. 1-8.

30.   Kelkar, R. R., and Godbole, R. V. The dependence of long-wave radiation on cloudiness, water vapor content and temperature. *Indian J. Meteorol. a. Geophys.,* 1970, Vol. 21, No. 4, p. 613-622.

31.   Kondratyev, K. Ya. Radiation in the atmosphere. Intern. Geophys. Ser. Vol. 12. N.Y., London, Acad. Press, 1969, 912 p.

32.   Kuhn, P.M., and Suomi, V.E. Infrared radiometer soundings on a synoptic scale. *J. Geophys. Res.,* 1960, Vol. 65, No. 11, p. 3369-3677.

33.   Kuhn, P. M., Suomi, V. E., and Darkow, G. L. Soundings of terrestrial radiation flux over Wisconsin. Mon. Wea. Rev., 1959, Vol. 87, No. 4, p. 129-135.

34.   McKee, T. B., and Klehr, J. T. Radiative effects of cloud geometry. In: Proc. of the Symposium on Radiation in the Atmosphere. Garmisch-Partenkirchen, FRG, 19-20 August 1976, publ. 1977, p. 217-219.

35.   McKee, T. B., and Klehr, J. T. Effects of cloud shape on scattered solar radiation. Mon. Wea. Rev., 1978, Vol. 106, No. 3, p. 399-404.

Chapter 14

1.   Zaitseva, N. A., and Feigel'son, E. M. Features of radiative heat transfer in the tropics according to actinometric-radiosounding data from the *TROPEKS-74* expedition (ATEP). Izv. Akad. Nauk SSSR, *Fiz. Atmos. Okeana,* 15 (2): 154, 1979.

2.   Gorodetskii, A. K., et al. The emissivity of clouds. Izv. Akad. Nauk SSSR, *Fiz. Atmos. Okeana,* 13 (4): 424, 1977.

3.   Kondrat'ev, K. Ya., Niilisk, Kh. Yu., and Noorma, R. Yu. The spectral distribution of atmospheric thermal radiation under conditions of stratus cloudiness. *Trudy GGO,* No. 275, p. 49, 1972.

4.   Reshetnikova, I. S., and Popov, O. I. The emissivity of low-level clouds in the 8-12 µm transmission window. Izv. Akad. Nauk SSSR, *Fiz. Atmos. Okeana,* 6 (6): 639, 1970.

5.   Feigel'son, E. M., and Tsvang, L. R. (eds.). Heat transfer in the atmosphere. A collection of papers. Nauka, Moscow, 1972.

6.   Feigel'son, E. M. Radiation processes in stratiform clouds. Nauka, Moscow, 1964.

7.   Feigel'son, E. M. Radiant heat transfer in a cloudy atmosphere. Israel Program for Scientific Translations, Jerusalem, 1973 (English translation).

8.   Shifrin, K. S. Transfer of thermal radiation in clouds. *Trudy GGO,* No. 46 (108), 1955.

9.   Cox, S. Radiative effects of cirrus clouds. *J. Atm. Sci.,* 1974, Vol. 31, No. 8, p. 2182-2188.

10.   Cox, S. Observations of cloud infrared effective emissivity. *J. Atm. Sci.,* 1976, Vol. 33, No. 2, p. 287-289.

11.   Griffith, K., and Cox, S. Infrared radiative properties of typical cirrus clouds inferred from broad band measurements. Atm. Sci. Pap. Dep. Atm. Sci. Colo. State Univ., 1977, No. 289, p. 103.

12.   Kuhn, P. Measured effective long-wave emissivity of clouds. Mon. Wea. Rev., 1963, Vol. 91, No. 10–12, p. 635–640.

13.   Stephens, G. L. Radiation profiles in extended water clouds II. *J. Atm. Sci.*, 1978, Vol. 35, No. 11, p. 2123–2129.

14.   Yamamoto, G., Tanaka, M., and Asano, S. Radiative transfer in water clouds in the infrared region. *J. Atm. Sci.*, 1970, Vol. 27, No. 2, p. 282–292.

15.   Yamamoto, G., Tanaka, M., and Asano, S. Radiative heat transfer in water clouds by infrared radiation. *J. Quant. Spectroscop. Radiat. Transf.*, 1971, Vol. 11, No. 6, p. 697–708.

Chapter 15

1.   Burova, L.P., and Voskresenskii, A. I. Moisture content and moisture transfer in the atmosphere over the North-Pole region. *Trudy AANII*, <u>323</u>: 25, 1976.

2.   Voskresenskii, A. I. The moisture content of the atmosphere over Antarctica. *Trudy AANII*, <u>327</u>: 56, 1976.

3.   Zaitseva, N. A., and Kostyanoi, G. N. The field of long-wave radiation in the free atmosphere (reference data). Gidrometeoizdat, Moscow, 1974.

4.   Zaitseva, N. A., and Shlyakhov, V. I. Results of many-year studies of long-wave radiation in the antarctic atmosphere. In: Antarctica, No. 17, p. 75. Nauka, Moscow, 1978.

5.   Zvereva, S. V. The attenuation of solar radiation in the polar regions. *Trudy AANII*, <u>287</u>: 171, 1969.

6.   Koptev, A. P., and Voskresenskii, A. I. Radiation properties of cloudiness. *Trudy AANII*, <u>239</u>: 39, 1962.

7.   Marshunova, M. S. Calculating the balance of long-wave radiation for an overcast arctic sky. *Trudy AANII*, 226: 109, 1959.

8.   Marshunova, M. S. The main regularities of the radiation balance of the underlying surface and atmosphere in the Arctic. *Trudy AANII*, <u>229</u>: 5, 1961.

9.   Marshunova, M. S. Formation conditions and properties of

radiation climate of Antarctica. Gidrometeoizdat, Leningrad, 1980.

10.   Pivovarova, Z. I. Radiation characteristics of the climate of the USSR. Gidrometeoizadat, Leningrad, 1977.

11.   Kondrat'ev, K. Ya. (ed.). Radiation characteristics of the atmosphere and the Earth's surface. Gidrometeoizdat, Leningrad, 1969.

12.   Timerev, A. A. Total radiation and albedo according to aircraft observations in the Arctic. *Trudy AANII*, <u>273</u>: 170, 1965.

13.   Shlyakhov, V. I. Results of actinometric sounding of the atmosphere in the Antarctic. In: Antarctica (Reports of Commission for 1962), p. 125. Izv. Akad. Nauk SSSR, Moscow, 1963.

Chapter 16

1.   Borovikov, A. M., Lavrent'ev, E. V., and Mazin, I. P. Morphological features of clouds in the eastern tropical Atlantic. In: *TROPEKS-74*, Vol. 1. The atmosphere, p. 468. Gidrometeoizdat, Leningrad, 1976.

2.   Kondrat'ev, K. Ya., et al. The effect of the aerosol on the radiation energetics of the tropical atmosphere. *Trudy GGO*, No. 415, p. 3, 1979.

3.   Ginzburg, A. S. Calculation of fluxes of thermal radiation on the basis of AETP data. In: Meteorological studies according to the program of the international tropical experiment, p. 15. Nauka, Moscow, 1977.

4.   Devyatova, V. A. Statistical characteristics of the vertical structure of single-layer stratiform cloudiness and temperature and humidity fields. *Trudy GMTs SSSR*, No. 165, p. 107, 1975.

5.   Egorov, B. N., and Kirillova, T. V. Attenuation of the total radiation by clouds of various kinds according to data of the ATEP-74 expedition. Izv. Akad. Nauk SSSR, *Fiz. Atmos. Okeana*, <u>15</u> (9): 987, 1979.

6.   Zaitseva, N. A., and Kostyanoi, G. N. The field of long-wave radiation in the free atmosphere (reference data). Gidrometeoizdat, Moscow, 1974.

7.   Zaitseva, N. A., and Krasnova, T. M. Variability of long-wave radiation in ATEP test area. In: *TROPEKS-74*, Vol. 1. The atmosphere, p. 542. Gidrometeoizdat, Leningrad, 1976.

8.   Zaitseva, N. A., and Feigel'son, E. M. Features of radiative heat transfer in the tropics according to actinometric-radiosounding data from the *TROPEKS-74* expedition (ATEP). Izv. Akad. Nauk SSSR, *Fiz. Atmos. Okeana,* <u>15</u> (2): 154, 1979.

9.   Kirillova, T. V., Timanovskaya, R. G., and Gudimenko, A. V. The statistical structure of the total radiation reaching the ocean surface under conditions of cumulus cloudiness. *Trudy GGO,* No. 388, p. 93, 1977.

10.   Kislov, A. V. Circulation of the atmosphere and formation of the transmission field over the region of the Atlantic tropical experiment. *Okeanologiya,* No. 4, p. 620, 1979.

11.   Abramov, R. V., et al. Convective clouds in the equatorial Atlantic according to instrumental and visual observations. In: *TROPEKS-74*, Vol. 1. The atmosphere, p. 474. Gidrometeoizdat, Leningrad, 1976.

12.   Kostyanoi, G. N. The accuracy of measuring long-wave radiation with an actinometric radiosonde. *Meteorologiya i Gidrologiya,* No. 4, 1975.

13.   Dyubkin, I. A., et al. The transmission coefficient of a cloudy atmosphere. *Trudy GGO,* No. 388, p. 106, 1977.

14.   Malkevich, M. S. Optical studies of the atmosphere using satellites. Nauka, Moscow, 1973.

15.   Semenchenko, B. A., et al. Mathematical modeling of the arrival of total solar radiation at the Earth's surface as a method of climatological analysis. In: Fundamental problems of climatogenesis, p. 68. Izd. MGU, Moscow, 1979.

16.   Kuusk, A. E., et al. The variability of cloudiness and radiation fields during the 13th voyage of R/V *Akademik Korolev.* In: *TROPEKS-74*, Vol. 1. The atmosphere, p. 600. Gidrometeoizdat, Leningrad, 1976.

17.   Obukhov, A. M. Statistical orthogonal expansions of empirical functions. Izv. Akad. Nauk SSSR, *Ser. Geofiz.,* No. 3, p. 432, 1960.

18.   Zaitseva, N. A., et al. Fluxes of thermal radiation in the tropical atmosphere. Preprint, IFA Akad. Nauk SSSR and TsAO, Moscow, 1979.

19.   Semenchenko, B. A., Kislov, A. V., and Mart'yanova, G. N. The components of the radiation balance at the equator. In: *TROPEKS-74*, Vol. 1. The atmosphere, p. 568, Gidrometeoizdat,

Leningrad, 1976.

20.   Timanovskaya, R. G. The statistical structure of fluxes of direct and total radiation at the Earth's surface for cumulus cloudiness. *Trudy GGO*, No. 297, p. 142, 1973.

21.   Timanovskaya, R. G. The space-time structure of fluxes of short-wave radiation for cumulus clouds. *Trudy GGO*, No. 326, p. 152, 1975.

22.   Petrosyan', M. A. (ed.). *TROPEKS-74*, Vol. 1. The atmosphere, Gidrometeoizdat, Leningrad, 1976.

23.   Feigel'son, E. M. Radiant heat transfer in a cloudy atmosphere. Israel Program for Scientific Translations, Jerusalem, 1973 (English translation).

24.   Carlson, T. N. Large-scale distribution of turbidity over the Northern Equatorial Atlantic. In: Proc. symposium on radiation. FRG, 19-28 August 1976, Science Press, USA, 1977, p. 555.

25.   Carlson, T. N., and Prospero J. Saharian air outbreaks: meteorology, aerosols and radiation. Report of the U.S. GATE Central program workshop held by NCAR, 25 July-12 August, 1977, p. 57-58.

26.   Semenchenko, B. A., and Kislov, A. V. The factors of variability of radiation fluxes on the oceanic surfaces in the tropics. In: Proc. Int. Sci. Conf. "Energetics of the tropical atmosphere". Geneva, ICSU/WMO, 1978, p. 189-195.

<u>INDEX</u>